The Philosophy of Science
A Collection of Essays

Series Editor

Lawrence Sklar
University of Michigan

A GARLAND SERIES
READINGS IN PHILOSOPHY

ROBERT NOZICK, *ADVISOR*
HARVARD UNIVERSITY

Series Contents

Explanation, Law and Cause

Edited with an introduction by

Lawrence Sklar
University of Michigan

Routledge
Taylor & Francis Group

NEW YORK AND LONDON

First published by Garland Publishing, Inc
This edition published 2013 by Routledge

Routledge
Taylor & Francis Group
711 Third Avenue
New York, NY 10017

Routledge
Taylor & Francis Group
2 Park Square
Milton Park, Abingdon
Oxon OX14 4RN

Routledge is an imprint of the Taylor & Francis Group, an informa business

Library of Congress Cataloging-in-Publication Data

Philosophy of science : a collection of essays / edited with an
   introduction by Lawrence Sklar.
        p.   cm.
      "A Garland series"—Ser. t.p.
      Includes bibliographical references.
      Contents: 1. Explanation, law, and cause — 2. The nature of
scientific theory — 3. Theory reduction and theory change —
4. Probability and confirmation — 5. Bayesian and non-inductive
methods — 6. The philosophy of physics.
         ISBN 0-8153-2700-5 (v.1 : alk. paper) — ISBN 0-8153-2701-3
(v. 2 : alk. paper) — ISBN 0-8153-2702-1 (v. 3 : alk. paper) —
ISBN 0-8153-2703-X (v. 4 : alk. paper) — ISBN 0-8153-3492-3
(v. 5 : alk. paper) — ISBN 0-8153-3493-1 (v. 6 : alk. paper)
         1. Science—Philosophy. 2. Science—Methodology. 3. Physics—
Philosophy. I. Sklar, Lawrence.

Q175.P51227 1999
501—dc21   99-40012
                                                                           CIP

SET ISBN   9780815326991
POD ISBN   9780415870719
    Vol1   9780815327004
    Vol2   9780815327011
    Vol3   9780815327028
    Vol4   9780815327035
    Vol5   9780815334927
    Vol6   9780815334934

Contents

Introduction

Long before the existence of anything properly called science, humankind has demanded explanations. We want to know not only what happens in the world, but also to know why what happens does happen. We want to understand the course of events in the world, and we search for reasons to account for what occurs. But what is it to have an answer to a "Why?" question? More specifically, what kinds of answers to such questions can we expect the scientist to provide? And how do scientific explanations differ in their special nature from the kinds of answers we might expect outside of the scientific context?

The first systematic philosophical exploration of these questions was undertaken by Aristotle. He offered a famous doctrine of "causes." Changes in the world were to be understood in terms of the object undergoing the change and the kind of change undergone, but also in terms of what we would now call the cause of the change, and for Aristotle, in terms of the end or purpose of the change as well. Throughout the history of philosophy concepts such as "cause" and "mechanism" have played important roles in trying to analyze the notion of what we must discover and reveal when we answer a scientific "Why?" question. The issues raised by Aristotle in his notion of final causes, that is to say ends or purposes of happenings in the world, have also remained recurrent themes in the philosophy of science. Is there any place in science for such "final cause" explanations?

Several crucial episodes in the history of science, especially in the history of physics, have played important roles in driving philosophical discussions about the nature of scientific explanation. One such episode was the debate surrounding Newton's theory of gravity. An objection was raised against Newton's account of gravity, that although Newton had found the law describing gravity, he had not shown us the mechanism by which gravity operated, and had, therefore, not truly explained gravitational force. Newton himself was quite sympathetic to the claim that a full explanation of gravity would require unveiling some such causal mechanism for the force.

In the nineteenth century there was also an intense debate between "Atomists" and "Energeticists" about what was needed for an account to be a full explanation of the phenomena. One recurring issue was whether one had provided an explanation for some phenomenon when one had placed the phenomenon in a scheme of regular occurrences in the world, or whether some additional elements were needed in order that an account be explanatory. Whereas the Atomists looked for the explanations of the observable phenomena and the regularities that governed these phenomena in the existence and nature of "hidden" causal mechanisms in the world, the Energeticists took explanation to be only the fitting of the observable phenomena into ever more general regular patterns of behavior within the realm of the observable phenomena.

These debates, along with major contributions from the philosophical side, such as Hume's eighteenth-century analysis of causation, an analysis that took causation to be grounded solely in constant conjunction of features in the world combined with human expectation or habit, led to a model of explanation that took the fundamental structure of scientific explanation to be merely subsuming the events of the world under lawlike generalizations. To explain an event, according to this account, was to tie it to other events by means of true general statements of regularities in the world.

This way of thinking about explanations led to the so-called deductive-nomological model of explanation. According to this view, explanations of particular occurrences in the world were arguments, arguments whose premises were statements of lawlike regularities and statements of the occurrence of a number of explaining events, and whose derivable conclusion, a conclusion that followed from the premises by deductive logic, was a statement describing the event to be explained. The deductive-nomological model also allowed for the explanation of the truth of laws as well. An explanation of a law would, once again, be a deductive argument. The premises of this argument would be statements of more fundamental or more general laws than the law to be explained, and the conclusion of the argument would be a statement of the law that needed explanation. This account of explanation related explanation closely to the scientific desires for the means of predicting and controlling what occurs in the world.

But does one really have a full explanation of what has happened when one has placed what has occurred into a pattern of regular occurrences? Numerous examples seem to show that subsuming under a regularity is not by itself enough for explanation. For example, we can derive the height of a building from the length of its shadow, the position of the sun, and the laws of geometric optics, but such a derivation would hardly explain why the building had the height it had. But it does seem to be the case that deriving the length of the shadow from the height of the building, the position of the sun, and the laws of optics would count as a genuine explanation. What else might be needed for something like a deductive-nomological derivation to count as a genuine explanation?

Sometimes it is suggested that the answer given to a "Why" question that constitutes an explanation is highly context dependent, changing as the state of knowledge or ignorance of the questioner varies. Various "pragmatic" issues of what is lacking in one's knowledge, and how the gap might be filled, are then crucial to an analysis of explanation. Others have suggested that in order for an answer to be explanatory, it must serve to "unify" a multiplicity of otherwise unrelated phenomena in some essential way. Explanations, then, are theoretical unifiers that place into a single framework a diversity of phenomena to be explained.

Finally, there are those who would argue that in order for an answer to a "Why?" question to be fully explanatory, it must reveal to the person asking the question aspects of the causal structure of the world. Here we see a return to the intuitive idea that has functioned so deeply in the philosophical issue of explanation ever since Aristotle. That is the idea that events in the world are related to one another as cause and effect, and that a deep understanding of what we take to be explanatory in science requires a deep understanding of what might be meant by the causal relationships among the events to be explained.

Various notions of causation and mechanism have been invoked as essential components of anything that can truly be called an explanation in science. Indeed, the idea that causation, and not mere correlation, plays a crucial role in explanation has played roles outside of academic philosophy itself, being important, for example, in numerous discussions of the nature of legal responsibility.

All of these issues are taken up in Section A of this volume.

First in the social sciences, and in this century within physics itself, the regularities that joined together the events in the world into patterns often turned out to be probabilistic or statistical regularities. How does this impinge on our analysis of what it is to be a scientific explanation?

From the regularity point of view, many novel questions arise. Is to explain an event to show it highly probable relative to the occurrence of other events? Or, instead, do we explain an event when we show that it is more probable relative to the events introduced as explanatory than it was relative to the background of other events? That is, is high probability needed for explanation, or is raised probability enough? Again, the probability of an event can vary with the reference class of events we pay attention to in our explanatory constructions. This variation is much richer than anything that appears when explanation is looked at in the case of strict, exceptionless, regularities of nature. How does this affect our notions of probabilistic or statistical explanation?

From the causal point of view, does probabilistic explanation really consist in explaining an occurrence of an event by showing that it is the causal effect of some genuine probabilistic or chance-like state of the world? What could such states be like and how do they explain their effects.

These issues are discussed in the readings in Section B.

Nearly all models of scientific explanation invoke the use of those scientific generalizations we call laws of nature. But what is a law of nature? It is the statement of some general truth about the world. But, it is often argued, it is more than that. A mere accidental generalization only says what, in general, is true, but a law of nature says, in some sense, what "must be true as a matter of physical necessity."

How are we to understand that intuitive distinction? Often it is argued that laws of nature differ from merely accidental generalizations in their ability to support counter-factual assertions. Thus, the mere fact that all of the coins in my pocket are copper does not support the claim that if some coin in my hand (perhaps made of silver) were in my pocket it would be copper. But the law that all metals conduct electricity (assuming it true) would support the counterfactual-conditional claim that if a piece of chalk in my hand were a metal, it too would conduct electricity.

The discussion of the nature of laws of nature involves a number of elements. One is the philosophical program of understanding the nature of counter-factual conditionals. What kind of understanding can we provide of these curious conditional utterances that tell us what would be the case were the world other than it actually is? We also require more understanding of what the relationship is between generalizations and such conditionals that makes laws of nature so special. Another problem is in trying to understand the place of such lawlike generalizations in our general body of assertions about the world. Philosophers would like to know, for example, if the special nature of lawlike assertions can be understood in terms of how those laws are established, since there seems to be a close connection between a generalization's being lawlike and its being confirmable by induction. We also want to know how laws function in the general scheme of scientific explanations. Could lawlikeness be a function of the place of a generalization in the overall scientific scheme? That is to say, do we count a generalization as being lawlike only when it fits neatly into our overall hierarchy of generalizations that play important roles in our fundamental science?

Or is it the case, that wholly distinctive "lawlike facts" must be introduced into the world in order for us to understand what true lawlike assertions are claiming about the world. Might we think, for example, of their being some special relationship between kinds or "universals" that grounds the lawlikeness of generalizations about things of the kinds in question?

Such issues are the subject matter of the readings in Section C.

How are we to understand what causes and effects are, and what the relationship is between events that makes us speak of one as the cause of the other? Ever since Hume's important discussion of causation in the eighteenth century, this has been a problem area for the philosophy of science.

Is causation a "primitive" relation between individual events, or is it, as Hume thought, a matter of the constant conjunction between various kinds of events? If the latter, what regularities must hold between events in the world so that we speak of one event as the cause of the other, and of the second as the effect of the first? A special set of problems arises when we realize that the relationships among the statements characterizing events (the descriptions of causes and effects), such as their being derivable from one another given some other assertions, only indirectly characterize the relationships among the events themselves. Failing to make the important distinction between an event and the relevant description of the event has led to much confusion in the past.

In the background behind these questions is a very general question about the connection between explanation and causation. Should we think of causation as the fundamental notion, with explanation being accounted for as the revelation of what causes there are in the world? Or, on the other hand, should we think of explanation, with all of its psychological and pragmatic aspects, as the fundamental concept, with causation to be explicated as a kind of projection onto the world of our explanatory notions?

But it is no easy task to say what we mean by the very notion of an event, and, having characterized what events are, it is difficult to say exactly and precisely what relationships hold among events that ought to be spoken of as causal. If causation is a matter of lawlike regularity, how, exactly, are laws invoked in accounting for the truth conditions of statements about causation?

Finally, when one moves from laws of nature thought of as exceptionless regularities to allow for the possibility of lawlike probabilistic features of the world, new questions arise about our idea of causation. Can we think of a cause of some effect as being some event that makes the probability of the so-called effect event high? Or is a causal event an event that raises the probability of the effect event's occurrence above what that probability would have been had the causal event not taken place? Attempts at defining causation in terms of probabilistic relations among events have led to many perplexities. The probabilistic relation that one event can bear to another is highly dependent on the background against which the relationship between the events is being considered. But can causation be so relative in that way?

Alternatively, should probabilistic causation be analyzed in terms of some irreducible chance-like states of the world that act as primitive probabilistic causes of their effects? From this perspective there is an irreducible notion of there being in the world events related to one another by a kind of "chancy" primitive causal relationship.

These are the subjects taken up in the selections in Section D of this volume.

STUDIES IN THE LOGIC OF EXPLANATION

CARL G. HEMPEL AND PAUL OPPENHEIM[1]

§1. Introduction.

To explain the phenomena in the world of our experience, to answer the question "why?" rather than only the question "what?", is one of the foremost objectives of all rational inquiry; and especially, scientific research in its various branches strives to go beyond a mere description of its subject matter by providing an explanation of the phenomena it investigates. While there is rather general agreement about this chief objective of science, there exists considerable difference of opinion as to the function and the essential characteristics of scientific explanation. In the present essay, an attempt will be made to shed some light on these issues by means of an elementary survey of the basic pattern of scientific explanation and a subsequent more rigorous analysis of the concept of law and of the logical structure of explanatory arguments.

The elementary survey is presented in Part I of this article; Part II contains an analysis of the concept of emergence; in Part III, an attempt is made to exhibit and to clarify in a more rigorous manner some of the peculiar and perplexing logical problems to which the familiar elementary analysis of explanation gives rise. Part IV, finally, is devoted to an examination of the idea of explanatory power of a theory; an explicit definition, and, based on it, a formal theory of this concept are developed for the case of a scientific language of simple logical structure.

PART I. ELEMENTARY SURVEY OF SCIENTIFIC EXPLANATION

§2. Some illustrations.

A mercury thermometer is rapidly immersed in hot water; there occurs a temporary drop of the mercury column, which is then followed by a swift rise. How is this phenomenon to be explained? The increase in temperature affects at first only the glass tube of the thermometer; it expands and thus provides a larger space for the mercury inside, whose surface therefore drops. As soon as by heat conduction the rise in temperature reaches the mercury, however, the latter expands, and as its coefficient of expansion is considerably larger than that of

[1] This paper represents the outcome of a series of discussions among the authors; their individual contributions cannot be separated in detail. The technical developments contained in Part IV, however, are due to the first author, who also put the article into its final form.

Some of the ideas presented in Part II were suggested by our common friend, Kurt Grelling, who, together with his wife, became a victim of Nazi terror during the war. Those ideas were developed by Grelling, in a discussion by correspondence with the present authors, of emergence and related concepts. By including at least some of that material, which is indicated in the text, in the present paper, we feel that we are realizing the hope expressed by Grelling that his contributions might not entirely fall into oblivion.

We wish to express our thanks to Dr. Rudolf Carnap, Dr. Herbert Feigl, Dr. Nelson Goodman, and Dr. W. V. Quine for stimulating discussions and constructive criticism.

1

glass, a rise of the mercury level results.—This account consists of statements of two kinds. Those of the first kind indicate certain conditions which are realized prior to, or at the same time as, the phenomenon to be explained; we shall refer to them briefly as antecedent conditions. In our illustration, the antecedent conditions include, among others, the fact that the thermometer consists of a glass tube which is partly filled with mercury, and that it is immersed into hot water. The statements of the second kind express certain general laws; in our case, these include the laws of the thermic expansion of mercury and of glass, and a statement about the small thermic conductivity of glass. The two sets of statements, if adequately and completely formulated, explain the phenomenon under consideration: They entail the consequence that the mercury will first drop, then rise. Thus, the event under discussion is explained by subsuming it under general laws, i.e., by showing that it occurred in accordance with those laws, by virtue of the realization of certain specified antecedent conditions.

Consider another illustration. To an observer in a row boat, that part of an oar which is under water appears to be bent upwards. The phenomenon is explained by means of general laws—mainly the law of refraction and the law that water is an optically denser medium than air—and by reference to certain antecedent conditions—especially the facts that part of the oar is in the water, part in the air, and that the oar is practically a straight piece of wood.—Thus, here again, the question "*Why* does the phenomenon happen?" is construed as meaning "according to what general laws, and by virtue of what antecedent conditions does the phenomenon occur?"

So far, we have considered exclusively the explanation of particular events occurring at a certain time and place. But the question "Why?" may be raised also in regard to general laws. Thus, in our last illustration, the question might be asked: Why does the propagation of light conform to the law of refraction? Classical physics answers in terms of the undulatory theory of light, i.e. by stating that the propagation of light is a wave phenomenon of a certain general type, and that all wave phenomena of that type satisfy the law of refraction. Thus, the explanation of a general regularity consists in subsuming it under another, more comprehensive regularity, under a more general law.—Similarly, the validity of Galileo's law for the free fall of bodies near the earth's surface can be explained by deducing it from a more comprehensive set of laws, namely Newton's laws of motion and his law of gravitation, together with some statements about particular facts, namely the mass and the radius of the earth.

§3. The basic pattern of scientific explanation.

From the preceding sample cases let us now abstract some general characteristics of scientific explanation. We divide an explanation into two major constituents, the explanandum and the explanans[2]. By the explanandum, we

---

[2] These two expressions, derived from the Latin *explanare*, were adopted in preference to the perhaps more customary terms "explicandum" and "explicans" in order to reserve the latter for use in the context of explication of meaning, or analysis. On explication in this sense, cf. Carnap, [Concepts], p. 513.—Abbreviated titles in brackets refer to the bibliography at the end of this article.

understand the sentence describing the phenomenon to be explained (not that phenomenon itself); by the explanans, the class of those sentences which are adduced to account for the phenomenon. As was noted before, the explanans falls into two subclasses; one of these contains certain sentences $C_1, C_2, \cdots, C_k$ which state specific antecedent conditions; the other is a set of sentences $L_1, L_2, \cdots L_r$ which represent general laws.

If a proposed explanation is to be sound, its constituents have to satisfy certain conditions of adequacy, which may be divided into logical and empirical conditions. For the following discussion, it will be sufficient to formulate these requirements in a slightly vague manner; in Part III, a more rigorous anlysis and a more precise restatement of these criteria will be presented.

I. *Logical conditions of adequacy.*

(R1) The explanandum must be a logical consequence of the explanans; in other words, the explanandum must be logically deducible from the information contained in the explanans, for otherwise, the explanans would not constitute adequate grounds for the explanandum.

(R2) The explanans must contain general laws, and these must actually be required for the derivation of the explanandum.—We shall not make it a necessary condition for a sound explanation, however, that the explanans must contain at least one statement which is not a law; for, to mention just one reason, we would surely want to consider as an explanation the derivation of the general regularities governing the motion of double stars from the laws of celestial mechanics, even though all the statements in the explanans are general laws.

(R3) The explanans must have empirical content; i.e., it must be capable, at least in principle, of test by experiment or observation.—This condition is implicit in (R1); for since the explanandum is assumed to describe some empirical phenomenon, it follows from (R1) that the explanans entails at least one consequence of empirical character, and this fact confers upon it testability and empirical content. But the point deserves special mention because, as will be seen in §4, certain arguments which have been offered as explanations in the natural and in the social sciences violate this requirement.

II. *Empirical condition of adequacy.*

(R4) The sentences constituting the explanans must be true.
That in a sound explanation, the statements constituting the explanans have to satisfy some condition of factual correctness is obvious. But it might seem more appropriate to stipulate that the explanans has to be highly confirmed by all the relevant evidence available rather than that it should be true. This stipulation however, leads to awkward consequences. Suppose that a certain phenomenon was explained at an earlier stage of science, by means of an explanans which was well supported by the evidence then at hand, but which had been highly disconfirmed by more recent empirical findings. In such a case, we

3

would have to say that originally the explanatory account was a correct explanation, but that it ceased to be one later, when unfavorable evidence was discovered. This does not appear to accord with sound common usage, which directs us to say that on the basis of the limited initial evidence, the truth of the explanans, and thus the soundness of the explanation, had been quite probable, but that the ampler evidence now available made it highly probable that the explanans was not true, and hence that the account in question was not—and had never been—a correct explanation. (A similar point will be made and illustrated, with respect to the requirement of truth for laws, in the beginning of §6.)

Some of the characteristics of an explanation which have been indicated so far may be summarized in the following schema:

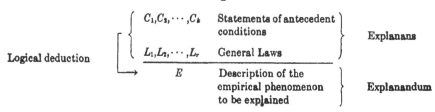

Let us note here that the same formal analysis, including the four necessary conditions, applies to scientific prediction as well as to explanation. The difference between the two is of a pragmatic character. If E is given, i.e. if we know that the phenomenon described by E has occurred, and a suitable set of statements C_1, C_2, $\cdots$, C_k, L_1, L_2, $\cdots$, L_r is provided afterwards, we speak of an explanation of the phenomenon in question. If the latter statements are given and E is derived prior to the occurrence of the phenomenon it describes, we speak of a prediction. It may be said, therefore, that an explanation is not fully adequate unless its explanans, if taken account of in time, could have served as a basis for predicting the phenomenon under consideration.[2a]—Consequently, whatever will be said in this article concerning the logical characteristics of explanation or prediction will be applicable to either, even if only one of them should be mentioned.

It is this potential predictive force which gives scientific explanation its importance: only to the extent that we are able to explain empirical facts can we attain the major objective of scientific research, namely not merely to record the phenomena of our experience, but to learn from them, by basing upon them theoretical generalizations which enable us to anticipate new occurrences and to control, at least to some extent, the changes in our environment.

Many explanations which are customarily offered, especially in pre-scientific discourse, lack this predictive character, however. Thus, it may be explained

---

[2a] The logical similarity of explanation and prediction, and the fact that one is directed towards past occurrences, the other towards future ones, is well expressed in the terms "post-dictability" and "predictability" used by Reichenbach in [Quantum Mechanics], p. 13.

that a car turned over on the road "because" one of its tires blew out while the car was travelling at high speed.   Clearly, on the basis of just this information, the accident could not have been predicted, for the explanans provides no explicit general laws by means of which the prediction might be effected, nor does it state adequately the antecedent conditions which would be needed for the prediction.—The same point may be illustrated by reference to W. S. Jevons's view that every explanation consists in pointing out a resemblance between facts, and that in some cases this process may require no reference to laws at all and "may involve nothing more than a single identity, as when we explain the appearance of shooting stars by showing that they are identical with portions of a comet".[3]   But clearly, this identity does not provide an explanation of the phenomenon of shooting stars unless we presuppose the laws governing the development of heat and light as the effect of friction.   The observation of similarities has explanatory value only if it involves at least tacit reference to general laws.

In some cases, incomplete explanatory arguments of the kind here illustrated suppress parts of the explanans simply as "obvious"; in other cases, they seem to involve the assumption that while the missing parts are not obvious, the incomplete explanans could at least, with appropriate effort, be so supplemented as to make a strict derivation of the explanandum possible.   This assumption may be justifiable in some cases, as when we say that a lump of sugar disappeared "because" it was put into hot tea, but it is surely not satisfied in many other cases.   Thus, when certain peculiarities in the work of an artist are explained as outgrowths of a specific type of neurosis, this observation may contain significant clues, but in general it does not afford a sufficient basis for a potential prediction of those peculiarities.   In cases of this kind, an incomplete explanation may at best be considered as indicating some positive correlation between the antecedent conditions adduced and the type of phenomenon to be explained, and as pointing out a direction in which further research might be carried on in order to complete the explanatory account.

The type of explanation which has been considered here so far is often referred to as causal explanation.   If E describes a particular event, then the antecedent circumstances described in the sentences C_1, C_2, $\cdots$, C_k may be said jointly to "cause" that event, in the sense that there are certain empirical regularities, expressed by the laws L_1, L_2, $\cdots$, L_r, which imply that whenever conditions of the kind indicated by C_1, C_2, $\cdots$, C_k occur, an event of the kind described in E will take place.   Statements such as L_1, L_2, $\cdots$, L_r, which assert general and unexceptional connections between specified characteristics of events, are customarily called causal, or deterministic, laws.   They are to be distinguished from the so-called statistical laws which assert that in the long run, an explicitly stated percentage of all cases satisfying a given set of conditions are accompanied by an event of a certain specified kind.   Certain cases of scientific explanation involve "subsumption" of the explanandum under a set of laws of which at least some are statistical in character.   Analysis of the peculiar logical structure

[3] [Principles], p. 533.

of that type of subsumption involves difficult special problems. The present essay will be restricted to an examination of the causal type of explanation, which has retained its significance in large segments of contemporary science, and even in some areas where a more adequate account calls for reference to statistical laws.[4]

§4. *Explanation in the non-physical sciences. Motivational and teleological approaches.*

Our characterization of scientific explanation is so far based on a study of cases taken from the physical sciences. But the general principles thus obtained apply also outside this area.[5] Thus, various types of behavior in laboratory animals and in human subjects are explained in psychology by subsumption under laws or even general theories of learning or conditioning; and while frequently, the regularities invoked cannot be stated with the same generality and precision as in physics or chemistry, it is clear, at least, that the general character of those explanations conforms to our earlier characterization.

Let us now consider an illustration involving sociological and economic factors. In the fall of 1946, there occurred at the cotton exchanges of the United States a price drop which was so severe that the exchanges in New York, New Orleans, and Chicago had to suspend their activities temporarily. In an attempt to explain this occurrence, newspapers traced it back to a large-scale speculator in New Orleans who had feared his holdings were too large and had therefore begun to liquidate his stocks; smaller speculators had then followed his example

---

[4] The account given above of the general characteristics of explanation and prediction in science is by no means novel; it merely summarizes and states explicitly some fundamental points which have been recognized by many scientists and methodologists.

Thus, e.g., Mill says: "An individual fact is said to be explained by pointing out its cause, that is, by stating the law or laws of causation of which its production is an instance", and "a law of uniformity in nature is said to be explained when another law or laws are pointed out, of which that law itself is but a case, and from which it could be deduced." ([Logic], Book III, Chapter XII, section 1). Similarly, Jevons, whose general characterization of explanation was critically discussed above, stresses that "the most important process of explanation consists in showing that an observed fact is one case of a general law or tendency." ([Principles], p. 533). Ducasse states the same point as follows: "Explanation essentially consists in the offering of a hypothesis of fact, standing to the fact to be explained as case of antecedent to case of consequent of some already known law of connection." ([Explanation], pp. 150–51). A lucid analysis of the fundamental structure of explanation and prediction was given by Popper in [Forschung], section 12, and, in an improved version, in his work [Society], especially in Chapter 25 and in note 7 referring to that chapter.—For a recent characterization of explanation as subsumption under general theories, cf., for example, Hull's concise discussion in [Principles], chapter I. A clear elementary examination of certain aspects of explanation is given in Hospers, [Explanation], and a concise survey of many of the essentials of scientific explanation which are considered in the first two parts of the present study may be found in Feigl, [Operationism], pp. 284 ff.

[5] On the subject of explanation in the social sciences, especially in history, cf. also the following publications, which may serve to supplement and amplify the brief discussion to be presented here: Hempel, [Laws]; Popper, [Society]; White, [Explanation]; and the articles *Cause* and *Understanding* in Beard and Hook, [Terminology].

in a panic and had thus touched off the critical decline. Without attempting to assess the merits of the argument, let us note that the explanation here suggested again involves statements about antecedent conditions and the assumption of general regularities. The former include the facts that the first speculator had large stocks of cotton, that there were smaller speculators with considerable holdings, that there existed the institution of the cotton exchanges with their specific mode of operation, etc. The general regularities referred to are—as often in semi-popular explanations—not explicitly mentioned; but there is obviously implied some form of the law of supply and demand to account for the drop in cotton prices in terms of the greatly increased supply under conditions of practically unchanged demand; besides, reliance is necessary on certain regularities in the behavior of individuals who are trying to preserve or improve their economic position. Such laws cannot be formulated at present with satisfactory precision and generality, and therefore, the suggested explanation is surely incomplete, but its intention is unmistakably to account for the phenomenon by integrating it into a general pattern of economic and socio-psychological regularities.

We turn to an explanatory argument taken from the field of linguistics.[6] In Northern France, there exist a large variety of words synonymous with the English "bee," whereas in Southern France, essentially only one such word is in existence. For this discrepancy, the explanation has been suggested that in the Latin epoch, the South of France used the word "apicula", the North the word "apis". The latter, because of a process of phonologic decay in Northern France, became the monosyllabic word "ê"; and monosyllables tend to be eliminated, especially if they contain few consonantic elements, for they are apt to give rise to misunderstandings. Thus, to avoid confusion, other words were selected. But "apicula", which was reduced to "abelho", remained clear enough and was retained, and finally it even entered into the standard language, in the form "abbeille". While the explanation here described is incomplete in the sense characterized in the previous section, it clearly exhibits reference to specific antecedent conditions as well as to general laws.[7]

While illustrations of this kind tend to support the view that explanation in biology, psychology, and the social sciences has the same structure as in the physical sciences, the opinion is rather widely held that in many instances, the causal type of explanation is essentially inadequate in fields other than physics and chemistry, and especially in the study of purposive behavior. Let us ex-

---

[6] The illustration is taken from Bonfante, [Semantics], section 3.

[7] While in each of the last two illustrations, certain regularities are unquestionably relied upon in the explanatory argument, it is not possible to argue convincingly that the intended laws, which at present cannot all be stated explicitly, are of a causal rather than a statistical character. It is quite possible that most or all of the regularities which will be discovered as sociology develops will be of a statistical type. Cf., on this point, the suggestive observations by Zilsel in [Empiricism] section 8, and [Laws]. This issue does not affect, however, the main point we wish to make here, namely that in the social no less than in the physical sciences, subsumption under general regularities is indispensable for the explanation and the theoretical understanding of any phenomenon.

amine briefly some of the reasons which have been adduced in support of this view.

One of the most familiar among them is the idea that events involving the activities of humans singly or in groups have a peculiar uniqueness and irrepeatability which makes them inaccessible to causal explanation because the latter, which its reliance upon uniformities, presupposes repeatability of the phenomena under consideration. This argument which, incidentally, has also been used in support of the contention that the experimental method is inapplicable in psychology and the social sciences, involves a misunderstanding of the logical character of causal explanation. Every individual event, in the physical sciences no less than in psychology or the social sciences, is unique in the sense that it, with all its peculiar characteristics, does not repeat itself. Nevertheless, individual events may conform to, and thus be explainable by means of, general laws of the causal type. For all that a causal law asserts is that any event of a specified kind, i.e. any event having certain specified characteristics, is accompanied by another event which in turn has certain specified characteristics; for example, that in any event involving friction, heat is developed. And all that is needed for the testability and applicability of such laws is the recurrence of events with the antecedent characteristics, i.e. the repetition of those characteristics, but not of their individual instances. Thus, the argument is inconclusive. It gives occasion, however, to emphasize an important point concerning our earlier analysis: When we spoke of the explanation of a single event, the term "event" referred to the occurrence of some more or less complex characteristic in a specific spatio-temporal location or in a certain individual object, and not to *all* the characteristics of that object, or to all that goes on in that space-time region.

A second argument that should be mentioned here[3] contends that the establishment of scientific generalizations—and thus of explanatory principles—for human behavior is impossible because the reactions of an individual in a given situation depend not only upon that situation, but also upon the previous history of the individual.—But surely, there is no *a priori* reason why generalizations should not be attainable which take into account this dependence of behavior on the past history of the agent. That indeed the given argument "proves" too much, and is therefore a *non sequitur*, is made evident by the existence of certain physical phenomena, such as magnetic hysteresis and elastic fatigue, in which the magnitude of a specific physical effect depends upon the past history of the system involved, and for which nevertheless certain general regularities have been established.

A third argument insists that the explanation of any phenomenon involving purposive behavior calls for reference to motivations and thus for teleological rather than causal analysis. Thus, for example, a fuller statement of the suggested explanation for the break in the cotton prices would have to indicate the large-scale speculator's motivations as one of the factors determining the event

[3] Cf., for example, F. H. Knight's presentation of this argument in [Limitations], pp. 251–52.

in question. Thus, we have to refer to goals sought, and this, so the argument runs, introduces a type of explanation alien to the physical sciences. Unquestionably, many of the—frequently incomplete—explanations which are offered for human actions involve reference to goals and motives; but does this make them essentially different from the causal explanations of physics and chemistry? One difference which suggests itself lies in the circumstance that in motivated behavior, the future appears to affect the present in a manner which is not found in the causal explanations of the physical sciences. But clearly, when the action of a person is motivated, say, by the desire to reach a certain objective, then it is not the as yet unrealized future event of attaining that goal which can be said to determine his present behavior, for indeed the goal may never be actually reached; rather—to put it in crude terms—it is (a) his desire, present before the action, to attain that particular objective, and (b) his belief, likewise present before the action, that such and such a course of action is most likely to have the desired effect. The determining motives and beliefs, therefore, have to be classified among the antecedent conditions of a motivational explanation, and there is no formal difference on this account between motivational and causal explanation.

Neither does the fact that motives are not accessible to direct observation by an outside observer constitute an essential difference between the two kinds of explanation; for also the determining factors adduced in physical explanations are very frequently inaccessible to direct observation. This is the case, for instance, when opposite electric charges are adduced in explanation of the mutual attraction of two metal spheres. The presence of those charges, while eluding all direct observation, can be ascertained by various kinds of indirect test, and that is sufficient to guarantee the empirical character of the explanatory statement. Similarly, the presence of certain motivations may be ascertainable only by indirect methods, which may include reference to linguistic utterances of the subject in question, slips of the pen or of the tongue, etc.; but as long as these methods are "operationally determined" with reasonable clarity and precision, there is no essential difference in this respect between motivational explanation and causal explanation in physics.

A potential danger of explanation by motives lies in the fact that the method lends itself to the facile construction of ex-post-facto accounts without predictive force. It is a widespread tendency to "explain" an action by ascribing it to motives conjectured only after the action has taken place. While this procedure is not in itself objectionable, its soundness requires that (1) the motivational assumptions in question be capable of test, and (2) that suitable general laws be available to lend explanatory power to the assumed motives. Disregard of these requirements frequently deprives alleged motivational explanations of their cognitive significance.

The explanation of an action in terms of the motives of the agent is sometimes considered as a special kind of teleological explanation. As was pointed out above, motivational explanation, if adequately formulated, conforms to the conditions for causal explanation, so that the term "teleological" is a misnomer if it is meant to imply either a non-causal character of the explanation or a peculiar determination of the present by the future. If this is borne in mind, however, the term "teleological" may be viewed, in this context, as referring to causal explanations in which some of the antecedent conditions are motives of the agent whose actions are to be explained.[9]

Teleological explanations of this kind have to be distinguished from a much more sweeping type, which has been claimed by certain schools of thought to be indispensable especially in biology. It consists in explaining characteristics of an organism by reference to certain ends or purposes which the characteristics are said to serve. In contradistinction to the cases examined before, the ends are not assumed here to be consciously or subconsciously pursued by the organism in question. Thus, for the phenomenon of mimicry, the explanation is sometimes offered that it serves the purpose of protecting the animals endowed with it from detection by its pursuers and thus tends to preserve the species. —Before teleological hypotheses of this kind can be appraised as to their potential explanatory power, their meaning has to be clarified. If they are intended somehow to express the idea that the purposes they refer to are inherent in the design of the universe, then clearly they are not capable of empirical test and thus violate the requirement (R3) stated in §3. In certain cases, however, assertions about the purposes of biological characteristics may be translatable into statements in non-teleological terminology which assert that those characteristics function in a specific manner which is essential to keeping the organism alive or to preserving the species.[10] An attempt to state precisely what is meant by this latter assertion—or by the similar one that without those characteristics, and other things being equal, the organism or the species would not survive—encounters considerable difficulties. But these need not be discussed here. For even if we assume that biological statements in teleological form can be adequately translated into descriptive statements about the life-preserving function of certain biological characteristics, it is clear that (1) the use of the concept of purpose is not essential in these contexts, since the term "purpose" can be completely eliminated from the statements in question, and (2) teleological assumptions, while now endowed with empirical content, cannot serve as explanatory principles in the customary contexts. Thus, e.g., the fact that a

---

[9] For a detailed logical analysis of the character and the function of the motivation concept in psychological theory, see Koch, [Motivation].—A stimulating discussion of teleological behavior from the standpoint of contemporary physics and biology is contained in the article [Teleology] by Rosenblueth, Wiener and Bigelow. The authors propose an interpretation of the concept of purpose which is free from metaphysical connotations, and they stress the importance of the concept thus obtained for a behavioristic analysis of machines and living organisms. While our formulations above intentionally use the crude terminology frequently applied in philosophical arguments concerning the applicability of causal explanation to purposive behavior, the analysis presented in the article referred to is couched in behavioristic terms and avoids reference to "motives" and the like.

[10] An analysis of teleological statements in biology along these lines may be found in Woodger, [Principles], especially pp. 432 ff; essentially the same interpretation is advocated by Kaufmann in [Methodology], chapter 8.

given species of butterflies displays a particular kind of coloring cannot be inferred from—and therefore cannot be explained by means of—the statement that this type of coloring has the effect of protecting the butterflies from detection by pursuing birds, nor can the presence of red corpuscles in the human blood be inferred from the statement that those corpuscles have a specific function in assimilating oxygen and that this function is essential for the maintenance of life.

One of the reasons for the perseverance of teleological considerations in biology probably lies in the fruitfulness of the teleological approach as a heuristic device: Biological research which was psychologically motivated by a teleological orientation, by an interest in purposes in nature, has frequently led to important results which can be stated in non-teleological terminology and which increase our scientific knowledge of the causal connections between biological phenomena.

Another aspect that lends appeal to teleological considerations is their anthropomorphic character. A teleological explanation tends to make us feel that we really "understand" the phenomenon in question, because it is accounted for in terms of purposes, with which we are familiar from our own experience of purposive behavior. But it is important to distinguish here understanding in the psychological sense of a feeling of empathic familiarity from understanding in the theoretical, or cognitive, sense of exhibiting the phenomenon to be explained as a special case of some general regularity. The frequent insistence that explanation means the reduction of something unfamiliar to ideas or experiences already familiar to us is indeed misleading. For while some scientific explanations do have this psychological effect, it is by no means universal: The free fall of a physical body may well be said to be a more familiar phenomenon than the law of gravitation, by means of which it can be explained; and surely the basic ideas of the theory of relativity will appear to many to be far less familiar than the phenomena for which the theory accounts.

"Familiarity" of the explicans is not only not necessary for a sound explanation—as we have just tried to show—, but it is not sufficient either. This is shown by the many cases in which a proposed explicans sounds suggestively familiar, but upon closer inspection proves to be a mere metaphor, or an account lacking testability, or a set of statements which includes no general laws and therefore lacks explanatory power. A case in point is the neovitalistic attempt to explain biological phenomena by reference to an entelechy or vital force. The crucial point here is not—as it is sometimes made out to be—that entelechies cannot be seen or otherwise directly observed; for that is true also of gravitational fields, and yet, reference to such fields is essential in the explanation of various physical phenomena. The decisive difference between the two cases is that the physical explanation provides (1) methods of testing, albeit indirectly, assertions about gravitational fields, and (2) general laws concerning the strength of gravitational fields, and the behavior of objects moving in them. Explanations by entelechies satisfy the analogue of neither of these two conditions. Failure to satisfy the first condition represents a violation of (R3); it renders all statements about entelechies inaccessible to empirical test and thus devoid of empirical meaning. Failure to comply with the second condition involves a

11

violation of (R2). It deprives the concept of entelechy of all explanatory import; for explanatory power never resides in a concept, but always in the general laws in which it functions. Therefore, notwithstanding the flavor of familiarity of the metaphor it invokes, the neovitalistic approach cannot provide theoretical understanding.

The preceding observations about familiarity and understanding can be applied, in a similar manner, to the view held by some scholars that the explanation, or the understanding, of human actions requires an empathic understanding of the personalities of the agents[11]. This understanding of another person in terms of one's own psychological functioning may prove a useful heuristic device in the search for general psychological principles which might provide a theoretical explanation; but the existence of empathy on the part of the scientist is neither a necessary nor a sufficient condition for the explanation, or the scientific understanding, of any human action. It is not necessary, for the behavior of psychotics or of people belonging to a culture very different from that of the scientist may sometimes be explainable and predictable in terms of general principles even though the scientist who establishes or applies those principles may not be able to understand his subjects empathically. And empathy is not sufficient to guarantee a sound explanation, for a strong feeling of empathy may exist even in cases where we completely misjudge a given personality. Moreover, as the late Dr. Zilsel has pointed out, empathy leads with ease to incompatible results; thus, when the population of a town has long been subjected to heavy bombing attacks, we can understand, in the empathic sense, that its morale should have broken down completely, but we can understand with the same ease also that it should have developed a defiant spirit of resistance. Arguments of this kind often appear quite convincing; but they are of an *ex post facto* character and lack cognitive significance unless they are supplemented by testable explanatory principles in the form of laws or theories.

Familiarity of the explanans, therefore, no matter whether it is achieved through the use of teleological terminology, through neovitalistic metaphors, or through other means, is no indication of the cognitive import and the predictive force of a proposed explanation. Besides, the extent to which an idea will be considered as familiar varies from person to person and from time to time, and a psychological factor of this kind certainly cannot serve as a standard in assessing the worth of a proposed explanation. The decisive requirement for every sound explanation remains that it subsume the explanandum under general laws.

PART II. ON THE IDEA OF EMERGENCE

§5. Levels of Explanation. Analysis of Emergence.

As has been shown above, a phenomenon may often be explained by sets of laws of different degrees of generality. The changing positions of a planet, for example, may be explained by subsumption under Kepler's laws, or by deriva-

---

[11] For a more detailed discussion of this view on the basis of the general principles outlined above, cf. Zilsel, [Empiricism], sections 7 and 8, and Hempel, [Laws], section 6.

tion from the far more comprehensive general law of gravitation in combination with the laws of motion, or finally by deduction from the general theory of relativity, which explains—and slightly modifies—the preceding set of laws. Similarly, the expansion of a gas with rising temperature at constant pressure may be explained by means of the Gas Law or by the more comprehensive kinetic theory of heat. The latter explains the Gas Law, and thus indirectly the phenomenon just mentioned, by means of (1) certain assumptions concerning the micro-behavior of gases (more specifically, the distributions of locations and speeds of the gas molecules) and (2) certain macro-micro principles, which connect such macro-characteristics of a gas as its temperature, pressure and volume with the micro-characteristics just mentioned.

In the sense of these illustrations, a distinction is frequently made between various levels of explanation[12]. Subsumption of a phenomenon under a general law directly connecting observable characteristics represents the first level; higher levels require the use of more or less abstract theoretical constructs which function in the context of some comprehensive theory. As the preceding illustrations show, the concept of higher-level explanation covers procedures of rather different character; one of the most important among them consists in explaining a class of phenomena by means of a theory concerning their micro-structure. The kinetic theory of heat, the atomic theory of matter, the electromagnetic as well as the quantum theory of light, and the gene theory of heredity are examples of this method. It is often felt that only the discovery of a micro-theory affords real scientific understanding of any type of phenomenon, because only it gives us insight into the inner mechanism of the phenomenon, so to speak. Consequently, classes of events for which no micro-theory was available have frequently been viewed as not actually understood; and concern with the theoretical status of phenomena which are unexplained in this sense may be considered as a theoretical root of the doctrine of emergence.

Generally speaking, the concept of emergence has been used to characterize certain phenomena as "novel", and this not merely in the psychological sense of being unexpected[13], but in the theoretical sense of being unexplainable, or unpredictable, on the basis of information concerning the spatial parts or other constituents of the systems in which the phenomena occur, and which in this context are often referred to as wholes. Thus, e.g., such characteristics of water as its transparence and liquidity at room temperature and atmospheric pressure, or its ability to quench thirst have been considered as emergent on the ground that they could not possibly have been predicted from a knowledge of the properties of its chemical constituents, hydrogen and oxygen. The weight of the compound, on the contrary, has been said not to be emergent because it is a mere "resultant" of its components and could have been predicted by simple addition even before the compound had been formed. The conceptions of ex-

[12] For a lucid brief exposition of this idea, see Feigl, [Operationism], pp. 284–288.

[13] Concerning the concept of novelty in its logical and psychological meanings, see also Stace, [Novelty].

planation and prediction which underly this idea of emergence call for various critical observations, and for corresponding changes in the concept of emergence.

(1) First, the question whether a given characteristic of a "whole", w, is emergent or not cannot be significantly raised until it has been stated what is to be understood by the parts or constituents of w. The volume of a brick wall, for example, may be inferable by addition from the volumes of its parts if the latter are understood to be the component bricks, but it is not so inferable from the volumes of the molecular components of the wall. Before we can significantly ask whether a characteristic W of an object w is emergent, we shall therefore have to state the intended meaning of the term "part of". This can be done by defining a specific relation Pt and stipulating that those and only those objects which stand in Pt to w count as parts or constituents of w. 'Pt' might be defined as meaning "constituent brick of" (with respect to buildings), or "molecule contained in" (for any physical object), or "chemical element contained in" (with respect to chemical compounds, or with respect to any material object), or "cell of" (with respect to organisms), etc. The term "whole" will be used here without any of its various connotations, merely as referring to any object w to which others stand in the specified relation Pt. In order to emphasize the dependence of the concept of part upon the definition of the relation Pt in each case, we shall sometimes speak of Pt-parts, to refer to parts as determined by the particular relation Pt under consideration.

(2) We turn to a second point of criticism. If a characteristic of a whole is to be qualified as emergent only if its occurrence cannot be inferred from a knowledge of all the properties of its parts, then, as Grelling has pointed out, no whole can have any emergent characteristics. Thus, to illustrate by reference to our earlier example, the properties of hydrogen include that of forming, if suitably combined with oxygen, a compound which is liquid, transparent, etc. Hence the liquidity, transparence, etc. of water can be inferred from certain properties of its chemical constituents. If the concept of emergence is not to be vacuous, therefore, it will be necessary to specify in every case a class G of attributes and to call a characteristic W of an object w emergent relatively to G and Pt if the occurrence of W in w cannot be inferred from a complete characterization of all the Pt-parts with respect to the attributes contained in G, i.e. from a statement which indicates, for every attribute in G, to which of the parts of w it applies. —Evidently, the occurrence of a characteristic may be emergent with respect to one class of attributes and not emergent with respect to another. The classes of attributes which the emergentists have in mind, and which are usually not explicitly indicated, will have to be construed as non-trivial, i.e. as not logically entailing the property of each constituent of forming, together with the other constituents, a whole with the characteristics under investigations. —Some fairly simple cases of emergence in the sense so far specified arise when the class G is restricted to certain simple properties of the parts, to the exclusion of spatial or other relations among them. Thus, the electromotive force of a system of several electric batteries cannot be inferred from the electromotive forces of its

14

constituents alone without a description, in terms of relational concepts, of the way in which the batteries are connected with each other.[14]

(3) Finally, the predictability of a given characteristic of an object on the basis of specified information concerning its parts will obviously depend on what general laws or theories are available.[15] Thus, the flow of an electric current in a wire connecting a piece of copper and a piece of zinc which are partly immersed in sulfuric acid is unexplainable, on the basis of information concerning any non-trivial set of attributes of copper, zinc and sulfuric acid, and the particular structure of the system under consideration, unless the theory available contains certain general laws concerning the functioning of batteries, or even more comprehensive principles of physical chemistry. If the theory includes such laws, on the other hand, then the occurrence of the current is predictable. Another illustration, which at the same time provides a good example for the point made under (2) above, is afforded by the optical activity of certain substances. The optical activity of sarco-lactic acid, for example, i.e. the fact that in solution it rotates the plane of polarization of plane-polarized light, cannot be predicted on the basis of the chemical characteristics of its constituent elements; rather, certain facts about the relations of the atoms constituting a molecule of sarco-lactic acid have to be known. The essential point is that the molecule in question contains an asymmetric carbon atom, i.e. one that holds four different atoms or groups, and if this piece of relational information is provided, the optical activity of the solution can be predicted provided that furthermore the theory available for the purpose embodies the law that the presence of one asymmetric carbon atom in a molecule implies optical activity of the solution; if the theory does not include this micro-macro law, then the phenomenon is emergent with respect to that theory.

An argument is sometimes advanced to the effect that phenomena such as the

[14] This observation connects the present discussion with a basic issue in Gestalt theory. Thus, e.g., the insistence that "a whole is more than the sum of its parts" may be construed as referring to characteristics of wholes whose prediction requires knowledge of certain structural relations among the parts. For a further examination of this point, see Grelling and Oppenheim, [Gestaltbegriff] and [Functional Whole].

[15] Logical analyses of emergence which make reference to the theories available have been propounded by Grelling and recently, in a very explicit form, by Henle in [Emergence]. In effect, Henle's definition characterizes a phenomenon as emergent if it cannot be predicted, by means of the theories accepted at the time, on the basis of the data available before its occurrence. In this interpretation of emergence, no reference is made to characteristics of parts or constitutents. Henle's concept of predictability differs from the one implicit in our discussion (and made explicit in Part III of this article) in that it implies derivability from the "simplest" hypothesis which can be formed on the basis of the data and theories available at the time. A number of suggestive observations on the idea of emergence and on Henle's analysis of it are contained in Bergmann's article [Emergence].— The idea that the concept of emergence, at least in some of its applications, is meant to refer to unpredictability by means of "simple" laws was advanced also by Grelling in the correspondence mentioned in note (1). Reliance on the motion of simplicity of hypotheses, however, involves considerable difficulties; in fact, no satisfactory definition of that concept is available at present.

flow of the current, or the optical activity, in our last examples, are absolutely emergent at least in the sense that they could not possibly have been predicted before they had been observed for the first time; in other words, that the laws requisite for their prediction could not have been arrived at on the basis of information available before their first observed occurrence.[16] This view is untenable, however. On the strength of data available at a given time, science often establishes generalizations by means of which it can forecast the occurrence of events the like of which have never before been encountered. Thus, generalizations based upon periodicities exhibited by the characteristics of chemical elements then known, enabled Mendeleeff in 1871 to predict the existence of a certain new element and to state correctly various properties of that element as well as of several of its compounds; the element in question, germanium, was not discovered until 1886.—A more recent illustration of the same point is provided by the development of the atomic bomb and the prediction, based on theoretical principles established prior to the event, of its explosion under specified conditions, and of its devastating release of energy.

As Grelling has stressed, the observation that the predictability of the occurrence of any characteristic depends upon the theoretical knowledge available, applies even to those cases in which, in the language of some emergentists, the characteristic of the whole is a mere resultant of the corresponding characteristics of the parts and can be obtained from the latter by addition. Thus, even the weight of a water molecule cannot be derived from the weights of its atomic constituents without the aid of a law which expresses the former as some specific mathematical function of the latter. That this function should be the sum is by no means self-evident; it is an empirical generalization, and at that not a strictly correct one, as relativistic physics has shown.

Failure to realize that the question of the predictability of a phenomenon cannot be significantly raised unless the theories available for the prediction have been specified has encouraged the misconception that certain phenomena have a mysterious quality of absolute unexplainability, and that their emergent status has to be accepted with "natural piety", as F. L. Morgan put it. The observations presented in the preceding discussion strip the idea of emergence of these unfounded connotations: emergence of a characteristic is not an ontological trait inherent in some phenomena; rather it is indicative of the scope of our knowl‐

[16] C. D. Broad, who in chapter 2 of his book, [Mind], gives a clear presentation and critical discussion of the essentials of emergentism, emphasizes the importance of "laws of composition" in predicting the characteristics of a whole on the basis of those of its parts. (cf. [Mind], pp. 61ff.); but he subscribes to the view characterized above and illustrates it specifically by the assertion that "if we want to know the chemical (and many of the physical) properties of a chemical compound, such as silver-chloride, it is absolutely necessary to study samples of *that particular compound*. . . . The essential point is that it would also be useless to study chemical compounds in general and to compare their properties with those of their elements in the hope of discovering a *general* law of composition by which the properties of *any* chemical compound could be foretold when the properties of its separate elements were known." (Ibid., p. 64)—That an achievement of precisely this sort has been possible on the basis of the periodic system of the elements is pointed out above.

edge at a given time; thus it has no absolute, but a relative character; and what is emergent with respect to the theories available today may lose its emergent status tomorrow.

The preceding considerations suggest the following redefinition of emergence: The occurrence of a characteristic W in an object w is emergent relatively to a theory T, a part relation Pt, and a class G of attributes if that occurrence cannot be deduced by means of T from a characterization of the Pt-parts of w with respect to all the attributes in G.

This formulation explicates the meaning of emergence with respect to *events* of a certain kind, namely the occurrence of some characteristic W in an object w. Frequently, emergence is attributed to *characteristics* rather than to events; this use of the concept of emergence may be interpreted as follows: A characteristic W is emergent relatively to T, Pt, and G if its occurrence in *any* object is emergent in the sense just indicated.

As far as its cognitive content is concerned, the emergentist assertion that the phenomena of life are emergent may now be construed, roughly, as an elliptic formulation of the following statement: Certain specifiable biological phenomena cannot be explained, by means of contemporary physico-chemical theories, on the basis of data concerning the physical and chemical characteristics of the atomic and molecular constituents of organisms. Similarly, the so-called emergent status of mind reduces to the assertion that present-day physical, chemical and biological theories do not suffice to explain all psychological phenomena on the basis of data concerning the physical, chemical, and biological characteristics of the cells or of the molecules or atoms constituting the organisms in question. But in this interpretation, the emergent character of biological and psychological phenomena becomes trivial; for the description of various biological phenomena requires terms which are not contained in the vocabulary of present day physics and chemistry; hence we cannot expect that all specifically biological phenomena are explainable, i.e. deductively inferable, by means of present day physico-chemical theories on the basis of initial conditions which themselves are described in exclusively physico-chemical terms. In order to obtain a less trivial interpretation of the assertion that the phenomena of life are emergent, we have therefore to include in the explanatory theory all those laws known at present which connect the physico-chemical with the biological "level", i.e., which contain, on the one hand, certain physical and chemical terms, including those required for the description of molecular structures, and on the other hand, certain concepts of biology. An analogous observation applies to the case of psychology. If the assertion that life and mind have an emergent status is interpreted in this sense, then its import can be summarized approximately by the statement that no explanation, in terms of micro-structure theories, is available at present for large classes of phenomena studied in biology and psychology.[17]

[17] The following passage from Tolman, [Behavior], may serve to support this interpretation: ". . . 'behavior-acts', though no doubt in complete one-to-one correspondence with the underlying molecular facts of physics and physiology, have, as 'molar' wholes, certain emergent properties of their own. . . . Further, these molar properties of behavior-acts

Assertions of this type, then, appear to represent the rational core of the doctrine of emergence. In its revised form, the idea of emergence no longer carries with it the connotation of absolute unpredictability—a notion which is objectionable not only because it involves and perpetuates certain logical misunderstandings, but also because, not unlike the ideas of neo-vitalism, it encourages an attitude of resignation which is stifling for scientific research. No doubt it is this characteristic, together with its theoretical sterility, which accounts for the rejection, by the majority of contemporary scientists, of the classical absolutistic doctrine of emergence.[18]

PART III. LOGICAL ANALYSIS OF LAW AND EXPLANATION

§6. Problems of the concept of general law.

From our general survey of the characteristics of scientific explanation, we now turn to a closer examination of its logical structure. The explanation of a phenomenon, we noted, consists in its subsumption under laws or under a theory. But what is a law, what is a theory? While the meaning of these concepts seems intuitively clear, an attempt to construct adequate explicit definitions for them encounters considerable difficulties. In the present section, some basic problems of the concept of law will be described and analyzed; in the next section, we intend to propose, on the basis of the suggestions thus obtained, definitions of law and of explanation for a formalized model language of a simple logical structure.

The concept of law will be construed here so as to apply to true statements only. The apparently plausible alternative procedure of requiring high confirmation rather than truth of a law seems to be inadequate: It would lead to a relativized concept of law, which would be expressed by the phrase "sentence S is a law relatively to the evidence E". This does not seem to accord with the meaning customarily assigned to the concept of law in science and in methodological inquiry. Thus, for example, we would not say that Bode's general formula for the distance of the planets from the sun was a law relatively to the astronomical evidence available in the 1770s, when Bode propounded it, and that it ceased to be a law after the discovery of Neptune and the determination of its distance from the sun; rather, we would say that the limited original evidence had given a high probability to the assumption that the formula was a law, whereas more recent additional information reduced that probability so much as to make it practically certain that Bode's formula is not generally true, and hence not a law.[18a]

---

cannot in the present state of our knowledge, i.e., prior to the working-out of many empirical correlations between behavior and its physiological correlates, be known even inferentially from a mere knowledge of the underlying, molecular, facts of physics and physiology." (l. c., pp. 7–8).—In a similar manner, Hull uses the distinction between molar and molecular theories and points out that theories of the latter type are not at present available in psychology. Cf. [Principles], pp. 19ff.; [Variables], p. 275.

[18] This attitude of the scientist is voiced, for example, by Hull in [Principles], pp. 24–28.

[18a] The requirement of truth for laws has the consequence that a given empirical statement S can never be definitely known to be a law; for the sentence affirming the truth of S

Apart from being true, a law will have to satisfy a number of additional conditions. These can be studied independently of the factual requirement of truth, for they refer, as it were, to all logically possible laws, no matter whether factually true or false. Adopting a convenient term proposed by Goodman[19], we will say that a sentence is lawlike if it has all the characteristics of a general law, with the possible exception of truth. Hence, every law is a lawlike sentence, but not conversely.

Our problem of analyzing the concept of law thus reduces to that of explicating the meaning of "lawlike sentence". We shall construe the class of lawlike sentences as including analytic general statements, such as "A rose is a rose", as well as the lawlike sentences of empirical science, which have empirical content.[20] It will not be necessary to require that each lawlike sentence permissible in explanatory contexts be of the second kind; rather, our definition of explanation will be so constructed as to guarantee the factual character of the totality of the laws—though not of every single one of them—which function in an explanation of an empirical fact.

What are the characteristics of lawlike sentences? First of all, lawlike sentences are statements of universal form, such as "All robins' eggs are greenish-blue", "All metals are conductors of electricity", "At constant pressure, any gas expands with increasing temperature". As these examples illustrate, a lawlike sentence usually is not only of universal, but also of conditional form; it makes an assertion to the effect that universally, if a certain set of conditions, C, is realized, then another specified set of conditions, E, is realized as well. The standard form for the symbolic expression of a lawlike sentence is therefore the universal conditional. However, since any conditional statement can be transformed into a non-conditional one, conditional form will not be considered as essential for a lawlike sentence, while universal character will be held indispensable.

But the requirement of universal form is not sufficient to characterize lawlike sentences. Suppose, for example, that a certain basket, b, contains at a certain time t a number of red apples and nothing else.[21] Then the statement

(S_1) Every apple in basket b at time t is red

is both true and of universal form. Yet the sentence does not qualify as a law; we would refuse, for example, to explain by subsumption under it the fact

---

is logically equivalent with S and is therefore capable only of acquiring a more or less high probability, or degree of confirmation, relatively to the experimental evidence available at any given time. On this point, cf. Carnap, [Remarks].—For an excellent non-technical exposition of the semantical concept of truth, which is here applied, the reader is referred to Tarski, [Truth].

[19] [Counterfactuals]. p. 125.

[20] This procedure was suggested by Goodman's approach in [Counterfactuals].—Reichenbach, in a detailed examination of the concept of law, similarly construes his concept of nomological statement as including both analytic and synthetic sentences; cf. [Logic], chapter VIII.

[21] The difficulty illustrated by this example was stated concisely by Langford ([Review]), who referred to it as the problem of distinguishing between universals of fact and causal universals. For further discussion and illustration of this point, see also Chisholm [Conditional], especially pp. 301f.—A systematic analysis of the problem was given by Goodman

that a particular apple chosen at random from the basket is red. What distinguishes S_1 from a lawlike sentence? Two points suggest themselves, which will be considered in turn, namely, finite scope, and reference to a specified object.

First, the sentence S_1 makes, in effect, an assertion about a finite number of objects only, and this seems irreconcilable with the claim to universality which is commonly associated with the notion of law.[22]   But are not Kepler's laws considered as lawlike although they refer to a finite set of planets only? And might we not even be willing to consider as lawlike a sentence such as the following?

(S_2) All the sixteen ice cubes in the freezing tray of this refrigerator have a temperature of less than 10 degrees centigrade.

This point might well be granted; but there is an essential difference between S_1 on the one hand and Kepler's laws as well as S_2 on the other: The latter, while finite in scope, are known to be consequences of more comprehensive laws whose scope is not limited, while for S_1 this is not the case.

Adopting a procedure recently suggested by Reichenbach[23], we will therefore distinguish between fundamental and derivative laws. A statement will be called a derivative law if it is of universal character and follows from some fundamental laws. The concept of fundamental law requires further clarification; so far, we may say that fundamental laws, and similarly fundamental lawlike sentences, should satisfy a certain condition of non-limitation of scope.

It would be excessive, however, to deny the status of fundamental lawlike sentence to all statements which, in effect, make an assertion about a finite class of objects only, for that would rule out also a sentence such as "All robins' eggs are greenish-blue", since presumably the class of all robins' eggs—past, present, and future—is finite. But again, there is an essential difference between this sentence and, say, S_1. It requires empirical knowledge to establish the finiteness of the class of robins' eggs, whereas, when the sentence S_1 is construed in a manner which renders it intuitively unlawlike, the terms "basket b" and "apple" are understood so as to imply finiteness of the class of apples in the basket at time t. Thus, so to speak, the meaning of its constitutive terms alone—without additional factual information—entails that S_1 has a finite scope.—Fundamental laws, then, will have to be construed so as to satisfy what we have called a condition of non-limited scope; our formulation of that condition however, which refers to what is entailed by "the meaning" of certain expressions, is too vague and will have to be revised later. Let us note in passing that the stipulation here envisaged would bar from the class of fundamental lawlike sentences also such undesirable candidates as "All uranic objects are spherical", where "uranic" means the property

in [Counterfactuals], especially part III.—While not concerned with the specific point under discussion, the detailed examination of counterfactual conditionals and their relation to laws of nature, in Chapter VIII of Lewis's work [Analysis], contains important observations on several of the issues raised in the present section.

[22] The view that laws should be construed as not being limited to a finite domain has been expressed, among others, by Popper ([Forschung], section 13) and by Reichenbach ([Logic], p. 369).

[23] [Logic], p. 361.—Our terminology as well as the definitions to be proposed later for the two types of law do not coincide with Reichenbach's, however.

of being the planet Uranus; indeed, while this sentence has universal form, it fails to satisfy the condition of non-limited scope.

In our search for a general characterization of lawlike sentences, we now turn to a second clue which is provided by the sentence S_1. In addition to violating the condition of non-limited scope, this sentence has the peculiarity of making reference to a particular object, the basket b; and this, too, seems to violate the universal character of a law.[24] The restriction which seems indicated here, should however again be applied to fundamental lawlike sentences only; for a true general statement about the free fall of physical bodies on the moon, while referring to a particular object, would still constitute a law, albeit a derivative one.

It seems reasonable to stipulate, therefore, that a fundamental lawlike sentence must be of universal form and must contain no essential—i.e., uneliminable —occurrences of designations for particular objects. But this is not sufficient; indeed, just at this point, a particularly serious difficulty presents itself. Consider the sentence

(S_3) Everything that is either an apple in basket b at time t or a sample of ferric oxide is red.

If we use a special expression, say "x is ferple", as synonymous with "x is either an apple in b at t or a sample of ferric oxide", then the content of S_3 can be expressed in the form

(S_4) Everything that is ferple is red.

The statement thus obtained is of universal form and contains no designations of particular objects, and it also satisfies the condition of non-limited scope; yet clearly, S_4 can qualify as a fundamental lawlike sentence no more than can S_3.

As long as "ferple" is a defined term of our language, the difficulty can readily be met by stipulating that after elimination of defined terms, a fundamental lawlike sentence must not contain essential occurrences of designations for particular objects. But this way out is of no avail when "ferple", or another term of the kind illustrated by it, is a primitive predicate of the language under consideration. This reflection indicates that certain restrictions have to be imposed upon those predicates—i.e., terms for properties or relations,—which may occur in fundamental lawlike sentences.[25]

---

[24] In physics, the idea that a law should not refer to any particular object has found its expression in the maxim that the general laws of physics should contain no reference to specific space-time points, and that spatio-temporal coordinates should occur in them only in the form of differences or differentials.

[25] The point illustrated by the sentences S_3 and S_4 above was made by Goodman, who has also emphasized the need to impose certain restrictions upon the predicates whose occurrence is to be permissible in lawlike sentences. These predicates are essentially the same as those which Goodman calls projectible. Goodman has suggested that the problems of establishing precise criteria for projectibility, of interpreting counterfactual conditionals, and of defining the concept of law are so intimately related as to be virtually aspects of a single problem. (Cf. his articles [Query] and [Counterfactuals].) One suggestion for an analysis of projectibility has recently been made by Carnap in [Application]. Goodman's note [Infirmities] contains critical observations on Carnap's proposals.

More specifically, the idea suggests itself of permitting a predicate in a fundamental lawlike sentence only if it is purely universal, or, as we shall say, purely qualitative, in character; in other words, if a statement of its meaning does not require reference to any one particular object or spatio-temporal location. Thus, the terms "soft", "green", "warmer than", "as long as", "liquid", "electrically charged", "female", "father of" are purely qualitative predicates, while "taller than the Eiffel Tower", "medieval", "lunar", "arctic", "Ming" are not.[26]

Exclusion from fundamental lawlike sentences of predicates which are not purely qualitative would at the same time ensure satisfaction of the condition of non-limited scope; for the meaning of a purely qualitative predicate does not require a finite extension; and indeed, all the sentences considered above which violate the condition of non-limited scope make explicit or implicit reference to specific objects.

The stipulation just proposed suffers, however, from the vagueness of the concept of purely qualitative predicate. The question whether indication of the meaning of a given predicate in English does or does not require reference to some one specific object does not always permit an unequivocal answer since English as a natural language does not provide explicit definitions or other clear explications of meaning for its terms. It seems therefore reasonable to attempt definition of the concept of law not with respect to English or any other natural language, but rather with respect to a formalized language—let us call it a model language, L,—which is governed by a well-determined system of logical rules, and in which every term either is characterized as primitive or is introduced by an explicit definition in terms of the primitives.

This reference to a well-determined system is customary in logical research and is indeed quite natural in the context of any attempt to develop precise criteria for certain logical distinctions. But it does not by itself suffice to overcome the specific difficulty under discussion. For while it is now readily possible to characterize as not purely qualitative all those among the defined predicates in L whose definiens contains an essential occurrence of some individual name, our problem remains open for the primitives of the language, whose meanings are not determined by definitions within the language, but rather by semantical rules of interpretation. For we want to permit the interpretation of the primitives of L by means of such attributes as blue, hard, solid, warmer, but

[26] That laws, in addition to being of universal form, must contain only purely universal predicates was clearly argued by Popper ([Forschung], sections 14, 15).—Our alternative expression "purely qualitative predicate" was chosen in analogy to Carnap's term "purely qualitative property" (cf. [Application]).—The above characterization of purely universal predicates seems preferable to a simpler and perhaps more customary one, to the effect that a statement of the meaning of the predicate must require no reference to particular objects. For this formulation might be too exclusive since it could be argued that stating the meaning of such purely qualitative terms as "blue" or "hot" requires illustrative reference to some particular object which has the quality in question. The essential point is that no one specific object has to be chosen; any one in the logically unlimited set of blue or of hot objects will do. In explicating the meaning of "taller than the Eiffel Tower", "being an apple in basket b at time t", "medieval", etc., however, reference has to be made to one specific object or to some one in a limited set of objects.

not by the properties of being a descendant of Napoleon, or an arctic animal, or a Greek statue; and the difficulty is precisely that of stating rigorous criteria for the distinction between the permissible and the non-permissible interpretations. Thus the problem of setting up an adequate definition for purely qualitative attributes now arises again; namely for the concepts of the metalanguage in which the semantical interpretation of the primitives is formulated. We may postpone an encounter with the difficulty by presupposing formalization of the semantical meta-language, the meta-meta-language, and so forth; but somewhere, we will have to stop at a non-formalized meta-language, and for it a characterization of purely qualitative predicates will be needed and will present much the same problems as non-formalized English, with which we began. The characterization of a purely qualitative predicate as one whose meaning can be made explicit without reference to any one particular object points to the intended meaning but does not explicate it precisely, and the problem of an adequate definition of purely qualitative predicates remains open.

There can be little doubt, however, that there exists a large number of property and relation terms which would be rather generally recognized as purely qualitative in the sense here pointed out, and as permissible in the formulation of fundamental lawlike sentences; some examples have been given above, and the list could be readily enlarged. When we speak of purely qualitative predicates, we shall henceforth have in mind predicates of this kind.

In the following section, a model language L of a rather simple logical structure will be described, whose primitives will be assumed to be qualitative in the sense just indicated. For this language, the concepts of law and explanation will then be defined in a manner which takes into account the general observations set forth in the present section.

§7. *Definition of law and explanation for a model language.*

Concerning the syntax of our model language L, we make the following assumptions:

L has the syntactical structure of the lower functional calculus without identity sign. In addition to the signs of alternation (disjunction), conjunction, and implication (conditional), and the symbols of universal and existential quantification with respect to individual variables, the vocabulary of L contains individual constants ('a', 'b', $\cdots$), individual variables ('x', 'y', $\cdots$), and predicates of any desired finite degree; the latter may include, in particular, predicates of degree 1 ('P', 'Q', $\cdots$), which express properties of individuals, and predicates of degree 2 ('R', 'S', $\cdots$), which express dyadic relations among individuals.

For simplicity, we assume that all predicates are primitive, i.e., undefined in L, or else that before the criteria subsequently to be developed are applied to a sentence, all defined predicates which it contains are eliminated in favor of primitives.

The syntactical rules for the formation of sentences and for logical inference in L are those of the lower functional calculus. No sentence may contain free variables, so that generality is always expressed by universal quantification.

For later reference, we now define, in purely syntactical terms, a number of auxiliary concepts. In the following definitions, S is always understood to be a sentence in L.

(7.1a) S is formally true (formally false) in L if S (the denial of S) can be proved in L, i.e. by means of the formal rules of logical inference for L. If two sentences are mutually derivable from each other in L, they will be called equivalent.

(7.1b) S is said to be a singular, or alternatively, a molecular sentence if S contains no variables. A singular sentence which contains no statement connectives is also called atomic. Illustrations: The sentences '$R(a, b) \supset (P(a) \cdot \sim Q(a))$', '$\sim Q(a)$', '$(R(a, b)$', '$P(a)$' are all singular, or molecular; the last two are atomic.

(7.1c) S is said to be a generalized sentence if it consists of one or more quantifiers followed by an expression which contains no quantifiers. S is said to be of universal form if it is a generalized sentence and all the quantifiers occurring in it are universal. S is called purely generalized (purely universal) if S is a generalized sentence (is of universal form) and contains no individual constants. S is said to be essentially universal if it is of universal form and not equivalent to a singular sentence. S is called essentially generalized if it is not equivalent to a singular sentence.

Illustrations: '$(x)(P(x) \supset Q(x))$', '$(x)R(a, x)$', '$(x)(P(x) \vee P(a))$', '$(x)(P(x) \vee \sim P(x))$', '$(Ex)(P(x) \cdot \sim Q(x))$', '$(Ex)(y)(R(a,x) \cdot S(a,y))$' are all generalized sentences; the first four are of universal form, the first and fourth are purely universal; the first and second are essentially universal, the third being equivalent to the singular sentence '$P(a)$', and the fourth to '$P(a) \vee \sim P(a)$'. All sentences except the third and fourth are essentially generalized.

Concerning the semantical interpretation of L, we lay down the following two stipulations:

(7.2a)   The primitive predicates of L are all purely qualitative.

(7.2b)   The universe of discourse of L, i.e., the domain of objects covered by the quantifiers, consists of all physical objects, or of all spatio-temporal locations.

A linguistic framework of the kind here characterized is not sufficient for the formulation of scientific theories since it contains no functors and does not provide the means for dealing with real numbers. Besides, the question is open at present whether a constitution system can be constructed in which all of the concepts of empirical science are reduced, by chains of explicit definitions, to a basis of primitives of a purely qualitative character. Nevertheless, we consider it worthwhile to study the problems at hand for the simplified type of language just described because the analysis of law and explanation is far from trivial even for our model language L, and because that analysis sheds light on the logical character of the concepts under investigation also in their application to more complex contexts.

In accordance with the considerations developed in section 6, we now define:

(7.3a) S is a fundamental lawlike sentence in L if S is purely universal; S is a fundamental law in L if S is purely universal and true.

(7.3b) S is a derivative law in L if (1) S is essentially, but not purely, universal and (2) there exists a set of fundamental laws in L which has S as a consequence.

(7.3c) S is a law in L if it is a fundamental or a derivative law in L.

The fundamental laws as here defined obviously include, besides general statements of empirical character, all those statements of purely universal form which are true on purely logical grounds; i.e. those which are formally true in L, such as '$(x)(P(x)\mathbf{v} \sim P(x))$', and those whose truth derives exclusively from the interpretation given to its constituents, as is the case with '$(x)(P(x) \supset Q(x))$', if 'P' is interpreted as meaning the property of being a father, and 'Q' that of being male.—The derivative laws, on the other hand, include neither of these categories; indeed, no fundamental law is also a derivative one.

As the primitives of L are purely qualitative, all the statements of universal form in L also satisfy the requirement of non-limited scope, and thus it is readily seen that the concept of law as defined above satisfies all the conditions suggested in section 6.[27]

The explanation of a phenomenon may involve generalized sentences which are not of universal form. We shall use the term "theory" to refer to such sentences, and we define this term by the following chain of definitions:

(7.4a) S is a fundamental theory if S is purely generalized and true.

(7.4b) S is a derivative theory in L if (1) S is essentially, but not purely, generalized and (2) there exists a set of fundamental theories in L which has S as a consequence.

(7.4c) S is a theory in L if it is a fundamental or a derivative theory in L.

By virtue of the above definitions, every law is also a theory, and every theory is true.

With the help of the concepts thus defined, we will now reformulate more precisely our earlier characterization of scientific explanation with specific reference to our model language L. It will be convenient to state our criteria for a sound explanation in the form of a definition for the expression "the ordered couple of sentences, (T, C), constitutes an explanans for the sentence E." Our analysis will be restricted to the explanation of particular events, i.e., to the case where the explanandum, E, is a singular sentence.[28]

---

[27] As defined above, fundamental laws include universal conditional statements with vacuous antecedents, such as "All mermaids are brunettes". This point does not appear to lead to undesirable consequences in the definition of explanation to be proposed later.— For an illuminating analysis of universal conditionals with vacuous antecedents, see Chapter VIII in Reichenbach's [Logic].

[28] This is not a matter of free choice: The precise rational reconstruction of explanation as applied to general regularities presents peculiar problems for which we can offer no solution at present. The core of the difficulty can be indicated briefly by reference to an example: Kepler's laws, K, may be conjoined with Boyle's law, B, to a stronger law $K.B$; but derivation of K from the latter would not be considered as an explanation of the regularities stated in Kepler's laws; rather, it would be viewed as representing, in effect, a pointless "explanation" of Kepler's laws by themselves. The derivation of Kepler's laws from Newton's laws of motion and of gravitation, on the other hand, would be recognized as a genuine explanation in terms of more comprehensive regularities, or so-called higher-level laws. The problem therefore arises of setting up clear-cut criteria for the distinction of levels of explanation or for a comparison of generalized sentences as to their comprehensiveness. The establishment of adequate criteria for this purpose is as yet an open problem.

In analogy to the concept of lawlike sentence, which need not satisfy a requirement of truth, we will first introduce an auxiliary concept of potential explanans, which is not subject to a requirement of truth; the notion of explanans will then be defined with the help of this auxiliary concept.—The considerations presented in Part I suggest the following initial stipulations:

(7.5)   An ordered couple of sentences, (T, C), constitutes a potential explanans for a singular sentence E only if

    (1) T is essentially generalized and C is singular

    (2) E is derivable in L from T and C jointly, but not from C alone.

(7.6)   An ordered couple of sentences, (T, C), constitutes an explanans for a singular sentence E if and only if

    (1) (T, C) is a potential explanans for E

    (2) T is a theory and C is true.

(7.6) is an explicit definition of explanation in terms of the concept of potential explanation.[29] On the other hand, (7.5) is not suggested as a definition, but as a statement of necessary conditions of potential explanation. These conditions will presently be shown not be sufficient, and additional requirements will be discussed by which (7.5) has to be supplemented in order to provide a definition of potential explanation.

Before we turn to this point, some remarks are called for concerning the formulation of (7.5). The analysis presented in Part I suggests that an explanans for a singular sentence consists of a class of generalized sentences and a class of singular ones. In (7.5), the elements of each of these classes separately are assumed to be conjoined to one sentence. This provision will simplify our formulations, and in the case of generalized sentences, it serves an additional purpose: A class of essentially generalized sentences may be equivalent to a singular sentence; thus, the class $\{'P(a)\mathbf{v}(x)Q(x)', \, 'P(a)\mathbf{v} \sim (x)Q(x)'\}$ is equivalent with the sentence '$P(a)$'. Since scientific explanation makes essential use of generalized sentences, sets of laws of this kind have to be ruled out; this is achieved above by combining all the generalized sentences in the explanans into one conjunction, T, and stipulating that T has to be essential generalized. —Again, since scientific explanation makes essential use of generalized sentences, E must not be a consequence of C alone: The law of gravitation, combined with the singular sentence "Mary is blonde and blue-eyed" does not constitute an explanans for "Mary is blonde". The last stipulation in (7.5) introduces the requisite restriction and thus prohibits complete self-explanation of the explanandum, i.e., the derivation of E from some singular sentence which has E as a consequence.—The same restriction also dispenses with the need for a special requirement to the effect that T has to have factual content if (T, C) is to be a potential explanans for an empirical sentence E. For if E is factual, then, since E is a consequence of T and C jointly, but not of C alone, T must be factual, too.

---

[29] It is necessary to stipulate, in (7.6) (2), that T be a theory rather than merely that T be true, for as was shown in section 6, the generalized sentences occurring in an explanans have to constitute a theory, and not every essentially generalized sentence which is true is actually a theory, i.e., a consequence of a set of purely generalized true sentences.

Our stipulations in (7.5) do not preclude, however, what might be termed partial self-explanation of the explanandum. Consider the sentences $T_1 =$ '$(x)(P(x) \supset Q(x))$', $C_1 =$ '$R(a, b) \cdot P(a) \cdot U(b)$', $E_1 =$ '$Q(a) \cdot R(a, b)$'. They satisfy all the requirements laid down in (7.5), but it seems counterintuitive to say that (T_1, C_1) potentially explains E_1, because the occurrence of the component '$R(a, b)$' of C_1 in the sentence E_1 amounts to a partial explanation of the explanandum by itself. Is it not possible to rule out, by an additional stipulation, all those cases in which E shares part of its content with C, i.e. where C and E have a common consequence which is not formally true in L? This stipulation would be tantamount to the requirement that C and E have to be exhaustive alternatives in the sense that their alternation is formally true, for the content which any two sentences have in common is expressed by their alternation. The proposed restriction, however, would be very severe. For if E does not share even part of its content with C, then C is altogether unnecessary for the derivation of E from T and C, i.e., E can be inferred from T alone. Therefore, in every potential explanation in which the singular component of the explanans is not dispensable, the explanandum is partly explained by itself. Take, for example, the potential explanation of $E_2 =$ '$Q(a)$' by $T_2 =$ '$(x)(P(x) \supset Q(x))$' and $C_2 =$ '$P(a)$', which satisfies (7.5), and which surely is intuitively unobjectionable. Its three components may be equivalently expressed by the following sentences: $T_2' =$ '$(x)(\sim P(x) \mathbf{v} Q(x))$'; $C_2' =$ '$(P(a)\mathbf{v}Q(a)) \cdot (P(a)\mathbf{v} \sim Q(a))$'; $E_2' =$ '$(P(a)\mathbf{v}Q(a)) \cdot (\sim P(a)\mathbf{v}Q(a))$'. This reformulation shows that part of the content of the explanandum is contained in the content of the singular component of the explanans and is, in this sense, explained by itself.

Our analysis has reached a point here where the customary intuitive idea of explanation becomes too vague to provide further guidance for rational reconstruction. Indeed, the last illustration strongly suggests that there may be no sharp boundary line which separates the intuitively permissible from the counterintuitive types of partial self-explanation; for even the potential explanation just considered, which is acceptable in its original formulation, might be judged unacceptable on intuitive grounds when transformed into the equivalent version given above.

The point illustrated by the last example is stated more explicitly in the following theorem, which we formulate here without proof.

(7.7) *Theorem.* Let (T, C) be a potential explanans for the singular sentence E. Then there exist three singular sentences, E_1, E_2, and C_1 in L such that E is equivalent to the conjunction $E_1 \cdot E_2$, C is equivalent to the conjunction $C_1 \cdot E_1$, and E_2 can be derived in L from T alone.[30]

In more intuitive terms, this means that if we represent the deductive structure

---

[30] In the formulation of the above theorem and subsequently, statement connective symbols are used not only as signs *in* L, but also autonomously in speaking *about* compound expressions of L. Thus, when 'S' and 'T' are names or name variables for sentences in L, their conjunction and disjunction will be designated by '$S.T$' and 'SvT', respectively; the conditional which has S as antecedent and T as consequent will be designated by '$S \supset T$', and the denial of S by '$\sim S$'. (Incidentally, this convention has already been used, tacitly, at one place in note 28).

of the given potential explanation by the schema $\{T, C\} \rightarrow E$, then this schema can be restated in the form $\{T, C_1 \cdot E_1\} \rightarrow E_1 \cdot E_2$, where E_2 follows from T alone, so that C_1 is entirely unnecessary as a premise; hence, the deductive schema under consideration can be reduced to $\{T, E_1\} \rightarrow E_1 \cdot E_2$, which can be decomposed into the two deductive schemata $\{T\} \rightarrow E_2$ and $\{E_1\} \rightarrow E_1$. The former of these might be called a purely theoretical explanation of E_2 by T, the latter a complete self-explanation of E_1. Theorem (7.7) shows, in other words, that every explanation whose explanandum is a singular sentence can be decomposed into a purely theoretical explanation and a complete self-explanation; and any explanation of this kind in which the singular constituent of the explanans is not completely unnecessary involves a partial self-explanation of the explanandum.[31]

To prohibit partial self-explanation altogether would therefore mean limitation of explanation to purely theoretical explanation. This measure seems too severely restrictive. On the other hand, an attempt to delimit, by some special rule, the permissible degree of self-explanation does not appear to be warranted because, as we saw, customary usage provides no guidance for such a delimitation, and because no systematic advantage seems to be gained by drawing some arbitrary dividing line. For these reasons, we refrain from laying down stipulations prohibiting partial self-explanation.

The conditions laid down in (7.5) fail to preclude yet another unacceptable type of explanatory argument, which is closely related to complete self-explanation, and which will have to be ruled out by an additional stipulation. The point is, briefly, that if we were to accept (7.5) as a definition, rather than merely as a statement of necessary conditions, for potential explanation, then, as a consequence of (7.6), any given particular fact could be explained by means of any true lawlike sentence whatsoever. More explicitly, if E is a true singular sentence—say, "Mt. Everest is snowcapped",—and T is a law—say, "All metals are good conductors of heat",—then there exists always a true singular sentence C such that E is derivable from T and C, but not from C alone; in other words, such that (7.5) is satisfied. Indeed, let T_* be some arbitrarily chosen particular instance of T, such as "If the Eiffel Tower is metal, it is a good conductor of heat". Now since E is true, so is the conditional $T_* \supset E$, and if the latter is chosen as the sentence C, then T, C, E satisfy the conditions laid down in (7.5).

In order to isolate the distinctive characteristic of this specious type of explanation, let us examine an especially simple case of the objectionable kind.

[31] The characteristic here referred to as partial self-explanation has to be distinguished from what is sometimes called the circularity of scientific explanation. The latter phrase has been used to cover two entirely different ideas. (a) One of these is the contention that the explanatory principles adduced in accounting for a specific phenomenon are inferred from that phenomenon, so that the entire explanatory process is circular. This belief is false, since general laws cannot be inferred from singular sentences. (b) It has also been argued that in a sound explanation the content of the explanandum is contained in that of the explanans. That is correct since the explanandum is a logical consequence of the explanans; but this peculiarity does not make scientific explanation trivially circular since the general laws occurring in the explanans go far beyond the content of the specific explanandum. For a fuller discussion of the circularity objection, see Feigl, [Operationism], pp. 286 ff, where this issue is dealt with very clearly.

Let $T_1 = $ '$(x)P(x)$' and $E_1 = $ '$R(a, b)$'; then the sentence $C_1 = $ '$P(a) \supset R(a, b)$' is formed in accordance with the preceding instructions, and T_1, C_1, E_1 satisfy the conditions (7.5). Yet, as the preceding example illustrates, we would not say that (T_1, C_1) constitutes a potential explanans for E_1. The rationale for the verdict may be stated as follows: If the theory T_1 on which the explanation rests, is actually true, then the sentence C_1, which can also be put into the form '$\sim P(a) \mathbf{v} R(a, b)$', can be verified, or shown to be true, only by verifying '$R(a, b)$', i.e., E_1. In this broader sense, E_1 is here explained by itself. And indeed, the peculiarity just pointed out clearly deprives the proposed potential explanation for E_1 of the predictive import which, as was noted in Part I, is essential for scientific explanation: E_1 could not possibly be predicted on the basis of T_1 and C_1 since the truth of C_1 cannot be ascertained in any manner which does not include verification of E_1. (7.5) should therefore be supplemented by a stipulation to the effect that if (T, C) is to be a potential explanans for E, then the assumption that T is true must not imply that verification of C necessitates verification of E.[32]

How can this idea be stated more precisely? Study of an illustration will suggest a definition of verification for molecular sentences. The sentence $M = $ '$(\sim P(a) \cdot Q(a)) \mathbf{v} R(a, b)$' may be verified in two different ways, either by ascertaining the truth of the two sentences '$\sim P(a)$' and '$Q(a)$', which jointly have M as a consequence, or by establishing the truth of the sentence '$R(a, b)$', which, again, has M as a consequence. Let us say that S is a basic sentence in L if S is either an atomic sentence or the denial of an atomic sentence in L. Verification of a molecular sentence S may then be defined generally as establishment of the truth of some class of basic sentences which has S as a consequence. Hence, the intended additional stipulation may be restated: The assumption that T is true must not imply that every class of true basic sentences which has C as a consequence also has E as a consequence.

As brief reflection shows, this stipulation may be expressed in the following form, which avoids reference to truth: T must be compatible in L with at least one class of basic sentences which has C but not E as a consequence; or, equivalently: There must exist at least one class of basic sentences which has C, but neither $\sim T$ nor E as a consequence in L.

If this requirement is met, then surely E cannot be a consequence of C, for otherwise there could be no class of basic sentences which has C but not E as a consequence; hence, supplementation of (7.5) by the new condition renders the second stipulation in (7.5) (2) superfluous.—We now define potential explanation as follows:

(7.8) An ordered couple of sentences, (T, C), constitutes a potential explanans for a singular sentence E if and only if the following conditions are satisfied:

(1) T is essentially generalized and C is singular

---

[32] It is important to distinguish clearly between the following two cases: (a) If T is true then C cannot be true without E being true; and (b) If T is true, C cannot be verified without E being verified.—Condition (a) must be satisfied by any potential explanation; the much more restictive condition (b) must not be satisfied if (T,C) is to be a potential explanans for E.

(2) E is derivable in L from T and C jointly

(3) T is compatible with at least one class of basic sentences which has C but not E as a consequence.

The definition of the concept of explanans by means of that of potential explanans as formulated in (7.6) remains unchanged.

In terms of our concept of explanans, we can give the following interpretation to the frequently used phrase "this fact is explainable by means of that theory": (7.9)   A singular sentence E is explainable by a theory T if there exists a singular sentence C such that (T, C) constitutes an explanans for E.

The concept of causal explanation, which has been examined here, is capable of various generalizations. One of these consists in permitting T to include statistical laws. This requires, however, a previous strengthening of the means of expression available in L, or the use of a complex theoretical apparatus in the metalanguage.—On the other hand, and independently of the admission of statistical laws among the explanatory principles, we may replace the strictly deductive requirement that E has to be a consequence of T and C jointly by the more liberal inductive one that E has to have a high degree of confirmation relatively to the conjunction of T and C. Both of these extensions of the concept of explanation open important prospects and raise a variety of new problems. In the present essay, however, these issues will not be further pursued.

PART IV. THE SYSTEMATIC POWER OF A THEORY

§8. Explication of the concept of systematic power.

Scientific laws and theories have the function of establishing systematic connections among the data of our experience, so as to make possible the derivation of some of those data from others. According as, at the time of the derivation, the derived data are, or are not yet, known to have occurred, the derivation is referred to as explanation or as prediction. Now it seems sometimes possible to compare different theories, at least in an intuitive manner, in regard to their explanatory or predictive powers: Some theories seem powerful in the sense of permitting the derivation of many data from a small amount of initial information, others seem less powerful, demanding comparatively more initial data, or yielding fewer results. Is it possible to give a precise interpretation to comparisons of this kind by defining, in a completely general manner, a numerical measure for the explanatory or predictive power of a theory? In the present section, we shall develop such a definition and examine some of its implications; in the following section, the definition will be expanded and a general theory of the concept under consideration will be outlined.

Since explanation and prediction have the same logical structure, namely that of a deductive systematization, we shall use the neutral term "systematic power" to refer to the intended concept. As is suggested by the preceding intuitive characterization, the systematic power of a theory T will be reflected in the ratio of the amount of information derivable by means of T to the amount of initial information required for that derivation. This ratio will obviously depend on the particular set of data, or of information, to which T is applied, and we shall therefore relativize our concept accordingly. Our aim, then, is to construct a definition for $s(T, K)$, the systematic power of a theory T with respect to a finite class K of data, or the degree to which T deductively systematizes the information contained in K.

Our concepts will be constructed again with specific reference to the language L. Any singular sentence in L will be said to express a potential datum, and K will accordingly be construed as a finite class of singular sentences[33]. T will be construed in a much broader sense than in the preceding sections; it may be any sentence in L, no matter whether essentially generalized or not. This liberal convention is adopted in the interest of the generality and simplicity of the definitions and theorems now to be developed.

To obtain values between 0 and 1 inclusive, we might now try to identify $s(T, K)$ with the percentage of those sentences in K which are derivable from the remainder by means of T. Thus, if $K_1 = \{$ '$P(a)$', '$Q(a)$', '$\sim P(b)$', '$\sim Q(b)$', '$Q(c)$', '$\sim P(d)$'$\}$, and $T_1 = $ '$(x)(P(x) \supset Q(x))$', then exactly the second and third sentence in K_1 are derivable by means of T_1 from the remainder, in fact from the first and fourth sentence. We might therefore consider setting $s(T_1, K_1) = 2/6 = 1/3$. But then, for the class $K_2 = \{$ '$P(a) \cdot Q(a)$', '$\sim P(b) \cdot \sim Q(b)$', '$Q(c)$', '$\sim P(d)$'$\}$, the same T_1 would have the s-value 0, although K_2 contains exactly the same information as K_1; again, for yet another formulation of that information, namely, $K_3 = \{$ '$P(a) \cdot \sim Q(b)$', '$Q(a) \cdot \sim P(b)$', '$Q(c)$', '$\sim P(d)$'$\}$, T_1 would have the s-value 1/4, and so on. But what we seek is a measure of the degree to which a given theory deductively systematizes a given body of factual information, i.e., a certain content, irrespective of the particular structure and grouping of the sentences in which that content happens to be expressed. We shall therefore make use of a method which represents the contents of any singular sentence or class of singular sentences as composed of certain uniquely determined smallest bits of information. By applying our general idea to these bits, we shall obtain a measure for the systematic power of T in K which is independent of the way in which the content of K is formulated. The sentences expressing those smallest bits of information will be called minimal sentences, and an exact formulation of the proposed procedure will be made possible by an explicit definition of this auxiliary concept. To this point we now turn.

If, as will be assumed here, the vocabulary of L contains fixed finite numbers of individual constants and of predicate constants, then only a certain finite number, say n, of different atomic sentences can be formulated in L. By a minimal

---

[33] As this stipulation shows, the term "datum" is here understood as covering actual as well as potential data. The convention that any singular sentence expresses a potential datum is plausible especially if the primitive predicates of L refer to attributes whose presence absence in specific instances can be ascertained by direct observation. In this case, each singular sentence in L may be considered as expressing a potential datum, in the sense of describing a logically possible state of affairs whose existence might be ascertained by direct observation.—The assumption that the primitives of L express directly observable attributes is, however, not essential for the definition and the formal theory of systematic power set forth in sections 8 and 9.

sentence in L, we will understand a disjunction of any number k $(0 \leqq k \leqq n)$ of different atomic sentences and the denials of the n-k remaining ones. Clearly, n atomic sentences determine 2^n minimal sentences. Thus, if a language L_1 contains exactly one individual constant, 'a', and exactly two primitive predicates, 'P' and 'Q', both of degree 1, then L_1 contains two atomic sentences, '$P(a)$' and '$Q(a)$', and four minimal sentences, namely, '$P(a)\mathbf{v}Q(a)$', '$P(a)\mathbf{v} \sim Q(a)$', '$\sim P(a)\mathbf{v}Q(a)$', '$\sim P(a)\mathbf{v} \sim Q(a)$'. If another language, L_2, contains in addition to the vocabulary of L_1 a second individual constant, 'b', and a predicate 'R' of degree 2, then L_2 contains eight atomic sentences and 256 minimal sentences, such as '$P(a)\mathbf{v}$ $P(b)\mathbf{v} \sim Q(a)\mathbf{v}$ $Q(b)\mathbf{v}$ $R(a, a)\mathbf{v}$ $R(a, b)\mathbf{v} \sim R(b, a)\mathbf{v}$ $\sim R(b, b)$'.

The term "minimal sentence" is to indicate that the statements in question are the singular sentences of smallest non-zero content in L, which means that every singular sentence in L which follows from a minimal sentence is either equivalent with that minimal sentence or formally true in L. However, minimal sentences do have consequences other than themselves which are not formally true in L, but these are not of singular form; '$(Ex)(P(x)\mathbf{v}Q(x))$' is such a consequence of '$P(a)\mathbf{v}Q(a)$' in L_1 above.

Furthermore, no two minimal sentences have any consequence in common which is not formally true in L; in other words, the contents of any two minimal sentences are mutually exclusive.

By virtue of the principles of the sentential calculus, every singular sentence which is not formally true in L can be transformed into a conjunction of uniquely determined minimal sentences; this conjunction will be called the minimal normal form of the sentence. Thus, e.g., in the language L_1 referred to above, the sentences '$P(a)$' and '$Q(a)$' have the minimal normal forms '$P(a)\mathbf{v}Q(a)) \cdot (P(a)\mathbf{v} \sim Q(a))$', and '$(P(a)\mathbf{v}Q(a)) \cdot (\sim P(a)\mathbf{v}Q(a))$', respectively; in L_2, the same sentences have minimal normal forms consisting of 128 conjoined minimal sentences each.—If a sentence is formally true in L, its content is zero, and it cannot be represented by a conjunction of minimal sentences. It will be convenient, however, to say that the minimal normal form of a formally true sentence in L is the vacuous conjunction of minimal sentences, which does not contain a single term.

As a consequence of the principle just mentioned, any class of singular sentences which are not all formally true can be represented by a sentence in minimal normal form. The basic idea outlined above for the explication of the concept of systematic power can now be expressed by the following definition:

(8.1) Let T be any sentence in L, and K any finite class of singular sentences in L which are not all formally true. If K' is the class of minimal sentences which occur in the minimal normal form of K, consider all divisions of K' into two mutually exclusive subclasses, K_1' and K_2', such that every sentence in K_2' is derivable from K_1' by means of T. Each division of this kind determines a ratio $n(K_2')/n(K')$, i.e. the number of minimal sentences in K_2' divided by the total number of minimal sentences in K'. Among the values of these ratios, there must be a largest one; $s(T, K)$ is to equal that maximum ratio. (Note that if all the elements of K were formally true, $n(K')$ would be 0 and the above ratio would not be defined.)

Illustration: Let L_1 contain only one individual constant, 'a', and only two predicates, 'P' and 'Q', both of degree 1. In L_1, let $T =$ '$(x)(P(x) \supset Q(x))$', $K = \{$'$P(a)$', '$Q(a)$'$\}$. Then we have $K' = \{$'$P(a)\mathbf{v}Q(a)$', '$P(a)\mathbf{v} \sim Q(a)$', '$\sim P(a)\mathbf{v}Q(a)$'$\}$. From the subclass K_1' consisting of the first two elements of K'—which together are equivalent to '$P(a)$'—we can derive, by means of T, the sentence '$Q(a)$', and from it, by pure logic, the third element of K'; it constitutes the only element of K_2'. No "better" systematization is possible, hence $s(T, K) = 1/3$.

Our definition leaves open, and is independent of, the question whether for a given K' there might not exist different divisions each of which would yield the maximum value for $n(K_2')/n(K')$. Actually, this can never happen: there exists always exactly one optimal subdivision of a given K'. This fact is a corollary of a general theorem, to which we now turn. It will be noticed that in the last illustration, K_2' can be derived from T alone, without the use of K_1' as a premise; indeed, '$\sim P(a)\mathbf{v}Q(a)$' is but a substitution instance of the sentence '$(x)(\sim P(x)\mathbf{v}Q(x))$', which is equivalent to T. The theorem now to be formulated, which might appear surprising at first, shows that this observation applies analogously in all other cases.

(8.2) Theorem. Let T be any sentence, K' a class of minimal sentences, and K_2' a subclass of K' such that every sentence in K_2' is derivable by means of T from the class $K - K_2'$; then every sentence in K_2' is derivable from T alone.

The proof, in outline, is as follows: Since the contents of any two different minimal sentences are mutually exclusive, so must be the contents of K_1' and K_2', which have not a single minimal sentence in common. But since the sentences of K_2' follow from K_1' and T jointly, they must therefore follow from T alone.

We note the following consequences of our theorem:

(8.2a) Theorem. In any class K' of minimal sentences, the largest subclass which is derivable from the remainder by means of a sentence T is identical with the class of those elements in K' which are derivable from T alone.

(8.2b) Theorem. Let T be any sentence, K a class of singular sentences which are not all formally true, K' the equivalent class of minimal sentences, and K_t' the class of those among the latter which are derivable from T alone. Then the concept s defined in (8.1) satisfies the following equation:

$$s(T, K) = n(K_t')/n(K')$$

§9. Systematic power and logical probability of a theory. Generalization of the concept of systematic power.

The concept of systematic power is closely related to that of degree of confirmation, or logical probability, of a theory. A study of this relationship will shed new light on the proposed definition of s, will suggest certain ways of generalizing it, and will finally lead to a general theory of systematic power which is formally analogous to that of logical probability.

The concept of logical probability, or degree of confirmation, is the central concept of inductive logic. Recently, different explicit definitions for this concept have been proposed, for languages of a structure similar to that of our model language, by Carnap[34] and by Helmer, Hempel, and Oppenheim[35].

While the definition of s proposed in the preceding section rests on the concept of minimal sentence, the basic concept in the construction of a measure for logical probability is that of state description, or, as we shall also say, of maximal sentence. A maximal sentence is the dual[36] of a minimal sentence in L; it is a conjunction of k ($0 \leq k \leq n$) different atomic sentences and of the denials of the remaining n-k atomic sentences. In a language with n atomic sentences, there exist 2^n state descriptions. Thus, e.g., the language L_1 repeatedly mentioned in §8 contains the following four maximal sentences: '$P(a) \cdot Q(a)$', '$P(a) \cdot \sim Q(a)$', '$\sim P(a) \cdot Q(a)$', '$\sim P(a) \cdot \sim Q(a)$'.

The term "maximal sentence" is to indicate that the sentences in question are the singular sentences of maximum non-universal content in L, which means that every singular sentence in L which has a maximal sentence as a consequence is either equivalent with that maximal sentence or formally false in L.

As we saw, every singular sentence can be represented in a conjunctive, or minimal, normal form, i.e., as a conjunction of certain uniquely determined minimal sentences; similarly, every singular sentence can be expressed also in a disjunctive, or maximal, normal form, i.e. as a disjunction of certain uniquely determined maximal sentences. In the language L_1, for example, '$P(a)$' has the minimal normal form '$(P(a) \lor Q(a)) \cdot (P(a) \lor \sim Q(a))$' and the maximal normal form '$(P(a) \cdot Q(a)) \lor (P(a) \cdot \sim Q(a))$'; the sentence '$P(a) \supset Q(a)$' has the minimal normal form '$\sim P(a) \lor Q(a)$' and the maximal normal form '$(P(a) \cdot Q(a)) \lor (\sim P(a) \cdot Q(a)) \lor (\sim P(a) \cdot \sim Q(a))$'; the minimal normal form of a formally true sentence is the vacuous conjunction, while its maximal normal form is the disjunction of all four state descriptions in L_1. The minimal normal form of any formally false sentence is the conjunction of all four minimal sentences in L_1, while its maximal normal form is the vacuous disjunction, as we shall say.

The minimal normal form of a singular sentence is well suited as an indicator of its content, for it represents the sentence as a conjunction of standard components whose contents are minimal and mutually exclusive. The maximal normal form of a sentence is suited as an indicator of its range, that is, intuitively speaking, of the variety of its different possible realizations, or of the variety of

[34] Cf. especially [Inductive Logic], [Concepts], [Application].

[35] See Helmer and Oppenheim, [Probability]; Hempel and Oppenheim, [Degree].—Certain general aspects of the relationship between the confirmation of a theory and its predictive or systematic success are examined in Hempel, [Studies], Part II, sections 7 and 8. The definition of s developed in the present essay establishes a quantitative counterpart of what, in that paper, is characterized, in non-numerical terms, as the prediction criterion of confirmation.

[36] For a definition and discussion of this concept, cf. Church, [Logic], p. 172.

those possible states of the world which, if realized, would make the statement true. Indeed, each maximal sentence may be considered as describing, as completely as is possible in L, one possible state of the world; and the state descriptions constituting the maximal normal form of a given singular sentence simply list those among the possible states which would make the sentence true.

Just as the contents of any two different minimal sentences, so also the ranges of any two maximal sentences are mutually exclusive: No possible state of the world can make two different maximal sentences true because any two maximal sentences are obviously incompatible with each other.[37]

Range and content of a sentence vary inversely. The more a sentence asserts, the smaller the variety of its possible realizations, and conversely. This relationship is reflected in the fact that the larger the number of constituents in the minimal normal form of a singular sentence, the smaller the number of constituents in its maximal normal form, and conversely. In fact, if the minimal normal form of a singular sentence U contains m_U of the $m = 2^n$ minimal sentences in L, then its maximal normal form contains $l_U = m - m_U$ of the m maximal sentences in L. This is illustrated by our last four examples, where $m = 4$, and $m_U = 2, 1, 0, 4$ respectively.

The preceding observations suggest that the content of any singular sentence U might be measured by the corresponding number m_U or by some magnitude proportional to it. Now it will prove convenient to restrict the values of the content measure function to the interval from 0 to 1, inclusive; and therefore, we define a measure, $g_1(U)$, for the content of any singular sentence in L by the formula

$$(9.1) \qquad\qquad g_1(U) = m_U/m$$

To any finite class K of singular sentences, we assign, as a measure $g_1(K)$ of its content, the value $g_1(S)$, where S is the conjunction of the elements of K.

By virtue of this definition, the equation in theorem (8.2b) may be rewritten:

$$s(T, K) = g_1(K_t')/g_1(K')$$

Here, K_t' is the class of all those minimal sentences in K' which are consequences of T. In the special case where T is a singular sentence, K_t' is therefore equivalent with $T \lor S$, where S is the conjunction of all the elements of K'. Hence, the preceding equation may then be transformed into

$$(9.2) \qquad\qquad s(T, S) = g_1(T \lor S)/g_1(S)$$

This formula holds when T and S are singular sentences, and S is not formally true. It bears a striking resemblance to the general schema for the definition of the logical probability of T in regard to S:

$$(9.3) \qquad\qquad p(T, S) = r(T \cdot S)/r(S)$$

[37] A more detailed discussion of the concept of range may be found in Carnap, [Inductive Logic], section 2, and in Carnap, [Semantics], sections 18 and 19, where the relation of range and content is examined at length.

Here, $r(U)$ is, for any sentence U in L, a measure of the range of U, T is any sentence in L, and S any sentence in L with $r(S) \neq 0$.

The several specific definitions which have been proposed for the concept of logical probability accord essentially with the pattern exhibited by (9.3)[38], but they differ in their choice of a specific measure function for ranges, i.e. in their definition of r. One idea which comes to mind is to assign, to any singular sentence U whose maximal normal form contains l_U maximal sentences, the range measure

$$(9.4) \qquad\qquad r_1(U) = l_U/m$$

which obviously is defined in strict analogy to the content measure g_1 for singular sentences as introduced in (9.1). For every singular sentence U, the two measures add up to unity:

$$(9.5) \qquad\qquad r_1(U) + g_1(U) = (l_U + m_U)/m = 1$$

As Carnap has shown, however, the range measure r_1 confers upon the corresponding concept of logical probability, i.e., upon the concept p_1 defined by means of it according to the schema (9.3), certain characteristics which are incompatible with the intended meaning of logical probability[39]; and Carnap as well as Helmer jointly with the present authors have suggested certain alternative measure functions for ranges, which lead to more satisfactory concepts of probability or of degree of confirmation. While we need not enter into details here, the following general remarks seem indicated to prepare the subsequent discussion.

The function r_1 measures the range of a singular sentence essentially by counting the number of maximal sentences in its maximal normal form; it thus gives equal weight to all maximal sentences (definition (9.1) deals analogously with minimal sentences). The alternative definitions just referred to are based on a different procedure. Carnap, in particular, lays down a rule which assigns a specific weight, i.e. a specific value of r, to each maximal sentence, but these weights are not the same for all maximal sentences. He then defines the range measure of any other singular sentence as the sum of the measures of its constituent maximal sentences. In terms of the function thus obtained—let us call it r_2—Carnap defines the corresponding concept of logical probability, which we shall call p_2, for singular sentences T, S in accordance with the schema (9.3): $p_2(T, S) = r_2(T. S)/r_2(S)$. The definitions of r_2 and p_2 are then extended, by means of certain limiting processes, to the cases where T and S are no longer both singular:[40]

[38] In Carnap's theory of logical probability, $p(T, S)$ is defined, for certain cases, as the limit which the function $r(T. S)/r(S)$ assumes under specified conditions (cf. Carnap, [Inductive Logic], p. 75); but we shall refrain here from considering this generalization of that type of definition which is represented by (9.3).

[39] [Inductive Logic], pp. 80–81.

[40] The alternative approach suggested by Helmer and the present authors involves use of a range measure function r_I which depends in a specified manner on the empirical information I available; hence, the range measure of any sentence U is determined only if a sentence

Now it can readily be seen that just as the function r_1 defined in (9.5) is but but one among an infinity of possible range measures, so the analogous function g_1 defined in (9.1) is but one among an infinity of possible content measures; and just as each range measure may serve to define, according to the schema (9.3), a corresponding measure of logical probability, so each content measure function may serve to define, by means of the schema illustrated by (9.2), a corresponding measure of systematic power. The method which suggests itself here for obtaining alternative content measure functions is to choose some range measure r other than r_1 and then to *define* a corresponding content measure g in terms of it by means of the formula

$$(9.6) \qquad\qquad g(U) = 1 - r(U)$$

so that g and r satisfy the analogue to (9.5) by definition. The function g thus defined will lead in turn, via a definition analogous to (9.2), to a corresponding concept s. Let us now consider this procedure a little more closely.

We assume that a function r is given which satisfies the customary requirements for range measures, namely:

(9.7) 1. $r(U)$ is uniquely determined for *all* sentences U in L.

2. $0 \leqq r(U) \leqq 1$ for every sentence U in L.

3. $r(U) = 1$ if the sentence U is formally true in L and thus has universal range.

4. $r(U_1 \mathbf{v} U_2) = r(U_1) + r(U_2)$ for any two sentences U_1, U_2 whose ranges are mutually exclusive, i.e., whose conjunction is formally false.

In terms of the given range measure let the corresponding content measure g be defined by means of (9.6). Then g can readily be shown to satisfy the following conditions:

(9.8) 1. $g(U)$ is uniquely determined for *all* sentences U in L.

2. $0 \leqq g(U) \leqq 1$ for every sentence U in L.

3. $g(U) = 1$ if the sentence U is formally false in L and thus has universal content.

I, expressing the available empirical information, is given. In terms of this range measure function, the concept of degree of confirmation, dc, can be defined by means of a formula similar to (9.3). The value of $dc(T, S)$ is not defined, however, in certain cases where S is generalized, as has been pointed out by McKinsey (cf. [Review]); also, the concept dc does not satisfy all the theorems of elementary probability theory (cf. the discussion of this point in the first two articles mentioned in note (35)); therefore, the degree of confirmation of a theory relatively to a given evidence is not a probability in the strict sense of the word. On the other hand, the definition of dc here referred to has certain methodologically desirable features, and it might therefore be of interest to construct a related concept of systematic power by means of the range measure function r_I. In the present paper, however, this question will not be pursued.

4. $g(U_1 \cdot U_2) = g(U_1) + g(U_2)$ for any two sentences U_1, U_2 whose contents are mutually exclusive, i.e., whose disjunction is formally true.

In analogy to (9.2), we next define, by means of g, a corresponding function s:

(9.9) $$s(T, S) = g(T \vee S)/g(S)$$

This function is determined for every sentence T, and for every sentence S with $g(S) \neq 0$, whereas the definition of systematic power given in §8 was restricted to those cases where S is singular and not formally true. Finally, our range measure r determines a corresponding probability function by virtue of the definition

(9.10) $$p(T, S) = r(T \cdot S)/r(S)$$

This formula determines the function p for any sentence T, and for any sentence S with $r(S) \neq 0$.

In this manner, every range measure r which satisfies (9.7) determines uniquely a corresponding content measure g which satisfies (9.8), a corresponding function s, defined by (9.9), and a corresponding function p, defined by (9.10). As a consequence of (9.7) and (9.10), the function p can be shown to satisfy the elementary laws of probability theory, especially those listed in (9.12) below; and by virtue of these, it is finally possible to establish a very simple relationship which obtains, for any given range measure r, between the corresponding concepts $p(T, S)$ and $s(T, S)$. Indeed, we have

(9.11)
$$
\begin{aligned}
s(T, S) &= g(T \vee S)/g(S) \\
&= (1 - r(T \vee S))/(1 - r(S)) \\
&= r(\sim (T \vee S))/r(\sim S) \\
&= r(\sim T \cdot \sim S)/r(\sim S) \\
&= p(\sim T, \sim S)
\end{aligned}
$$

We now list, without proof, some theorems concerning p and s which follow from our assumptions and definitions; they hold in all cases where the values of p and s referred to exist, i.e., where the r-value of the second arguments of p, and the g-value of the second arguments of s, is not 0.

(9.12) (1) a. $\qquad\qquad 0 \leqq p(T, S) \leqq 1$
$\qquad$ b. $\qquad\qquad 0 \leqq s(T, S) \leqq 1$
$\qquad$ (2) a. $\qquad p(\sim T, S) = 1 - p(T, S)$
$\qquad$ b. $\qquad s(\sim T, S) = 1 - s(T, S)$
$\qquad$ (3) a. $p(T_1 \vee T_2, S) = p(T_1, S) + p(T_2, S) - p(T_1 \cdot T_2, S)$
$\qquad$ b. $s(T_1 \cdot T_2, S) = s(T_1, S) + s(T_2, S) - s(T_1 \vee T_2, S)$
$\qquad$ (4) a. $p(T_1 \cdot T_2, S) = p(T_1, S) \cdot p(T_2, T_1 \cdot S)$
$\qquad$ b. $s(T_1 \vee T_2, S) = s(T_1, S) \cdot s(T_2, T_1 \vee S)$

In the above grouping, these theorems exemplify the relationship of dual correspondence which obtains between p and s. A general characterization of this correspondence is given in the following theorem, which can be proved on the basis of (9.11), and which is stated here in a slightly informal manner in order to avoid the tedium of lengthy formulations.

(*9.13*) *Dualism theorem.* From any demonstrable general formula expressing an equality or an inequality concerning p, a demonstrable formula concerning s is obtained if 'p' is replaced, throughout, by 's', and '$\cdot$' and 'v' are exchanged for each other. The same exchange, and replacement of 's' by 'p', conversely transforms any theorem expressing an equality or an inequality concerning s into a theorem about p.

We began our analysis of the systematic power of a theory in regard to a class of data by interpreting this concept, in §8, as a measure of the optimum ratio of those among the given data which are derivable from the remainder by means of the theory. Systematic elaboration of this idea has led to the definition, in the present section, of a more general concept of systematic power, which proved to be the dual counterpart of the concept of logical probability. This extension of our original interpretation yields a simpler and more comprehensive theory than would have been attainable on the basis of our initial definition.

But the theory of systematic power, in its narrower as well as in its generalized version, is, just like the theory of logical probability, purely formal in character, and a significant application of either theory in epistemology or the methodology of science requires the solution of certain fundamental problems which concern the logical structure of the language of science and the interpretation of its concepts. One urgent desideratum here is the further elucidation of the requirement of purely qualitative primitives in the language of science; another crucial problem is that of choosing, among an infinity of formal possibilities, an adequate range measure r. The complexity and difficulty of the issues which arise in these contexts has been brought to light by recent investigations[41]; it can only be hoped that recent developments in formal theory will soon be followed by progress in solving those open problems and thus clarifying the conditions for a sound application of the theories of logical probability and of systematic power.

Queens College, Flushing, N. Y.
Princeton, N. J.

BIBLIOGRAPHY
Throughout the article, the abbreviated titles in brackets are used for reference

Beard, Charles A., and Hook, Sidney. [Terminology] Problems of terminology in historical writing. Chapter IV of Theory and practice in historical study: A report of the Committee on Historiography. Social Science Research Council, New York, 1946.

Bergmann, Gustav. [Emergence] Holism, historicism, and emergence. *Philosophy of Science*, vol. 11 (1944), pp. 209–221.

Bonfante, G. [Semantics] Semantics, language. An article in P. L. Harriman, ed., The encyclopedia of psychology. Philosophical Library, New York, 1946.

Broad, C. D. [Mind] The mind and its place in nature. New York, 1925.

---

[41] Cf. especially Goodman, [Query], [Counterfactuals], [Infirmities], and Carnap, [Application]. See also notes (21) and (25).

Carnap, Rudolf. [Semantics] Introduction to semantics. Harvard University Press,
    1942.
――――. [Inductive Logic] On inductive logic. *Philosophy of science*, vol 12 (1945), pp.
    72–97.
――――. [Concepts] The two concepts of probability. *Philosophy and phenomenological
    research*, vol. 5 (1945), pp. 513–532.
――――. [Remarks] Remarks on induction and truth. *Philosophy and phenomenological
    research*, vol. 6 (1946), pp. 590–602.
――――. [Application] On the application of inductive logic. *Philosophy and phenomen-
    ological research*, vol. 8 (1947), pp. 133–147.
Chisholm, Roderick M. [Conditional] The contrary-to-fact conditional. *Mind*, vol. 55
    (1946), pp. 289–307.
Church, Alonzo. [Logic] Logic, formal. An article in Dagobert D. Runes, ed. The
    dictionary of philosophy. Philosophical Library, New York, 1942.
Ducasse, C. J. [Explanation] Explanation, mechanism, and teleology. *The journal of
    philosophy*, vol. 22 (1925), pp. 150–155.
Feigl, Herbert. [Operationism] Operationism and scientific method. *Psychological re-
    view*, vol. 52 (1945), pp. 250–259 and 284–288.
Goodman, Nelson. [Query] A query on confirmation. *The journal of philosophy*, vol.
    43 (1946), pp. 383–385.
――――. [Counterfactuals]. The problem of counterfactual conditionals. *The journal of
    philosophy*, vol. 44 (1947), pp. 113–128.
――――. [Infirmities] On infirmities of confirmation theory. *Philosophy and pheno-
    menological research*, vol. 8 (1947), pp. 149–151.
Grelling, Kurt and Oppenheim, Paul. [Gestaltbegriff] Der Gestaltbegriff im Lichte der
    neuen Logik. *Erkenntnis*, vol. 7 (1937–38), pp. 211–225 and 357–359.
Grelling, Kurt and Oppenheim, Paul. [Functional Whole] Logical Analysis of "Gestalt"
    as "Functional whole". Preprinted for distribution at Fifth Internat. Congress for
    the Unity of Science, Cambridge, Mass., 1939.
Helmer, Olaf and Oppenheim, Paul. [Probability] A syntactical definition of probability
    and of degree of confirmation. *The journal of symbolic logic*, vol. 10 (1945), pp. 25–60.
Hempel, Carl G. [Laws] The function of general laws in history. *The journal of phil-
    osophy*, vol. 39 (1942), pp. 35–48.
――――. [Studies] Studies in the logic of confirmation. *Mind*, vol. 54 (1945); Part I:
    pp. 1–26, Part II: pp. 97–121.
Hempel, Carl G. and Oppenheim, Paul. [Degree] A definition of "degree of confirmation".
    Philosophy of science, vol. 12 (1945), pp. 98–115.
Henle, Paul. [Emergence] The status of emergence. *The journal of philosophy*, vol. 39
    (1942), pp. 486–493.
Hospers, John. [Explanation] On explanation. *The journal of philosophy*, vol. 43 (1946),
    pp. 337–356.
Hull, Clark L. [Variables] The problem of intervening variables in molar behavior theory.
    Psychological review, vol. 50 (1943), pp. 273–291.
――――. [Principles] Principles of behavior. New York, 1943.
Jevons, W. Stanley. [Principles] The principles of science. London, 1924. (1st ed. 1874).
Kaufmann, Felix. [Methodology] Methodology of the social sciences. New York, 1944.
Knight, Frank H. [Limitations] The limitations of scientific method in economics. In
    Tugwell, R., ed., The trend of economics. New York, 1924.
Koch, Sigmund. [Motivation] The logical character of the motivation concept. *Psycho-
    logical review*, vol. 48 (1941). Part I: pp. 15–38, Part II: pp. 127–154.
Langford, C. H. [Review] Review in *The journal of symbolic logic*, vol. 6 (1941), pp. 67–68
Lewis, C. I. [Analysis] An analysis of knowledge and valuation. La Salle, Ill., 1946.
McKinsey, J. C. C. [Review] Review of Helmer and Oppenheim, [Probability]. *Mathe-
    matical reviews*, vol. 7 (1946), p. 45.

Mill, John Stuart. [Logic]   A system of Logic.

Morgan, C. Lloyd.   Emergent evolution, New York, 1923.

———.   The emergence of novelty.   New York, 1933.

Popper, Karl. [Forschung]   Logik der Forschung.   Wien, 1935.

———.   [Society]   The open society and its enemies.   London, 1945.

Reichenbach, Hans. [Logic]   Elements of symbolic logic.   New York, 1947.

———.   [Quantum mechanics]   Philosophic foundations of quantum mechanics.   University of California Press, 1944.

Rosenblueth, A., Wiener, N., and Bigelow, J. [Teleology]   Behavior, Purpose, and Teleology.   *Philosophy of science*, vol. 10 (1943), pp. 18–24.

Stace, W. T. [Novelty]   Novelty, indeterminism and emergence.   *Philosophical review*, vol. 48 (1939), pp. 296–310.

Tarski, Alfred. [Truth]   The semantical conception of truth, and the foundations of semantics.   *Philosophy and phenomenological research*, vol. 4 (1944), pp. 341–376.

Tolman, Edward Chase. [Behavior]   Purposive behavior in animals and men.   New York 1932.

White, Morton G. [Explanation]   Historical explanation. *Mind*, vol. 52 (1943), pp. 212–229.

Woodger, J. H. [Principles]   Biological principles.   New York, 1929.

Zilsel, Edgar. [Empiricism]   Problems of empiricism.   In *International encyclopedia of unified science*, vol. II, no. 8.   The University of Chicago Press, 1941.

———.   [Laws]   Physics and the problem of historico-sociological laws.   *Philosophy of science*, vol. 8 (1941), pp. 567–579.

AMERICAN PHILOSOPHICAL QUARTERLY
Volume 14, Number 2, April 1977

Paper presented at the 51st Annual Meeting of the
American Philosophical Association, Pacific Division,
Portland, Oregon, March 1977

VII. THE PRAGMATICS OF EXPLANATION

BAS C. VAN FRAASSEN

THERE are two problems about scientific explanation. The first is to describe it: when is something explained? The second is to show why (or in what sense) explanation is a virtue. Presumably we have no explanation unless we have a good theory; one which is independently worthy of acceptance. But what virtue is there in explanation over and above this? I believe that philosophical concern with the first problem has been led thoroughly astray by mistaken views on the second.

I. FALSE IDEALS

To begin I wish to dispute three ideas about explanation that seem to have a subliminal influence on the discussion. The first is that explanation is a relation simply between a theory or hypothesis and the phenomena or facts, just like truth for example. The second is that explanatory power cannot be logically separated from certain other virtues of a theory, notably truth or acceptability. And the third is that explanation is the overriding virtue, the end of scientific inquiry.

When is something explained? As a foil to the above three ideas, let me propose the simple answer: *when we have a theory which explains*. Note first that "have" is not "have on the books"; I cannot claim to have such a theory without implying that this theory is acceptable all told. Note also that both "have" and "explains" are tensed; and that I have allowed that we can have a theory which does not explain, or "have on the books" an unacceptable one that does. Newton's theory explained the tides but not the advance in the perihelion of mercury; we used to have an acceptable theory, provided by Newton, which bore (or bears timelessly?) the explanation relationship to some facts but not to all. My answer also implies

that we can intelligibly say that the theory explains, and not merely that people can explain by means of the theory. But this consequence is not very restrictive, because the former could be an ellipsis for the latter.

There are questions of usage here. I am happy to report that the history of science allows systematic use of both idioms. In Huygens and Young the typical phrasing seemed to be that phenomenon may be explained *by means of* principles, laws and hypotheses, or *according to* a view.[1] On the other hand, Fresnel writes to Arago in 1815 "Tous ces phénomènes ... sont réunis et expliqués par la même théorie des vibrations," and Lavoisier says that the oxygen hypothesis he proposes *explains* the phenomena of combustion.[2] Darwin also speaks in the latter idiom: "In scientific investigations it is permitted to invent any hypothesis, and if it explains various large and independent classes of facts it rises to the rank of a well-grounded theory"; though elsewhere he says that the facts of geographical distribution are *explicable on* the theory of migration.[3]

My answer did separate acceptance of the theory from its explanatory power. Of course, the second can be a reason for the first; but *that* requires their separation. Various philosophers have held that explanation logically requires true (or acceptable) theories as premises. Otherwise, they hold, we can at most mistakenly believe that we have an explanation.

This is also a question of usage, and again usage is quite clear. Lavoisier said of the phlogiston hypothesis that it is too vague and consequently "s'adapte a toutes les explications dans lesquelles on veut le faire entrer."[4] Darwin explicitly allows explanations by false theories when he says "It can hardly be supposed that a false theory would

---

[1] I owe these and following references to my student Mr. Paul Thagard. For instance see C. Huygens, *Treatise on Light*, tr. by S. P. Thompson (New York, 1962), pp. 19, 20, 22, 63; Thomas Young, *Miscellaneous Works*, ed. by George Peacock (London, 1855), Vol. I, pp. 168, 170.

[2] Augustin Fresnel, *Oeuvres Complètes* (Paris, 1866), Vol. I, p. 36 (see also pp. 254, 355); Antoine Lavoisier, *Oeuvres* (Paris, 1862), Vol. II, p. 233.

[3] Charles Darwin, *The Variation of Animals and Plants* (London, 1868), Vol. I, p. 9; *On the Origin of the Species* (Facs. of first edition, Cambridge, Mass., 1964), p. 408.

[4] Antoine Lavoisier, *op. cit.*, p. 640.

explain, in so satisfactory a manner as does the theory of natural selection, the several large classes of facts above specified."[5] More recently, Gilbert Harman has argued similarly: that a theory explains certain phenomena is part of the evidence that leads us to accept it. But that means that the explanation-relation is visible beforehand. Finally, we criticize theories selectively: a discussion of celestial mechanics around the turn of the century would surely contain the assertion that Newton's theory does explain many planetary phenomena, though not the advance in the perihelion of Mercury.

There is a third false ideal, which I consider worst: that explanation is the *summum bonum* and exact aim of science. A virtue could be overriding in one of two ways. The first is that it is a minimal criterion of acceptability. Such is consistency with the facts in the domain of application (though not necessarily with all data, if these are dubitable!). Explanation is not like that, or else a theory would not be acceptable at all unless it explained all facts in its domain. The second way in which a virtue may be overriding is that of being required when it can be had. This would mean that if two theories pass other tests (empirical adequacy, simplicity) equally well, then the one which explains more must be accepted. As I have argued elsewhere,[6] and as we shall see in connection with Salmon's views below, a precise formulation of this demand requires hidden variables for indeterministic theories. But of course, hidden variables are rejected in scientific practice as so much "metaphysical baggage" when they make no difference in empirical predictions.

II. A BIASED HISTORY

I will outline the attempts to characterize explanation of the past three decades, with no pretense of objectivity. On the contrary, the selection is meant to illustrate the diagnosis, and point to the solution, of the next section.

1. *Hempel*

In 1966, Hempel summarized his views by listing two main criteria for explanation. The first is the criterion of *explanatory relevance*: "the explanatory information adduced affords good grounds for

believing that the phenomenon to be explained did, or does, indeed occur."[7] That information has two components, one supplied by the scientific theory, the other consisting of auxiliary factual information. The relationship of providing good grounds is explicated as (a) implying (D–N case), or (b) conferring a high probability (I–S case), which is not lowered by the addition of other (available) evidence.

As Hempel points out, this criterion is not a sufficient condition for explanation: the red shift gives us good grounds for believing that distant galaxies are receding from us, but does not explain why they do. The classic case is the *barometer example*: the storm will come exactly if the barometers fall, which they do exactly if the atmospheric conditions are of the correct sort; yet only the last factor explains. Nor is the criterion a necessary condition; for this the classic case is the *paresis example*. We explain why the mayor, alone among the townsfolk, contracted paresis by his history of latent, contracted syphilis; yet such histories are followed by paresis in only a small percentage of cases.

The second criterion is the requirement of *testability*; but since all serious candidates for the role of scientific theory meet this, it cannot help to remove the noted defects.

2. *Beckner, Putnam, and Salmon*

The criterion of explanatory relevance was revised in one direction, informally by Beckner and Putnam and precisely by Salmon. Morton Beckner, in his discussion of evolution theory, pointed out that this often explains a phenomenon only by showing how it could have happened, given certain possible conditions.[8] Evolutionists do this by constructing models of processes which utilize only genetic and natural selection mechanisms, in which the outcome agrees with the actual phenomenon. Parallel conclusions were drawn by Hilary Putnam about the way in which celestial phenomena are explained by Newton's theory of gravity: celestial motions could indeed be as they are, given a certain possible (though not, known) distribution of masses in the universe.[9]

We may take the paresis example to be explained similarly. Mere consistency with the theory is of

[5] *Origin* (sixth ed., New York, 1962), p. 476.
[6] "Wilfrid Sellars on Scientific Realism," *Dialogue*, vol. 14 (1975), pp. 606–616.
[7] C. G. Hempel, *Philosophy of Natural Science* (Englewood Cliffs, New Jersey, 1966), p. 48.
[8] *The Biological Way of Thought* (Berkeley, 1968), p. 176; this was first published in 1959.
[9] In a paper of which a summary is found in Frederick Suppe (ed.), *The Structure of Scientific Theories* (Urbana, Ill., 1974).

course much too weak, since that is implied by logical irrelevance. Hence Wesley Salmon made this precise as follows: to explain is to exhibit (the) statistically relevant factors.[10] (I shall leave till later the qualifications about "screening off.") Since this sort of explication discards the talk about modelling and mechanisms of Beckner and Putnam, it may not capture enough. And indeed, I am not satisfied with Salmon's arguments that his criterion provides a sufficient condition. He gives the example of an equal mixture of Uranium 238 atoms and Polonium 214 atoms, which makes the Geiger counter click in interval $(t, t+m)$. This means that one of the atoms disintegrated. Why did it? The correct answer will be: because it was a Uranium 238 atom, if that is so—although the probability of its disintegration is much higher relative to the previous knowledge that the atom belonged to the described mixture.[11] The problem with this argument is that, on Salmon's criterion, we can explain not only why there was a disintegration, but also why *that* atom disintegrated *just then*. And surely that is exactly one of those facts which atomic physics leaves unexplained?

But there is a more serious general criticism. Whatever the phenomenon is, we can amass the statistically relevant factors, as long as the theory does not rule out the phenomenon altogether. "What more could one ask of an explanation?" Salmon inquires.[12] But in that case, as soon as we have an empirically adequate theory, we have an explanation of every fact in its domain. We may claim an explanation as soon as we have shown that the phenomenon can be embedded in some model allowed by the theory—that is, does not throw doubt on the theory's empirical adequacy.[13] But surely that is too sanguine?

3. *Global Properties*

Explanatory power cannot be identified with empirical adequacy; but it may still reside in the performance of the theory as a whole. This view is accompanied by the conviction that science does not explain individual facts but general regularities and was developed in different ways by Michael Friedman and James Greeno. Friedman says explicitly that in his view, "the kind of understanding provided by science is global rather than local" and consists in the simplification and unification imposed on our world picture.[14] That S_1 explains S_2 is a conjunction of two facts: S_1 implies S_2 relative to our background knowledge (and/or belief) K, *and* S_1 unifies and simplifies the set of its consequences relative to K. Friedman will no doubt wish to weaken the first condition in view of Salmon's work.

The precise explication Friedman gives of the second condition does not work, and is not likely to have a near variant that does.[15] But here we may look at Greeno's proposal.[16] His abstract and closing statement subscribe to the same general view as Friedman. But he takes as his model of a theory one which specifies a single probability space Q as the correct one, plus two partitions (or random variables) of which one is designated *explanandum* and the other *explanans*. An example: sociology cannot explain why Albert, who lives in San Francisco and whose father has a high income, steals a car. Nor is it meant to. But it does explain delinquency in terms of such other factors as residence and parental income. The degree of explanatory power is measured by an ingeniously devised quantity which measures the information I the theory provides of the explanandum variable M on the basis of explanans S. This measure takes its maximum value if all conditional probabilities $P(M_i/S_j)$ are zero or one (D–N case), and its minimum value zero if S and M are statistically independent.

Unfortunately, this way of measuring the unification imposed on our data abandons Friedman's insight that scientific understanding cannot be identified as a function of grounds for rational expectation. For if we let S and M describe the behavior of the barometer and coming storms, with $P(\text{barometer falls}) = P(\text{storm comes}) = 0.2$, $P(\text{storm comes/barometer falls}) = 1$, and $P(\text{storm comes/barometer does not fall}) = 0$, then the quantity I

---

[10] "Statistical Explanation," pp. 173–231 in R. G. Colodny (ed.), *The Nature and Function of Scientific Theories* (Pittsburgh, 1970); reprinted also in Salmon's book cited below.

[11] *Ibid.*, pp. 207–209. Nancy Cartwright has further, unpublished, counter-examples to the necessity and sufficiency of Salmon's criterion.

[12] *Ibid.*, p. 222.

[13] These concepts are discussed in my "To Save the Phenomena," *The Journal of Philosophy*, vol. 73 (1976), forthcoming.

[14] "Explanation and Scientific Understanding," *The Journal of Philosophy*, vol. 71 (1974), pp. 5–19.

[15] See Philip Kitcher, "Explanation, Conjunction, and Unification," *The Journal of Philosophy*, vol. 73 (1976), pp. 207–212.

[16] "Explanation and Information," pp. 89–103 in Wesley Salmon (ed.), *Statistical Explanation and Statistical Relevance* (Pittsburgh, 1971). This paper was originally published with a different title in *Philosophy of Science*, vol. 37 (1970), pp. 279–293.

takes its maximum value. Indeed, it does so whether we designate M or S as explanans.

It would seem that such asymmetries as exhibited by the red shift and barometer examples must necessarily remain recalcitrant for any attempt to strengthen Hempel's or Salmon's criteria by global restraints on theories alone.

4. The Major Difficulties

There are two main difficulties, illustrated by the old paresis and barometer examples, which none of the examined positions can handle. The first is that there are cases, clearly in a theory's domain, where the request for explanation is nevertheless rejected. We can explain why John, rather than his brothers contracted paresis, for he had syphilis; but not why he, among all those syphilitics, got paresis. Medical science is incomplete, and hopes to find the answer some day. But the example of the uranium atom disintegrating just then rather than later, is formally similar and we believe the theory to be complete. We also reject such questions as the Aristotelians asked the Galileans: why does a body free of impressed forces retain its velocity? The importance of this sort of case, and its pervasive character, has been repeatedly discussed by Adolf Grünbaum.

The second difficulty is the asymmetry revealed by the barometer: even if the theory implies that one condition obtains when and only when another does, it may be that it explains the one in terms of the other and not vice versa. An example which combines both the first and second difficulty is this: according to atomic physics, each chemical element has a characteristic atomic structure and a characteristic spectrum (of light emitted upon excitation). Yet the spectrum is explained by the atomic structure, and the question why a substance has that structure does not arise at all (except in the trivial sense that the questioner may need to have the terms explained to him).

5. Causality

Why are there no longer any Tasmanian natives? Well, they were a nuisance, so the white settlers just kept shooting them till there were none left. The request was not for population statistics, but for the story; though in some truncated way, the statistics "tell" the story.

In a later paper Salmon gives a primary place to causal mechanisms in explanation.[17] Events are

bound into causal chains by two relations: spatio-temporal continuity and statistical relevance. Explanation requires the exhibition of such chains. Salmon's point of departure is Reichenbach's *principle of the common cause*: every relation of statistical relevance ought to be explained by one of causal relevance. This means that a correlation of simultaneous values must be explained by a prior common cause. Salmon gives two statistical conditions that must be met by a common cause C of events A and B:

(a) $P(A \ \& \ B/C) = P(A/C)P(B/C)$

(b) $P(A/B \ \& \ C) = P(A/C)$ "C screens off B from A."

If $P(B/C) \neq 0$ these are equivalent, and symmetric in A and B.

Suppose that explanation is typically the demand for a common cause. Then we still have the problem: when does this arise? Atmospheric conditions explain the correlation between barometer and storm, say; but are still prior causes required to explain the correlation between atmospheric conditions and falling barometers?

In the quantum domain, Salmon says, causality is violated because "causal influence is not transmitted with spatio-temporal continuity." But the situation is worse. To assume Reichenbach's principle to be satisfiable, continuity aside, is to rule out all genuinely indeterministic theories. As example, let a theory say that C is invariably followed by one of the incompatible events A, B, or D, each with probability $1/3$. Let us suppose the theory complete, and its probabilities irreducible, with C the complete specification of state. Then we will find a correlation for which only C could be the common cause, but it is not. Assuming that A, B, D are always preceded by C and that they have low but equal prior probabilities, there is a statistical correlation between $\phi = (A$ or $D)$ and $\psi = (B$ or $D)$, for $P(\phi/\psi) = P(\psi/\phi) = 1/2 \neq P(\phi)$. But C, the only available candidate, does not screen off ϕ from ψ: $P(\phi/C \ \& \ \psi) = P(\phi/\psi) = 1/2 \neq P(\phi/C)$ which is $2/3$. Although this may sound complicated, the construction is so general that almost any irreducibly probabilistic situation will give a similar example. Thus Reichenbach's *principle of the common cause* is in fact a demand for hidden variables.

Yet we retain the feeling that Salmon has given an essential clue to the asymmetries of explanation. For surely the crucial point about the barometer is

[17] "Theoretical Explanation," pp. 118–145 in Stephan Körner (ed.), *Explanation* (Oxford, 1975).

THE PRAGMATICS OF EXPLANATION

that the atmospheric conditions screen off the barometer fall from the storm? The general point that the asymmetries are totally bound up with causality was argued in a provocative article by B. A. Brody.[18] Aristotle certainly discussed examples of asymmetries: the planets do not twinkle because they are near, yet they are near if and only if they do not twinkle (*Posterior Analytics*, I, 13). Not all explanations are causal, says Brody, but the others use a second Aristotelian notion, that of essence. The spectrum angle is a clear case: sodium has that spectrum because it has this atomic structure, which is its essence.

Brody's account has the further advantage that he can say when questions do not arise: other properties are explained in terms of essence, but the request for an explanation of the essence does not arise. However, I do not see how he would distinguish between the questions why the uranium atom disintegrated and why it disintegrated just then. In addition there is the problem that modern science is not formulated in terms of causes and essences, and it seems doubtful that these concepts can be redefined in terms which do occur there.

6. *Why-Questions*

A why-question is a request for explanation. Sylvain Bromberger called P the *presupposition* of the question *Why-P?* and restated the problem of explanation as that of giving the conditions under which proposition Q is a correct answer to a why-question with presupposition P.[19] However, Bengt Hannson has pointed out that "Why was it John who ate the apple?" and "Why was it the apple which John ate?: are different why-questions, although the comprised proposition is the same.[20] The difference can be indicated by such phrasing, or by emphasis ("Why did *John* . . .?") or by an auxiliary clause ("Why did John rather than . . .?"). Hannson says that an explanation is requested, not of a proposition or fact, but of an *aspect* of a proposition.

As is at least suggested by Hannson, we can cover all these cases by saying that we wish an explanation of why P is true in contrast to other members of a set X or propositions. This explains the tension in our reaction to the paresis-example. The question why the mayor, in contrast to other townfolk generally, contracted paresis *has* a true

correct answer: because of his latent syphilis. But the question why he did in contrast to the other syphilitics in his country club, has no true correct answer. Intuitively we may say: Q is a correct answer to *Why P in contrast to X?* only if Q gives reasons to expect that P, in contrast to the other members of X. Hannson's proposal for a precise criterion is: the probability of P given Q is higher than the average of the probabilities of R given Q, for members R of X.

Hannson points out that the set X of alternatives is often left tacit; the two questions about paresis might well be expressed by the same sentence in different contexts. The important point is that explanations are not requested of propositions, and consequently a distinction can be drawn between answered and rejected requests in a clear way. However, Hannson makes Q a correct answer to *Why P in contrast to X?* when Q is statistically irrelevant, when P is already more likely than the rest; or when Q implies P but not the others. I do not see how he can handle the barometer (or red shift, or spectrum) asymmetries. On his precise criterion, that the barometer fell is a correct answer to why it will storm as opposed to be calm. The difficulty is very deep: if P and R are necessarily equivalent, according to our accepted theories, how can *Why P in contrast to X?* be distinguished from *Why R in contrast to X?*

III. THE SOLUTION

1. *Prejudices*

Two convictions have prejudiced the discussion of explanation, one methodological and one substantive.

The first is that a philosophical account must aim to produce necessary and sufficient conditions for theory T explaining phenomenon E. A similar prejudice plagued the discussion of counter-factuals for twenty years, requiring the exact conditions under which, if A were the case, B would be. Stalnaker's liberating insight was that these conditions are largely determined by context and speaker's interest. This brings the central question to light: what *form* can these conditions take?

The second conviction is that explanatory power is a virtue of theories by themselves, or of their relation to the world, like simplicity, predictive

[18] "Towards an Aristotelian Theory of Scientific Explanation," *Philosophy of Science*, vol. 39 (1972), pp. 20–31.
[19] "Why-Questions," pp. 86–108 in R. G. Colodny (ed.), *Mind and Cosmos* (Pittsburgh, 1966).
[20] "Explanations—Of What?" (mimeographed: Stanford University, 1974).

strength, truth, empirical adequacy. There is again an analogy with counterfactuals: it used to be thought that science contains, or directly implies, counterfactuals. In all but limiting cases, however, the proposition expressed is highly context-dependent, and the implication is there at most relative to the determining contextual factors, such as speakers' interest.

2. Diagnosis

The earlier accounts lead us to the format: C explains E relative to theory T exactly if (a) T has certain global virtues, and (b) T implies a certain proposition $\phi(C, E)$ expressible in the language of logic and probability theory. Different accounts directed themselves to the specification of what should go into (a) and (b). We may add, following Beckner and Putnam, that T explains E exactly if there is a proposition C consistent with T (and presumably, background beliefs) such that C explains E relative to T.

The significant modifications were proposed by Hannson and Brody. The former pointed out that the explanadum E cannot be reified as a proposition: we request the explanation of something F in contrast to its alternatives X (the latter generally tacitly specified by context). This modification is absolutely necessary to handle some of our puzzles. It requires that in (b) above we replace "$\phi(C, E)$" by the formula form "$\psi(C, F, X)$." But the problem of asymmetries remains recalcitrant, because if T implies the necessary equivalence of F and F' (say, atomic structure and characteristic spectrum), then T will also imply $\psi(C, F', X)$ if and only if it implies $\psi(C, F, X)$.

The only account we have seen which grapples at all successfully with this, is Brody's. For Brody points out that even properties which we believe to be constantly conjoined in all possible circumstances, can be divided into essences and accidents, or related as cause and effect. In this sense, the asymmetries were no problem for Aristotle.

3. The logical problem

We have now seen exactly what logical problem is posed by the asymmetries. To put it in current terms: how can we distinguish propositions which are true in exactly the same possible worlds?

There are several known approaches that use impossible worlds. David Lewis, in his discussion of causality, suggests that we should look not only to the worlds theory T allows as possible, but also to those it rules out as impossible, and speaks of counterfactuals which are counterlegal. Relevant logic and entailment draw distinctions between logically equivalent sentences and their semantics devised by Routley and Meyer use both inconsistent and incomplete worlds. I believe such approaches to be totally inappropriate for the problem of explanation, for when we look at actual explanations of phenomena by theories, we do not see any detours through circumstances or events ruled out as impossible by the theory.

A further approach, developed by Rolf Schock, Romane Clark, and myself distinguishes sentences by the facts that make them true. The idea is simple. That it rains, that it does not rain, that it snows, and that it does not snow, are four distinct facts. The disjunction that it rains or does not rain is made true equally by the first and second, and not by the third or fourth, which distinguishes it from the logically equivalent disjunction that it snows or does not snow.[11] The distinction remains even if there is also a fact of its raining or not raining, distinct or identical with that of its snowing or not snowing.

This approach can work for the asymmetries of explanation. Such asymmetries are possible because, for example, the distinct facts that light is emitted with wavelengths λ, μ, ... conjointly make up the characteristic spectrum, while quite different facts conjoin to make up the atomic structure. So we have shown how such asymmetries *can* arise, in the way that Stalnaker showed how failures of transitivity in counterfactuals *can* arise. But while we have the distinct facts to classify asymmetrically, we still have the non-logical problem: whence comes the classification? The only suggestion so far is that it comes from Aristotle's concepts of cause and essence; but if so, modern science will not supply it.

4. The Aristotelian Sieve

I believe that we should return to Aristotle more thoroughly, and in two ways. To begin, I will state without argument how I understand Aristotle's theory of science. Scientific activity is divided into two parts, *demonstration* and *explanation*, the former treated mainly by the *Posterior Analytics* and the latter mainly by Book II of the *Physics*. Illustrations

---

[11] Cf. my "Facts and Tautological Entailments," *The Journal of Philosophy*, vol. 66 (1969), pp. 477-487 and in A. R. Anderson, *et al*, (ed.), *Entailment* (Princeton, 1975); and "Extension, Intension, and Comprehension" in Milton Munitz (ed.), *Logic and Ontology* (New York, 1973).

in the former are mainly examples of explanations in which the results of demonstration are *applied*; this is why the examples contain premises and conclusions which are not necessary and universal principles, although demonstration is only to and from such principles. Thus the division corresponds to our pure versus applied science. There is no reason to think that principles and demonstrations have such words as "cause" and "essence" in them, although looking at pure science from outside, Aristotle could say that its principles state causes and essences. In applications, the principles may be filtered through a conceptual sieve originating outside science.

The doctrine of the four "causes" (*aitiai*) allows for the systematic ambiguity or context-dependence of why-questions.[22] Aristotle's example (Physics II, 3; 195a) is of a lantern. In a modern example, the question why the porch light is on may be answered "because I flipped the switch" or "because we are expecting company," and the context determines which is appropriate. Probabilistic relations cannot distinguish these. Which factors are explanatory is decided not by features of the scientific theory but by concerns brought from outside. This is true even if we ask specifically for an "efficient cause," for how far back in the chain should we look, and which factors are merely auxiliary contributors?

Aristotle would not have agreed that essence is context-dependent. The essence is what the thing *is*, hence, its sum of classificatory properties. Realism has always asserted that ontological distinctions determine the "natural" classification. But which property is counted as explanatory and which as explained seems to me clearly context dependent. For consider Bromberger's flagpole example: the shadow is so long because the pole has this height, and not conversely. At first sight, no contextual factor could reverse this asymmetry, because the pole's height is a property it has in and by itself, and its shadow is a very accidental feature. The general principle linking the two is that its shadow is a function $f(x, t)$ of its height x and the time t (the latter determining the sun's elevation). But imagine the pole is the pointer on a giant sundial. Then the values of f have desired properties for each time t, and we appeal to these to explain why it is (had to be) such a tall pole. We may again draw a parallel to counterfactuals.

Professor Geach drew my attention to the following spurious argument: If John asked his father for money, then they would not have quarreled (because John is too proud to ask after a quarrel). Also if John asked and they hadn't quarreled, he would receive. By the usual logic of counterfactuals, it follows that if John asked his father for money, he would receive. But we know that he would not, because they have in fact quarreled. The fallacy is of equivocation, because "what was kept constant" changed in the middle of the monologue. (Or if you like, the aspects by which worlds are graded as more or less similar to this one.) Because science cannot dictate what speakers decide to "keep constant" it contains no counterfactuals. By exact parallel, *science contains no explanations.*

5. The Logic of Why-Questions

What remains of the problem of explanation is to study its logic, which is the logic of why-questions. This can be put to some extent, but not totally, in the general form developed by Harrah and Belnap and others.[23]

A question admits of three classes of response, *direct answers*, *corrections*, and *comments*. A *presupposition*, it has been held, is any proposition implied by all direct answers, or equivalently, denied by a correction. I believe we must add that the question "Why P, in contrast to X?" also presupposes that (a) P is a member of X, (b) P is true and the majority of X are not. This opens the door to the possibility that a question may not be uniquely determined by its set of direct answers. The question itself should decompose into factors which determine that set: the *topic* P, the *alternatives* X, and a *request specification* (of which the doctrine of the four "causes" is perhaps the first description).

We have seen that the propositions involved in question and answer must be individuated by something more than the set of possible worlds. I propose that we use the facts that make them true (see footnote 21). The context will determine an asymmetric relation among these facts, of *explanatory relevance*; it will also determine the theory or beliefs which determine which worlds are *possible*, and what is *probable* relative to what.

We must now determine what direct answers are and how they are evaluated. They must be made

[22] Cf. Julius Moravcik, "Aristotle on Adequate Explanations," *Synthese*, vol. 28 (1974), pp. 3–18.
[23] Cf. N. D. Belnap, Jr., "Questions: Their Presuppositions, and How They Can Fail to Arise," *The Logical Way of Doing Things*, ed. by Karel Lambert (New Haven, 1969), pp. 23–39.

true by facts (and only by facts forcing such) which are explanatorily relevant to those which make the topic true. Moreover, these facts must be statistically relevant, telling for the topic in contrast to the alternatives generally; this part I believe to be explicable by probabilities, combining Salmon's and Hannson's account. How strongly the answers count for the topic should be part of their evaluation as better or worse answers.

The main difference from such simple questions as "Which cat is on the mat?" lies in the relation of a why-question to its presuppositions. A why-question may fail to arise because it is ill-posed (P is false, or most of X is true), or because only question-begging answers tell probabilistically for P in contrast to X generally, or because none of the factors that do tell for P are explanatorily relevant in the question-context. Scientific theory enters mainly in the evaluation of possibilities and probabilities, which is only part of the process, and which it has in common with other applications such as prediction and control.

IV. SIMPLE PLEASURES

There are no explanations in science. How did philosophers come to mislocate explanation among semantic rather than pragmatic relations? This was certainly in part because the positivists tended to identify the pragmatic with subjective psychological features. They looked for measures by which to evaluate theories. Truth and empirical adequacy are such, but they are weak, being preserved when a theory is watered down. Some measure of "goodness of fit" was also needed, which did not reduce to a purely internal criterion such as simplicity, but concerned the theory's relation to the world. The studies of explanation have gone some way toward giving us such a measure, but it was a mistake to call this explanatory power. The fact that seemed to confirm this error was that we do not say that we *have* an explanation unless we have a theory which is acceptable, and victorious in its competition with alternatives, whereby we can explain. Theories are applied in explanation, but the peculiar and puzzling features of explanation are supplied by other factors involved. I shall now redescribe several familiar subjects from this point of view.

When a scientist campaigns on behalf of an advocated theory, he will point out how our situation will change if we accept it. Hitherto unsuspected factors become relevant, known relations are revealed to be strands of an intricate web, some terribly puzzling questions are laid to rest as not arising at all. We shall be in a much better position to explain. But equally, we shall be in a much better position to predict and control. The features of the theory that will make this possible are its empirical adequacy and logical strength, not special "explanatory power" and "control power." On the other hand, it is also a mistake to say explanatory power is nothing but those other features, for then we are defeated by asymmetries having no "objective" basis in science.

Why are *new* predictions so much more to the credit of a theory than agreement with the old? Because they tend to bring to light new phenomena which the older theories cannot explain. But of course, in doing so, they throw doubt on the empirical adequacy of the older theory: they show that a pre-condition for explanation is not met. As Boltzmann said of the radiometer, "the theories based on older hydrodynamic experience can never describe" these phenomena.[24] The failure in explanation is a by-product.

Scientific inference is inference to the best explanation. That does not rule at all for the supremacy of explanation among the virtues of theories. For we evaluate how good an explanation is given by how good a theory is used to give it, how close it fits to the empirical facts, how internally simple and coherent the explanation. There is a further evaluation in terms of a prior judgment of which kinds of factors are explanatorily relevant. If this further evaluation took precedence, overriding other considerations, explanation would be the peculiar virtue sought above all. But this is not so: instead, science schools our imagination so as to revise just those prior judgments of what satisfies and eliminates wonder.

Explanatory power is something we value and desire. But we are as ready, for the sake of scientific progress, to dismiss questions as not really arising at all. Explanation is indeed a virtue; but still, less a virtue than an anthropocentric pleasure.[25]

University of Toronto and University of Southern California   *Received September 1, 1976*

[24] Ludwig Boltzmann, *Lectures on Gas Theory*, tr. by S. G. Brush (Berkeley, 1964), p. 25.
[25] The author wishes to acknowledge helpful discussions and correspondence with Professors N. Cartwright, B. Hannson, K. Lambert, and W. Salmon, and the financial support of the Canada Council.

Philosophy of Science

December, 1981

EXPLANATORY UNIFICATION*

PHILIP KITCHER†

Department of Philosophy
University of Vermont

The official model of explanation proposed by the logical empiricists, the covering law model, is subject to familiar objections. The goal of the present paper is to explore an unofficial view of explanation which logical empiricists have sometimes suggested, the view of explanation as unification. I try to show that this view can be developed so as to provide insight into major episodes in the history of science, and that it can overcome some of the most serious difficulties besetting the covering law model.

1. The Decline and Fall of the Covering Law Model. One of the great apparent triumphs of logical empiricism was its official theory of explanation. In a series of lucid studies (Hempel 1965, Chapters 9, 10, 12; Hempel 1962; Hempel 1966), C. G. Hempel showed how to articulate precisely an idea which had received a hazy formulation from traditional empiricists such as Hume and Mill. The picture of explanation which Hempel presented, the *covering law model*, begins with the idea that explanation is derivation. When a scientist explains a phenomenon, he derives (deductively or inductively) a sentence describing that phenomenon (the *explanandum* sentence) from a set of sentences (the *explanans*) which must contain at least one general law.

*Received September 1980; Revised March 1981.

†A distant ancestor of this paper was read to the Dartmouth College Philosophy Colloquium in the Spring of 1977. I would like to thank those who participated, especially Merrie Bergmann and Jim Moor, for their helpful suggestions. I am also grateful to two anonymous referees for *Philosophy of Science* whose extremely constructive criticisms have led to substantial improvements. Finally, I want to acknowledge the amount I have learned from the writing and the teaching of Peter Hempel. The present essay is a token payment on an enormous debt.

Philosophy of Science, 48 (1981) pp. 507–531.

Today the model has fallen on hard times. Yet it was never the empiricists' whole story about explanation. Behind the official model stood an unofficial model, a view of explanation which was not treated precisely, but which sometimes emerged in discussions of theoretical explanation. In contrasting scientific explanation with the idea of reducing unfamiliar phenomena to familiar phenomena, Hempel suggests this unofficial view: "What scientific explanation, especially theoretical explanation, aims at is not [an] intuitive and highly subjective kind of understanding, but an objective kind of insight that is achieved by a systematic unification, by exhibiting the phenomena as manifestations of common, underlying structures and processes that conform to specific, testable, basic principles" (Hempel 1966, p. 83; see also Hempel 1965, pp. 345, 444). Herbert Feigl makes a similar point: "The aim of scientific explanation throughout the ages has been *unification*, i.e., the comprehending of a maximum of facts and regularities in terms of a minimum of theoretical concepts and assumptions" (Feigl 1970, p. 12).

This unofficial view, which regards explanation as unification, is, I think, more promising than the official view. My aim in this paper is to develop the view and to present its virtues. Since the picture of explanation which results is rather complex, my exposition will be programmatic, but I shall try to show that the unofficial view can avoid some prominent shortcomings of the covering law model.

Why should we want an account of scientific explanation? Two reasons present themselves. Firstly, we would like to understand and to evaluate the popular claim that the natural sciences do not merely pile up unrelated items of knowledge of more or less practical significance, but that they increase our understanding of the world. A theory of explanation should show us *how* scientific explanation advances our understanding. (Michael Friedman cogently presents this demand in his (1974)). Secondly, an account of explanation ought to enable us to comprehend and to arbitrate disputes in past and present science. Embryonic theories are often defended by appeal to their explanatory power. A theory of explanation should enable us to judge the adequacy of the defense.

The covering law model satisfies neither of these *desiderata*. Its difficulties stem from the fact that, when it is viewed as providing a set of necessary *and sufficient* conditions for explanation, it is far too liberal. Many derivations which are intuitively nonexplanatory meet the conditions of the model. Unable to make relatively gross distinctions, the model is quite powerless to adjudicate the more subtle considerations about explanatory adequacy which are the focus of scientific debate. Moreover, our ability to derive a description of a phenomenon from a set of premises *containing a law* seems quite tangential to our understanding of the phenomenon. Why should it be that exactly those derivations which employ laws advance our understanding?

The unofficial theory appears to do better. As Friedman points out, we can easily connect the notion of unification with that of understanding. (However, as I have argued in my (1976), Friedman's analysis of unification is faulty; the account of unification offered below is indirectly defended by my diagnosis of the problems for his approach.) Furthermore, as we shall see below, the acceptance of some major programs of scientific research—such as, the Newtonian program of eighteenth century physics and chemistry, and the Darwinian program of nineteenth century biology—depended on recognizing promises for unifying, and thereby explaining, the phenomena. Reasonable skepticism may protest at this point that the attractions of the unofficial view stem from its unclarity. Let us see.

2. Explanation: Some Pragmatic Issues. Our first task is to formulate the problem of scientific explanation clearly, filtering out a host of issues which need not concern us here. The most obvious way in which to categorize explanation is to view it as an activity. In this activity we answer the actual or anticipated questions of an actual or anticipated audience. We do so by presenting reasons. We draw on the beliefs we hold, frequently using or adapting arguments furnished to us by the sciences.

Recognizing the connection between explanations and arguments, proponents of the covering law model (and other writers on explanation) have identified explanations as special types of arguments. But although I shall follow the covering law model in employing the notion of argument to characterize that of explanation, I shall not adopt the ontological thesis that explanations are arguments. Following Peter Achinstein's thorough discussion of ontological issues concerning explanation in his (1977), I shall suppose that an explanation is an ordered pair consisting of a proposition and an act type.[1] The relevance of arguments to explanation resides in the fact that what makes an ordered pair $(p,$ explaining $q)$ an explanation is that a sentence expressing p bears an appropriate relation to a particular argument. (Achinstein shows how the central idea of the covering law model can be viewed in this way.) So I am supposing that there are acts of explanation which draw on arguments supplied by science, reformulating the traditional problem of explanation as the ques-

---

[1]Strictly speaking, this is one of two views which emerge from Achinstein's discussion and which he regards as equally satisfactory. As Achinstein goes on to point out, either of these ontological theses can be developed to capture the central idea of the covering law model.

tion: What features should a scientific argument have if it is to serve as the basis for an act of explanation?[2]

The complex relation between scientific explanation and scientific argument may be illuminated by a simple example. Imagine a mythical Galileo confronted by a mythical fusilier who wants to know why his gun attains maximum range when it is mounted on a flat plain, if the barrel is elevated at 45° to the horizontal. Galileo reformulates this question as the question of why an ideal projectile, projected with fixed velocity from a perfectly smooth horizontal plane and subject only to gravitational acceleration, attains maximum range when the angle of elevation of the projection is 45°. He defends this reformulation by arguing that the effects of air resistance in the case of the actual projectile, the cannonball, are insignificant, and that the curvature of the earth and the unevenness of the ground can be neglected. He then selects a kinematical argument which shows that, for fixed velocity, an ideal projectile attains maximum range when the angle of elevation is 45°. He adapts this argument by explaining to the fusilier some unfamiliar terms ('uniform acceleration', let us say), motivating some problematic principles (such as the law of composition of velocities), and by omitting some obvious computational steps. Both Galileo and the fusilier depart satisfied.

The most general problem of scientific explanation is to determine the conditions which must be met if science is to be used in answering an explanation-seeking question Q. I shall restrict my attention to explanation-seeking why-questions, and I shall attempt to determine the conditions under which an argument whose conclusion is S can be used to answer the question "Why is it the case that S?". More colloquially, my project will be that of deciding when an argument explains why its conclusion is true.[3]

---

[2]To pose the problem in this way we may still invite the charge that *arguments* should not be viewed as the bases for acts of explanation. Many of the criticisms levelled against the covering law model by Wesley Salmon in his seminal paper on statistical explanation (Salmon 1970) can be reformulated to support this charge. My discussion in section 7 will show how some of the difficulties raised by Salmon for the covering law model do not bedevil my account. However, I shall not respond directly to the points about statistical explanation and statistical inference advanced by Salmon and by Richard Jeffrey in his (1970). I believe that Peter Railton has shown how these specific difficulties concerning statistical explanation can be accommodated by an approach which takes explanations to be (or be based on) arguments (see Railton 1978), and that the account offered in section 4 of his paper can be adapted to complement my own.

[3]Of course, in restricting my attention to why-questions I am following the tradition of philosophical discussion of scientific explanation: as Bromberger notes in section IV of his (1966) not all explanations are directed at why-questions, but attempts to characterize explanatory responses to why-questions have a special interest for the philosophy of science because of the connection to a range of methodological issues. I believe that the account of explanation offered in the present paper could be extended to cover explanatory answers to some other kinds of questions (such as how-questions). But I do want to disavow the

We leave on one side a number of interesting, and difficult issues. So, for example, I shall not discuss the general relation between explanation-seeking questions and the arguments which can be used to answer them, nor the pragmatic conditions governing the idealization of questions and the adaptation of scientific arguments to the needs of the audience. (For illuminating discussions of some of these issues, see Bromberger 1962.) Given that so much is dismissed, does anything remain?

In a provocative article, (van Fraassen 1977) Bas van Fraassen denies, in effect, that there are any issues about scientific explanation other than the pragmatic questions I have just banished. After a survey of attempts to provide a theory of explanation he appears to conclude that the idea that explanatory power is a special virtue of theories is a myth. We accept scientific theories on the basis of their empirical adequacy and simplicity, and, having done so, we use the arguments with which they supply us to give explanations. This activity of applying scientific arguments in explanation accords with extra-scientific, "pragmatic", conditions. Moreover, our views about these extra-scientific factors are revised in the light of our acceptance of new theories: ". . . science schools our imag-ination so as to revise just those prior judgments of what satisfies and eliminates wonder" (van Fraassen 1977, p. 150). Thus there are no con-text-independent conditions, beyond those of simplicity and empirical adequacy which distinguish arguments for use in explanation.

van Fraassen's approach does not fit well with some examples from the history of science—such as the acceptance of Newtonian theory of matter and Darwin's theory of evolution—examples in which the ex-planatory promise of a theory was appreciated in advance of the articu-lation of a theory with predictive power. (See below pp. 512–14.) More-over, the account I shall offer provides an answer to skepticism that no "global constraints" (van Fraassen 1977, p. 146) on explanation can avoid the familiar problems of asymmetry and irrelevance, problems which bedevil the covering law model. I shall try to respond to van Fraas-sen's challenge by showing that there are certain context-independent fea-

---

claim that unification is relevant to all types of explanation. If one believes that expla-nations are sometimes offered in response to what-questions (for example), so that it is correct to talk of someone explaining what a gene is, then one should allow that some types of explanation can be characterized independently of the notions of unification or of argument. I ignore these kinds of explanation in part because they lack the methodo-logical significance of explanations directed at why-questions and in part because the prob-lem of characterizing explanatory answers to what-questions seems so much less recalci-trant than that of characterizing explanatory answers to why-questions (for a similar assessment, see Belnap and Steel 1976, pp. 86–7). Thus I would regard a full account of explanation as a heterogeneous affair, because the conditions required of adequate answers to different types of questions are rather different, and I intend the present essay to make a proposal about how *part* of this account (the most interesting part) should be developed.

tures of arguments which distinguish them for application in response to explanation-seeking why-questions, and that we can assess theories (including embryonic theories) by their ability to provide us with such arguments. Hence I think that it is possible to defend the thesis that historical appeals to the explanatory power of theories involve recognition of a virtue over and beyond considerations of simplicity and predictive power.

Resuming our main theme, we can use the example of Galileo and the fusilier to achieve a further refinement of our problem. Galileo selects and adapts an argument from his new kinematics—that is, he draws an argument from a set of arguments available for explanatory purposes, a set which I shall call the *explanatory store*. We may think of the sciences not as providing us with many unrelated individual arguments which can be used in individual acts of explanation, but as offering a reserve of explanatory arguments, which we may tap as need arises. Approaching the issue in this way, we shall be led to present our problem as that of specifying the conditions which must be met by the explanatory store.

The set of arguments which science supplies for adaptation in acts of explanation will change with our changing beliefs. Therefore the appropriate *analysandum* is the notion of the store of arguments relative to a set of accepted sentences. Suppose that, at the point in the history of inquiry which interests us, the set of accepted sentences is K. (I shall assume, for simplicity's sake, that K is consistent. Should our beliefs be inconsistent then it is more appropriate to regard K as some tidied version of our beliefs.) The general problem I have set is that of specifying $E(K)$, the *explanatory store over K*, which is the set of arguments acceptable as the basis for acts of explanation by those whose beliefs are exactly the members of K. (For the purposes of this paper I shall assume that, for each K there is exactly one $E(K)$.)

The unofficial view answers the problem: for each K, $E(K)$ is the set of arguments which best unifies K. My task is to articulate the answer. I begin by looking at two historical episodes in which the desire for unification played a crucial role. In both cases, we find three important features: (i) prior to the articulation of a theory with high predictive power, certain proposals for theory construction are favored on grounds of their explanatory promise; (ii) the explanatory power of embryonic theories is explicitly tied to the notion of unification; (iii) particular features of the theories are taken to support their claims to unification. Recognition of (i) and (ii) will illustrate points that have already been made, while (iii) will point towards an analysis of the concept of unification.

3. A Newtonian Program. Newton's achievements in dynamics, astronomy and optics inspired some of his successors to undertake an ambitious program which I shall call "dynamic corpuscularianism".[4] *Principia* had shown how to obtain the motions of bodies from a knowledge of the forces acting on them, and had also demonstrated the possibility of dealing with gravitational systems in a unified way. The next step would be to isolate a few basic force laws, akin to the law of universal gravitation, so that, applying the basic laws to specifications of the dispositions of the ultimate parts of bodies, all of the phenomena of nature could be derived. Chemical reactions, for example, might be understood in terms of the rearrangement of ultimate parts under the action of cohesive and repulsive forces. The phenomena of reflection, refraction and diffraction of light might be viewed as resulting from a special force of attraction between light corpuscles and ordinary matter. These speculations encouraged eighteenth century Newtonians to construct very general hypotheses about inter-atomic forces—even in the absence of any confirming evidence for the existence of such forces.

In the preface to *Principia*, Newton had already indicated that he took dynamic corpuscularianism to be a program deserving the attention of the scientific community:

> I wish we could derive the rest of the phenomena of Nature by the same kind of reasoning from mechanical principles, for I am induced by many reasons to suspect that they may all depend upon certain forces by which the particles of bodies, by some causes hitherto unknown, are either mutually impelled towards one another, and cohere in regular figures, or are repelled and recede from one another (Newton 1962, p. xviii. See also Newton 1952, pp. 401-2).

This, and other influential passages, inspired Newton's successors to try to complete the unification of science by finding further force laws analogous to the law of universal gravitation. Dynamic corpuscularianism remained popular so long as there was promise of significant unification. Its appeal began to fade only when repeated attempts to specify force laws were found to invoke so many different (apparently incompatible) attractive and repulsive forces that the goal of unification appeared unlikely. Yet that goal could still motivate renewed efforts to implement the program. In the second half of the eighteenth century Boscovich revived dynamic corpuscularian hopes by claiming that the whole of natural

---

[4]For illuminating accounts of Newton's influence on eighteenth century research see Cohen (1956) and Schofield (1969). I have simplified the discussion by considering only *one* of the programs which eighteenth century scientists derived from Newton's work. A more extended treatment would reveal the existence of several different approaches aimed at unifying science, and I believe that the theory of explanation proposed in this paper may help in the historical task of understanding the diverse aspirations of different Newtonians. (For the problems involved in this enterprise, see Heimann and McGuire 1971.)

philosophy can be reduced to "one law of forces existing in nature."[5]

The passage I have quoted from Newton suggests the nature of the unification that was being sought. *Principia* had exhibited how one style of argument, one "kind of reasoning from mechanical principles", could be used in the derivation of descriptions of many, diverse, phenomena. The unifying power of Newton's work consisted in its demonstration that one *pattern* of argument could be used again and again in the derivation of a wide range of accepted sentences. (I shall give a representation of the Newtonian pattern in Section 5.) In searching for force laws analogous to the law of universal gravitation, Newton's successors were trying to generalize the pattern of argument presented in *Principia*, so that one "kind of reasoning" would suffice to derive all phenomena of motion. If, furthermore, the facts studied by chemistry, optics, physiology and so forth, could be related to facts about particle motion, then one general pattern of argument would be used in the derivation of all phenomena. I suggest that this is the ideal of unification at which Newton's immediate successors aimed, which came to seem less likely to be attained as the eighteenth century wore on, and which Boscovich's work endeavored, with some success, to reinstate.

4. The Reception of Darwin's Evolutionary Theory. The picture of unification which emerges from the last section may be summarized quite simply: a theory unifies our beliefs when it provides one (or more generally, a few) pattern(s) of argument which can be used in the derivation of a large number of sentences which we accept. I shall try to develop this idea more precisely in later sections. But first I want to show how a different example suggests the same view of unification.

In several places, Darwin claims that his conclusion that species evolve through natural selection should be accepted because of its explanatory power, that ". . . the doctrine must sink or swim according as it groups and explains phenomena" (F. Darwin 1887; Vol. 2, p. 155, quoted in Hull 1974, p. 292). Yet, as he often laments, he is unable to provide any complete derivation of any biological phenomenon—our ignorance of the appropriate facts and regularities is "profound". How, then, can he contend that the primary virtue of the new theory is its explanatory power?

The answer lies in the fact that Darwin's evolutionary theory promises to unify a host of biological phenomena (C. Darwin 1964, pp. 243-4). The eventual unification would consist in derivations of descriptions of

---

[5]See Boscovich (1966) Part III, especially p. 134. For an introduction to Boscovich's work, see the essays by L. L. Whyte and Z. Markovic in Whyte (1961). For the influence of Boscovich on British science, see the essays of Pearce Williams and Schofield in the same volume, and Schofield (1969).

these phenomena which would instantiate a common pattern. When Darwin expounds his doctrine what he offers us is the pattern. Instead of detailed explanations of the presence of some particular trait in some particular species, Darwin presents two "imaginary examples" (C. Darwin 1964, pp. 90-96) and a diagram, which shows, in a general way, the evolution of species *represented by schematic letters* (1964, pp. 116-126). In doing so, he exhibits a pattern of argument, which, he maintains, can be instantiated, *in principle*, by a complete and rigorous derivation of descriptions of the characteristics of any current species. The derivation would employ the principle of natural selection—as well as premises describing ancestral forms and the nature of their environment and the (unknown) laws of variation and inheritance. In place of detailed evolutionary stories, Darwin offers *explanation-sketches*. By showing how a particular characteristic would be advantageous to a particular species, he indicates an explanation of the emergence of that characteristic in the species, suggesting the outline of an argument instantiating the general pattern.

From this perspective, much of Darwin's argumentation in the *Origin* (and in other works) becomes readily comprehensible. Darwin attempts to show how his pattern can be applied to a host of biological phenomena. He claims that, by using arguments which instantiate the pattern, we can account for analogous variations in kindred species, for the greater variability of specific (as opposed to generic) characteristics, for the facts about geographical distribution, and so forth. But he is also required to resist challenges that the pattern cannot be applied in some cases, that premises for arguments instantiating the pattern will not be forthcoming. So, for example, Darwin must show how evolutionary stories, fashioned after his pattern, can be told to account for the emergence of complex organs. In both aspects of his argument, whether he is responding to those who would limit the application of his pattern or whether he is campaigning for its use within a realm of biological phenomena, Darwin has the same goal. He aims to show that his theory should be accepted because it unifies and explains.

5. Argument Patterns. Our two historical examples[6] have led us to the conclusion that the notion of an argument pattern is central to that of explanatory unification. Quite different considerations could easily have pointed us in the same direction. If someone were to distinguish between the explanatory worth of two arguments instantiating a common pattern,

---

[6]The examples could easily be multiplied. I think it is possible to understand the structure and explanatory power of such theories as modern evolutionary theory, transmission genetics, plate tectonics, and sociobiology in the terms I develop here.

then we would regard that person as an explanatory deviant. To grasp the concept of explanation is to see that if one accepts an argument as explanatory, one is thereby committed to accepting as explanatory other arguments which instantiate the same pattern.

To say that members of a set of arguments instantiate a common pattern is to recognize that the arguments in the set are similar in some interesting way. With different interests, people may fasten on different similarities, and may arrive at different notions of argument pattern. Our enterprise is to characterize the concept of argument pattern which plays a role in the explanatory activity of scientists.

Formal logic, ancient and modern, is concerned in one obvious sense with patterns of argument. The logician proceeds by isolating a small set of expressions (the logical vocabulary), considers the schemata formed from sentences by replacing with dummy letters all expressions which do not belong to this set, and tries to specify which sequences of these schemata are valid patterns of argument. The pattern of argument which is taught to students of Newtonian dynamics is not a pattern of the kind which interests logicians. It has instantiations with different logical structures. (A rigorous derivation of the equations of motion of different dynamical systems would have a logical structure depending on the number of bodies involved and the mathematical details of the integration.) Moreover, an argument can only instantiate the Newtonian pattern if particular *non*logical terms, 'force', 'mass' and 'acceleration', occur in it in particular ways. However, the logician's approach can help us to isolate the notion of argument pattern which we require.

Let us say that a *schematic sentence* is an expression obtained by replacing some, but not necessarily all, the nonlogical expressions occurring in a sentence with dummy letters. A set of *filling instructions* for a schematic sentence is a set of directions for replacing the dummy letters of the schematic sentence, such that, for each dummy letter, there is a direction which tells us how it should be replaced. A *schematic argument* is a sequence of schematic sentences. A *classification* for a schematic argument is a set of sentences which describe the inferential characteristics of the schematic argument: its function is to tell us which terms in the sequence are to be regarded as premises, which are to be inferred from which, what rules of inference are to be used, and so forth.

We can use these ideas to define the concept of a *general argument pattern*. A general argument pattern is a triple consisting of a schematic argument, a set of sets of filling instructions containing one set of filling instructions for each term of the schematic argument, and a classification for the schematic argument. A sequence of sentences instantiates the general argument pattern just in case it meets the following conditions:

(i) The sequence has the same number of terms as the schematic argument of the general argument pattern.

(ii) Each sentence in the sequence is obtained from the corresponding schematic sentence in accordance with the appropriate set of filling instructions.

(iii) It is possible to construct a chain of reasoning which assigns to each sentence the status accorded to the corresponding schematic sentence by the classification.

We can make these definitions more intuitive by considering the way in which they apply to the Newtonian example. Restricting ourselves to the basic pattern used in treating systems which contain one body (such as the pendulum and the projectile) we may represent the schematic argument as follows:

(1) The force on α is β.
(2) The acceleration of α is γ.
(3) Force = mass·acceleration.
(4) (Mass of α)·(γ) = β
(5) $\delta = \theta$

The filling instructions tell us that all occurrences of 'α' are to be replaced by an expression referring to the body under investigation; occurrences of 'β' are to be replaced by an algebraic expression referring to a function of the variable coordinates and of time; 'γ' is to be replaced by an expression which gives the acceleration of the body as a function of its coordinates and their time-derivatives (thus, in the case of a one-dimensional motion along the x-axis of a Cartesian coordinate system, 'γ' would be replaced by the expression 'd^2x/dt^2'); 'δ' is to be replaced by an expression referring to the variable coordinates of the body, and 'θ' is to be replaced by an explicit function of time, (thus the sentences which instantiate (5) reveal the dependence of the variable coordinates on time, and so provide specifications of the positions of the body in question throughout the motion). The classification of the argument tells us that (1)-(3) have the status of premises, that (4) is obtained from them by substituting identicals, and that (5) follows from (4) using algebraic manipulation and the techniques of the calculus.

Although the argument patterns which interest logicians are general argument patterns in the sense just defined, our example exhibits clearly the features which distinguish the kinds of patterns which scientists are trained to use. Whereas logicians are concerned to display all the schematic premises which are employed and to specify exactly which rules of inference are used, our example allows for the use of premises (math-

ematical assumptions) which do not occur as terms of the schematic argument and it does not give a complete description of the way in which
the route from (4) to (5) is to go. Moreover, our pattern does not replace
all nonlogical expressions by dummy letters. Because some nonlogical
expressions remain, the pattern imposes special demands on arguments
which instantiate it. In a different way, restrictions are set by the instructions for replacing dummy letters. The patterns of logicians are very liberal in both these latter respects. The conditions for replacing dummy
letters in Aristotelian syllogisms, or first-order schemata, require only
that some letters be replaced with predicates, others with names.

Arguments may be similar either in terms of their logical structure or
in terms of the nonlogical vocabulary they employ at corresponding
places. I think that the notion of similarity (and the corresponding notion
of pattern) which is central to the explanatory activity of scientists results
from a compromise in demanding these two kinds of similarity. I propose
that scientists are interested in *stringent* patterns of argument, patterns
which contain some nonlogical expressions and which are fairly similar
in terms of logical structure. The Newtonian pattern cited above furnishes
a good example. Although arguments instantiating this pattern do not
have exactly the same logical structure, the classification imposes conditions which ensure that there will be similarities in logical structure
among such arguments. Moreover, the presence of the nonlogical terms
sets strict requirements on the instantiations and so ensures a different
type of kinship among them. Thus, without trying to provide an exact
analysis of the notion of stringency, we may suppose that the stringency
of a pattern is determined by two different constraints: (1) the conditions
on the substitution of expressions for dummy letters, jointly imposed by
the presence of nonlogical expressions in the pattern and by the filling
instructions; and, (2) the conditions on the logical structure, imposed by
the classification. If both conditions are relaxed completely then the notion of pattern degenerates so as to admit *any* argument. If both conditions
are simultaneously made as strict as possible, then we obtain another
degenerate case, a ''pattern'' which is its own unique instantiation. If
condition (2) is tightened at the total expense of (1), we produce the
logician's notion of pattern. The use of condition (1) requires that arguments instantiating a common pattern draw on a common nonlogical
vocabulary. We can glimpse here that ideal of unification through the use
of a few theoretical concepts which the remarks of Hempel and Feigl
suggest.

Ideally, we should develop a precise account of how these two kinds
of similarity are weighted against one another. The best strategy for obtaining such an account is to see how claims about stringency occur in
scientific discussions. But scientists do not make explicit assessments of

the stringency of argument patterns. Instead they evaluate the ability of a theory to explain and to unify. The way to a refined account of stringency lies through the notions of explanation and unification.

6. Explanation as Unification. As I have posed it, the problem of explanation is to specify which set of arguments we ought to accept for explanatory purposes given that we hold certain sentences to be true. Obviously this formulation can encourage confusion: we must not think of a scientific community as *first* deciding what sentences it will accept and *then* adopting the appropriate set of arguments. The Newtonian and Darwinian examples should convince us that the promise of explanatory power enters into the modification of our beliefs. So, in proposing that $E(K)$ is a function of K, I do not mean to suggest that the acceptance of K must be temporally prior to the adoption of $E(K)$.

$E(K)$ is to be that set of arguments which best unifies K. There are, of course, usually many ways of deriving some sentences in K from others. Let us call a set of arguments which derives some members of K from other members of K a *systematization* of K. We may then think of $E(K)$ as the best systematization of K.

Let us begin by making explicit an idealization which I have just made tacitly. A set of arguments will be said to be *acceptable relative to K* just in case every argument in the set consists of a sequence of steps which accord with elementary valid rules of inference (deductive or inductive) and if every premise of every argument in the set belongs to K. When we are considering ways of systematizing K we restrict our attention to those sets of arguments which are acceptable relative to K. This is an idealization because we sometimes use as the basis of acts of explanation arguments furnished by theories whose principles we no longer believe. I shall not investigate this practice nor the considerations which justify us in engaging in it. The most obvious way to extend my idealized picture to accommodate it is to regard the explanatory store over K, as I characterize it here, as being supplemented with an extra class of arguments meeting the following conditions: (a) from the perspective of K, the premises of these arguments are approximately true; (b) these arguments can be viewed as approximating the structure of (parts of) arguments in $E(K)$; (c) the arguments are simpler than the corresponding arguments in $E(K)$. Plainly, to spell out these conditions precisely would lead into issues which are tangential to my main goal in this paper.

The moral of the Newtonian and Darwinian examples is that unification is achieved by using similar arguments in the derivation of many accepted sentences. When we confront the set of possible systematizations of K we should therefore attend to the *patterns* of argument which are employed in each systematization. Let us introduce the notion of a *gener-*

ating set: if Σ is a set of arguments then a generating set for Σ is a set of argument patterns Π such that each argument in Σ is an instantiation of some pattern in Π. A generating set for Σ will be said to be *complete with respect to K* if and only if every argument which is acceptable relative to K and which instantiates a pattern in Π belongs to Σ. In determining the explanatory store $E(K)$ we first narrow our choice to those sets of arguments which are acceptable relative to K, the systematizations of K. Then we consider, for each such set of arguments, the various generating sets of argument patterns which are complete with respect to K. (The importance of the requirement of completeness is to debar explanatory deviants who use patterns selectively.) Among these latter sets we select that set with the greatest unifying power (according to criteria shortly to be indicated) and we call the selected set the *basis* of the set of arguments in question. The explanatory store over K is that systematization whose basis does best by the criteria of unifying power.

This complicated picture can be made clearer, perhaps, with the help of a diagram.

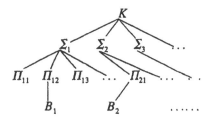

Systematizations, sets of arguments acceptable relative to K.

Complete generating sets. Π_{ij} is a generating set for Σ_i which is complete with respect to K.

Bases. B_i is the basis for Σ_i, and is selected as the best of the Π_{ij} on the basis of unifying power.

If B_k is the basis with the greatest unifying power then $E(K) = \Sigma_k$.

The task which confronts us is now formulated as that of specifying the factors which determine the unifying power of a set of argument patterns. Our Newtonian and Darwinian examples inspire an obvious suggestion: unifying power is achieved by generating a large number of accepted sentences as the conclusions of acceptable arguments which instantiate a few, stringent patterns. With this in mind, we define the *conclusion set* of a set of arguments Σ, $C(\Sigma)$, to be the set of sentences which occur as conclusions of some argument in Σ. So we might propose that the unifying power of a basis B_i with respect to K varies directly with the size of $C(\Sigma_i)$, varies directly with the stringency of the patterns which belong to B_i, and varies inversely with the number of members of B_i. This proposal is along the right lines, but it is, unfortunately, too simple.

The pattern of argument which derives a specification of the positions of bodies as explicit functions of time from a specification of the forces acting on those bodies is, indeed, central to Newtonian explanations. But not every argument used in Newtonian explanations instantiates this pattern. Some Newtonian derivations consist of an argument instantiating the pattern followed by further derivations from the conclusion. Thus, for example, when we explain why a pendulum has the period it does we may draw on an argument which *first* derives the equation of motion of the pendulum and *then* continues by deriving the period. Similarly, in explaining why projectiles projected with fixed velocity obtain maximum range when projected at 45° to the horizontal, we first show how the values of the horizontal and vertical coordinates can be found as functions of time and the angle of elevation, use our results to compute the horizontal distance travelled by the time the projectile returns to the horizontal, and then show how this distance is a maximum when the angle of elevation of projection is 45°. In both cases we take further steps beyond the computation of the explicit equations of motion—and the further steps in each case are different.

If we consider the entire range of arguments which Newtonian dynamics supplies for explanatory purposes, we find that these arguments instantiate a number of different patterns. Yet these patterns are not entirely distinct, for all of them proceed by using the computation of explicit equations of motion as a prelude to further derivation. It is natural to suggest that the pattern of computing equations of motion is the *core* pattern provided by Newtonian theory, and that the theory also shows how conclusions generated by arguments instantiating the core pattern can be used to derive further conclusions. In some Newtonian explanations, the core pattern is supplemented by a *problem-reducing pattern*, a pattern of argument which shows how to obtain a further type of conclusion from explicit equations of motion.

This suggests that our conditions on unifying power should be modified, so that, instead of merely counting the number of different patterns in a basis, we pay attention to similarities among them. All the patterns in the basis may contain a common core pattern, that is, each of them may contain some pattern as a subpattern. The unifying power of a basis is obviously increased if some (or all) of the patterns it contains share a common core pattern.

As I mentioned at the beginning of this paper, the account of explanation as unification is complicated. The explanatory store is determined on the basis of criteria which pull in different directions, and I shall make no attempt here to specify precisely the ways in which these criteria are to be balanced against one another. Instead, I shall show that some traditional problems of scientific explanation can be solved without more detailed specification of the conditions on unifying power. For the account I have indicated has two important corollaries.

(A) Let Σ, Σ' be sets of arguments which are acceptable relative to K and which meet the following conditions:

    (i) the basis of Σ' is as good as the basis of Σ in terms of the criteria of stringency of patterns, paucity of patterns, presence of core patterns, and so forth.

    (ii) $C(\Sigma)$ is a proper subset of $C(\Sigma')$.

    Then $\Sigma \neq E(K)$.

(B) Let Σ, Σ' be sets of arguments which are acceptable relative to K and which meet the following conditions:

    (i) $C(\Sigma) = C(\Sigma')$

    (ii) the basis of Σ' is a proper subset of the basis of Σ.

    Then $\Sigma \neq E(K)$

(A) and (B) tell us that sets of arguments which do equally well in terms of some of our conditions are to be ranked according to their relative ability to satisfy the rest. I shall try to show that (A) and (B) have interesting consequences.

7. Asymmetry, Irrelevance and Accidental Generalization.

Some familiar difficulties beset the covering law model. The *asymmetry problem* arises because some scientific laws have the logical form of equivalences. Such laws can be used "in either direction". Thus a law asserting that the satisfaction of a condition C_1 is equivalent to the satisfaction of a condition C_2 can be used in two different kinds of argument. From a premise asserting that an object meets C_1, we can use the law to infer that it meets C_2; conversely, from a premise asserting that an object meets C_2, we can use the law to infer that it meets C_1. The asymmetry problem is generated by noting that in many such cases one of these derivations can be used in giving explanations while the other cannot.

Consider a hoary example. (For further examples, see Bromberger 1966.) We can explain why a simple pendulum has the period it does by deriving a specification of the period from a specification of the length and the law which relates length and period. But we cannot explain the length of the pendulum by deriving a specification of the length from a specification of the period and the same law. What accounts for our different assessment of these two arguments? Why does it seem that one is explanatory while the other "gets things backwards"? The covering law model fails to distinguish the two, and thus fails to provide answers.

The *irrelevance problem* is equally vexing. The problem arises because we can sometimes find a lawlike connection between an accidental and irrelevant occurrence and an event or state which would have come about independently of that occurrence. Imagine that Milo the magician waves his hands over a sample of table salt, thereby "hexing" it. It is true (and

I shall suppose, lawlike) that all hexed samples of table salt dissolve when placed in water. Hence we can construct a derivation of the dissolving of Milo's hexed sample of salt by citing the circumstances of the hexing. Although this derivation fits the covering law model, it is, by our ordinary lights, nonexplanatory. (This example is given by Wesley Salmon in his (1970); Salmon attributes it to Henry Kyburg. For more examples, see Achinstein 1971.)

The covering law model explicitly debars a further type of derivation which any account of explanation ought to exclude. Arguments whose premises contain no laws, but which make essential use of accidental generalizations are intuitively nonexplanatory. Thus, if we derive the conclusion that Horace is bald from premises stating that Horace is a member of the Greenbury School Board and that all members of the Greenbury School Board are bald we do not thereby explain why Horace is bald. (See Hempel 1965, p. 339.) We shall have to show that our account does not admit as explanatory derivations of this kind.

I want to show that the account of explanation I have sketched contains sufficient resources to solve these problems.[7] In each case we shall pursue a common strategy. Faced with an argument we want to exclude from the explanatory store we endeavor to show that any set of arguments containing the unwanted argument could not provide the best unification of our beliefs. Specifically, we shall try to show either that any such set of arguments will be more limited than some other set with an equally satisfactory basis, or that the basis of the set must fare worse according to the criterion of using the smallest number of most stringent patterns. That is, we shall appeal to the corollaries (A) and (B) given above. In actual practice, this strategy for exclusion is less complicated than one might fear, and, as we shall see, its applications to the examples just discussed brings out what is intuitively wrong with the derivations we reject.

Consider first the irrelevance problem. Suppose that we were to accept as explanatory the argument which derives a description of the dissolving of the salt from a description of Milo's act of hexing. What will be our policy for explaining the dissolving of samples of salt which have not been hexed? If we offer the usual chemical arguments in these latter cases then we shall commit ourselves to an inflated basis for the set of arguments we accept as explanatory. For, unlike the person who explains *all* cases of dissolving of samples of salt by using the standard chemical

---

[7]More exactly, I shall try to show that my account can solve some of the principal versions of these difficulties which have been used to discredit the covering law model. I believe that it can also overcome more refined versions of the problems than I consider here, but to demonstrate that would require a more lengthy exposition.

pattern of argument, we shall be committed to the use of two different patterns of argument in covering such cases. Nor is the use of the extra pattern of argument offset by its applicability in explaining other phenomena. Our policy employs one extra pattern of argument without extending the range of things we can derive from our favored set of arguments. Conversely, if we eschew the standard chemical pattern of argument (just using the pattern which appeals to the hexing) we shall find ourselves unable to apply our favored pattern to cases in which the sample of salt dissolved has not been hexed. Moreover, the pattern we use will not fall under the more general patterns we employ to explain chemical phenomena such as solution, precipitation and so forth. Hence the unifying power of the basis for our preferred set of arguments will be less than that of the basis for the set of arguments we normally accept as explanatory.[8]

If we explain the dissolving of the sample of salt which Milo has hexed by appealing to the hexing then we are faced with the problems of explaining the dissolving of unhexed samples of salt. We have two options: (a) to adopt two patterns of argument corresponding to the two kinds of case; (b) to adopt one pattern of argument whose instantiations apply just to the cases of hexed salt. If we choose (a) then we shall be in conflict with [B], whereas choice of (b) will be ruled out by [A]. The general moral is that appeals to hexing fasten on a local and accidental feature of the cases of solution. By contrast our standard arguments instantiate a pattern which can be generally applied.[9]

[8] There is an objection to this line of reasoning. Can't we view the arguments $<(x)((Sx \& Hx) \rightarrow Dx), Sa \& Ha, Da>$, $<(x)((Sx \& \sim Hx) \rightarrow Dx), Sb \& \sim Hb, Db>$ as instantiating a common pattern? I reply that, insofar as we can view these arguments as instantiating a common pattern, the standard pair of comparable (low-level) derivations—$<(x)(Sx \rightarrow Dx), Sa, Da>$, $<(x)(Sx \rightarrow Dx), Sb, Db>$—share a more stringent common pattern. Hence, incorporating the deviant derivations in the explanatory store would give us an inferior basis. We can justify the claim that the pattern instantiated by the standard pair of derivations is more stringent than that shared by the deviant derivations, by noting that representation of the deviant pattern would compel us to broaden our conception of schematic sentence, and, even were we to do so, the deviant pattern would contain a "degree of freedom" which the standard pattern lacks. For a representation of the deviant "pattern" would take the form $<(x)((Sx \& \alpha Hx) \rightarrow Dx), Sa \& \alpha Ha, Da>$, where '$\alpha$' is to be replaced uniformly either with the null symbol or with '$\sim$'. Even if we waive my requirement that, in schematic sentences, we substitute for *non*logical vocabulary, it is evident that this "pattern" is more accommodating than the standard pattern.

[9] However, the strategy I have recommended will not avail with a different type of case. Suppose that a deviant wants to explain the dissolving of the salt by appealing to some property which holds universally. That is, the "explanatory" arguments are to begin from some premise such as "$(x)((x$ is a sample of salt $\& x$ does not violate conservation of energy$) \rightarrow x$ dissolves in water)" or "$(x)((x$ is a sample of salt $\& x = x) \rightarrow x$ dissolves in water)." I would handle these cases somewhat differently. If the deviant's explanatory store were to be as unified as our own, then it would contain arguments corresponding to ours in which a redundant conjunct systematically occurred, and I think it would be plausible to invoke a criterion of simplicity to advocate dropping that conjunct.

A similar strategy succeeds with the asymmetry problem. We have general ways of explaining why bodies have the dimensions they do. Our practice is to describe the circumstances leading to the formation of the object in question and then to show how it has since been modified. Let us call explanations of this kind "origin and development derivations". (In some cases, the details of the original formation of the object are more important; with other objects, features of its subsequent modification are crucial.) Suppose now that we admit as explanatory a derivation of the length of a simple pendulum from a specification of the period. Then we shall either have to explain the lengths of *non*swinging bodies by employing quite a different style of explanation (an origin and development derivation) or we shall have to forego explaining the lengths of such bodies. The situation is exactly parallel to that of the irrelevance problem. Admitting the argument which is intuitively nonexplanatory saddles us with a set of arguments which is less good at unifying our beliefs than the set we normally choose for explanatory purposes.

Our approach also solves a more refined version of the pendulum problem (given by Paul Teller in his (1974)). Many bodies which are not currently executing pendulum motion *could* be making small oscillations, and, were they to do so, the period of their motion would be functionally related to their dimensions. For such bodies we can specify the *dispositional period* as the period which the body would have if it were to execute small oscillations. Someone may now suggest that we can construct derivations of the dimensions of bodies from specifications of their dispositional periods, thereby generating an argument pattern which can be applied as generally as that instantiated in origin and development explanations. This suggestion is mistaken. There are some objects—such as the Earth and the Crab Nebula—which *could not* be pendulums, and for which the notion of a dispositional period makes no sense. Hence, the argument pattern proposed cannot entirely supplant our origin and development derivations, and, in consequence, acceptance of it would fail to achieve the best unification of our beliefs.

The problem posed by accidental generalizations can be handled in parallel fashion. We have a general pattern of argument, using principles of physiology, which we apply to explain cases of baldness. This pattern is generally applicable, whereas that which derives ascriptions of baldness using the principle that all members of the Greenbury School Board are bald is not. Hence, as in the other cases, sets which contain the unwanted derivation will be ruled out by one of the conditions (A), (B).

Of course, this does not show that an account of explanation along the lines I have suggested would sanction only derivations which satisfy the conditions imposed by the covering law model. For I have not argued that an explanatory derivation need contain *any* sentence of universal form. What *does* seem to follow from the account of explanation as unification is that explanatory arguments must not use accidental generalization, and, in this respect, the new account appears to underscore and generalize an important insight of the covering law model. Moreover, our success with the problems of asymmetry and irrelevance indicates that, even in the absence of a detailed account of the notion of stringency and of the way in which generality of the consequence set is weighed against paucity and stringency of the patterns in the basis, the view of explanation as unification has the resources to solve some traditional difficulties for theories of explanation.

8. Spurious Unification. Unfortunately there is a fly in the ointment. One of the most aggravating problems for the covering law model has been its failure to exclude certain types of self-explanation. (For a classic source of difficulties see Eberle, Kaplan and Montague 1961.) As it stands, the account of explanation as unification seems to be even more vulnerable on this score. The problem derives from a phenomenon which I shall call *spurious unification*.

Consider, first, a difficulty which Hempel and Oppenheim noted in a seminal article (Hempel 1965, Chapter 10). Suppose that we conjoin two laws. Then we can derive one of the laws from the conjunction, and the derivation conforms to the covering law model (unless, of course, the model is restricted to cover only the explanation of singular sentences; Hempel and Oppenheim do, in fact, make this restriction). To quote Hempel and Oppenheim:

> The core of the difficulty can be indicated briefly by reference to an example: Kepler's laws, K, may be conjoined with Boyle's law, B, to a stronger law $K \cdot B$; but derivation of K from the latter would not be considered as an explanation of the regularities stated in Kepler's laws; rather it would be viewed as representing, in effect, a pointless "explanation" of Kepler's laws by themselves. (Hempel 1965, p. 273 fn. 33.)

This problem is magnified for our account. For, why may we not unify our beliefs *completely* by deriving all of them using arguments which instantiate the one pattern?

$$\frac{\alpha \ \& \ B}{\alpha} \qquad [\text{`}\alpha\text{' is to be replaced by any sentence we accept.}]$$

Or, to make matters even more simple, why should we not unify our beliefs by using the most trivial pattern of self-derivation?

$$\alpha \qquad [\text{`}\alpha\text{' is to be replaced by any sentence we accept.}]$$

There is an obvious reply. The patterns just cited may succeed admirably in satisfying our criteria of using a few patterns of argument to generate many beliefs, but they fail dismally when judged by the criterion of stringency. Recall that the stringency of a pattern is assessed by adopting a compromise between two constraints: stringent patterns are not only to have instantiations with similar logical structures; their instantiations are also to contain similar nonlogical vocabulary at similar places. Now both of the above argument patterns are very lax in allowing any vocabulary whatever to appear in the place of 'α'. Hence we can argue that, according to our intuitive concept of stringency, they should be excluded as non-stringent.

Although this reply is promising, it does not entirely quash the objection. A defender of the unwanted argument patterns may *artificially* introduce restrictions on the pattern to make it more stringent. So, for example, if we suppose that one of *our* favorite patterns (such as the Newtonian pattern displayed above) is applied to generate conclusions meeting a particular condition C, the defender of the patterns just cited may propose that 'α' is to be replaced, not by any sentence, but by a sentence which meets C. He may then legitimately point out that his newly contrived pattern is as stringent as our favored pattern. Inspired by this partial success, he may adopt a general strategy. Wherever we use an argument pattern to generate a particular type of conclusion, he may use some argument pattern which involves self-derivation, placing an appropriate restriction on the sentences to be substituted for the dummy letters. In this way, he will mimic whatever unification we achieve. His "unification" is obviously spurious. How do we debar it?

The answer comes from recognizing the way in which the stringency of the unwanted patterns was produced. Any condition on the substitution of sentences for dummy letters would have done equally well, provided only that it imposed constraints comparable to those imposed by acceptable patterns. Thus the stringency of the restricted pattern seems accidental to it. This accidental quality is exposed when we notice that we can vary the filling instructions, while retaining the same syntactic structure, to obtain a host of other argument patterns with equally many instantiations. By contrast, the constraints imposed on the substitution of nonlogical vocabulary in the Newtonian pattern (for example) cannot be amended without destroying the stringency of the pattern or without depriving it of its ability to furnish us with many instantiations. Thus the constraints imposed in the Newtonian pattern are essential to its functioning; those imposed in the unwanted pattern are not.

Let us formulate this idea as an explicit requirement. If the filling instructions associated with a pattern P could be replaced by different filling instructions, allowing for the substitution of a class of expressions of the

71

same syntactic category, to yield a pattern P' and if P' would allow the derivation of *any* sentence, then the unification achieved by P is spurious. Consider, in this light, any of the patterns which we have been trying to debar. In each case, we can vary the filling instructions to produce an even more "successful" pattern. So, for example, given the pattern:

α ['α' is to be replaced by a sentence meeting condition C]

we can generalize the filling instructions to obtain

α ['α' is to be replaced by any sentence].

Thus, under our new requirement, the unification achieved by the original pattern is spurious.

In a moment I shall try to show how this requirement can be motivated, both by appealing to the intuition which underlies the view of explanation as unification and by recognizing the role that something like my requirement has played in the history of science. Before I do so, I want to examine a slightly different kind of example which initially appears to threaten my account. Imagine that a group of religious fanatics decides to argue for the explanatory power of some theological doctrines by claiming that these doctrines unify their beliefs about the world. They suggest that their beliefs can be systematized by using the following pattern:

God wants it to be the case that α.   ['α' is to be replaced by any
What God wants to be the case is   accepted sentence describing
the case.   the physical world]
_____
                      α

The new requirement will also identify as spurious the pattern just presented, and will thus block the claim that the theological doctrines that God exists and has the power to actualize his wishes have explanatory power. For it is easy to see that we can modify the filling instructions to obtain a pattern that will yield any sentence whatsoever.

Why should patterns whose filling instructions can be modified to accommodate any sentence be suspect? The answer is that, in such patterns, the nonlogical vocabulary which remains is idling. The presence of that nonlogical vocabulary imposes no constraints on the expressions we can substitute for the dummy symbols, so that, beyond the specification that a place be filled by expressions of a particular syntactic category, the structure we impose by means of filling instructions is quite incidental. Thus the patterns in question do not genuinely reflect the contents of our beliefs. The explanatory store should present the order of natural phenomena which is exposed by what we think we know. To do so, it must exhibit connections among our beliefs beyond those which could be found among any beliefs. Patterns of self-derivation and the type of pattern exemplified in the example of the theological community merely provide trivial, omnipresent connections, and, in consequence, the unification they offer is spurious.

My requirement obviously has some kinship with the requirement that the principles put forward in giving explanations be testable. As previous writers have insisted that genuine explanatory theories should not be able to cater to all possible evidence, I am demanding that genuinely unifying patterns should not be able to accommodate all conclusions. The requirement that I have proposed accords well with some of the issues which scientists have addressed in discussing the explanatory merits of particular theories. Thus several of Darwin's opponents complain that the explanatory benefits claimed for the embryonic theory of evolution are illusory, on the grounds that the style of reasoning suggested could be adapted to any conclusion. (For a particularly acute statement of the complaint, see the review by Fleeming Jenkin, printed in Hull 1974, especially p. 342.) Similarly, Lavoisier denied that the explanatory power of the phlogiston theory was genuine, accusing that theory of using a type of reasoning which could adapt itself to any conclusion (Lavoisier 1862, Volume II p. 233). Hence I suggest that some problems of spurious unification can be solved in the way I have indicated, and that the solution conforms both to our intuitions about explanatory unification and to the considerations which are used in scientific debate.

However, I do not wish to claim that my requirement will debar all types of spurious unification. It may be possible to find other unwanted patterns which circumvent my requirement. A full characterization of the notion of a stringent argument pattern should provide a criterion for excluding the unwanted patterns. My claim in this section is that it will do so by counting as spurious the unification achieved by patterns which adapt themselves to any conclusion and by patterns which accidentally restrict such universally hospitable patterns. I have also tried to show how this claim can be developed to block the most obvious cases of spurious unification.

9. Conclusions. I have sketched an account of explanation as unification, attempting to show that such an account has the resources to provide insight into episodes in the history of science and to overcome some traditional problems for the covering law model. In conclusion, let me indicate very briefly how my view of explanation as unification suggests how scientific explanation yields understanding. By using a few patterns of argument in the derivation of many beliefs we minimize the number of *types* of premises we must take as underived. That is, we reduce, in so far as possible, the number of types of facts we must accept as brute.

Hence we can endorse something close to Friedman's view of the merits of explanatory unification (Friedman 1974, pp. 18-19).

Quite evidently, I have only *sketched* an account of explanation. To provide precise analyses of the notions I have introduced, the basic approach to explanation offered here must be refined against concrete examples of scientific practice. What needs to be done is to look closely at the argument patterns favored by scientists and attempt to understand what characteristics they share. If I am right, the scientific search for explanation is governed by a maxim, once formulated succinctly by E. M. Forster. Only connect.

REFERENCES

Achinstein, P. (1971), *Law and Explanation*. Oxford: Oxford University Press.
Achinstein, P. (1977), "What is an Explanation?", *American Philosophical Quarterly* 14: pp. 1–15.
Belnap, N. and Steel, T. B. (1976), *The Logic of Questions and Answers*. New Haven: Yale University Press.
Boscovich, R. J. (1966), *A Theory of Natural Philosophy* (trans. J. M. Child). Cambridge: M.I.T. Press.
Bromberger, S. (1962), "An Approach to Explanation", in R. J. Butler (ed.), *Analytical Philosophy* (First Series). Oxford: Blackwell.
Bromberger, S. (1966), "Why-Questions", in R. Colodny (ed.), *Mind and Cosmos*. Pittsburgh: University of Pittsburgh Press.
Cohen, I. B. (1956), *Franklin and Newton*. Philadelphia: American Philosophical Society.
Darwin, C. (1964), *On the Origin of Species*, Facsimile of the First Edition, edited by E. Mayr. Cambridge: Harvard University Press.
Darwin, F. (1887), *The Life and Letters of Charles Darwin*. London: John Murray.
Eberle, R., Kaplan, D., and Montague, R. (1961), "Hempel and Oppenheim on Explanation", *Philosophy of Science* 28: pp. 418–28.
Feigl, H. (1970), "The 'Orthodox' View of Theories: Remarks in Defense as well as Critique", in M. Radner and S. Winokur (eds.), *Minnesota Studies in the Philosophy of Science*, Volume IV. Minneapolis: University of Minnesota Press.
Friedman, M. (1974), "Explanation and Scientific Understanding", *Journal of Philosophy LXXI*: pp. 5–19.
Heimann, P., and McGuire, J. E. (1971), "Newtonian Forces and Lockean Powers", *Historical Studies in the Physical Sciences 3*: pp. 233–306.
Hempel, C. G. (1965), *Aspects of Scientific Explanation*. New York: The Free Press.
Hempel, C. G. (1962), "Deductive-Nonlogical vs. Statistical Explanation", in H. Feigl and G. Maxwell (eds.) *Minnesota Studies in the Philosophy of Science*, Volume III. Minneapolis: University of Minnesota Press.
Hempel, C. G. (1966), *Philosophy of Natural Science*. Englewood Cliffs: Prentice-Hall.
Hull, D. (ed.) (1974), *Darwin and his Critics*. Cambridge: Harvard University Press.
Jeffrey, R. (1970), "Statistical Explanation vs. Statistical Inference", in N. Rescher (ed.), *Essays in Honor of Carl G. Hempel*. Dordrecht: D. Reidel.
Kitcher, P. S. (1976), "Explanation, Conjunction and Unification", *Journal of Philosophy*, LXXIII: pp. 207–12.
Lavoisier, A. (1862), *Oeuvres*. Paris.
Newton, I. (1962), *The Mathematical Principles of Natural Philosophy* (trans. A. Motte and F. Cajori). Berkeley: University of California Press.
Newton, I. (1952), *Opticks*. New York: Dover.
Railton, P. (1978), "A Deductive-Nomological Model of Probabilistic Explanation", *Philosophy of Science 45*: pp. 206–226.
Salmon, W. (1970), "Statistical Explanation", in R. Colodny (ed.), *The Nature and Func-*

tion of Scientific Theories. Pittsburgh: University of Pittsburgh Press.

Schofield, R. E. (1969), *Mechanism and Materialism*. Princeton: Princeton University Press.

Teller, P. (1974), "On Why-Questions", *Noûs* VIII: pp. 371–80.

van Fraassen, B. (1977), "The Pragmatics of Explanation", *American Philosophical Quarterly 14*: pp. 143–50.

Whyte, L. L. (ed.) (1961), *Roger Joseph Boscovich*. London: Allen and Unwin.

THE JOURNAL OF PHILOSOPHY

VOLUME LXXI, NO. 1, JANUARY 17, 1974

EXPLANATION AND SCIENTIFIC UNDERSTANDING *

W HY does water turn to steam when heated? Why do the planets obey Kepler's laws? Why is light refracted by a prism? These are typical of the questions science tries to answer. Consider, for example, the answer to the first question: Water is made of tiny molecules in a state of constant motion. Between these molecules are intermolecular forces, which, at normal temperatures, are sufficient to hold them together. If the water is heated, however, the energy, and consequently the motion, of the molecules increases. If the water is heated sufficiently the molecules acquire enough energy to overcome the intermolecular forces—they fly apart and escape into the atmosphere. Thus, the water gives off steam. This account answers our question. Our little story seems to give us understanding of the process by which water turns to steam. The phenomenon is now more intelligible or comprehensible. How does this work? What is it about our little story, and scientific explanations generally, that gives us understanding of the world—what is it for a phenomenon to be scientifically understandable?

Two aspects of our example are of special interest. First, what is explained is a general regularity or pattern of behavior—a law, if you like—i.e., that water turns to steam when heated. Although most of the philosophical literature on explanation deals with the explanation of particular events, the type of explanation illustrated by the above account seems much more typical of the physical sciences. Explanations of particular events are comparatively rare— found only perhaps in geology and astronomy. Second, our little story explains one phenomenon, the changing of water into steam,

* An earlier version of this paper was read at the University of Massachusetts at Amherst. I am indebted to members of the philosophy department there, especially Fred Feldman, for helpful comments. I would also like to thank Hartry Field, Clark Glymour, and David Hills for valuable criticism and conversation.

by relating it to another phenomenon, the behavior of the molecules of which water is composed. This relation is commonly described as *reduction:* the explained phenomenon is said to be reduced to the explaining phenomenon; e.g., the behavior of water is reduced to the behavior of molecules. Thus, the central problem for the theory of scientific explanation comes down to this: what is the relation between phenomena in virtue of which one phenomenon can constitute an explanation of another, and what is it about this relation that gives understanding of the explained phenomenon?

When I ask that a theory of scientific explanation tell us what it is about the explanation relation that produces understanding, I do not suppose that 'scientific understanding' is a clear notion. Nor do I suppose that it is possible to say what scientific understanding is in advance of giving a theory of explanation. It is not reasonable to require that a theory of explanation proceed by first defining 'scientific understanding' and then showing how its reconstruction of the explanation relation produces scientific understanding. We can find out what scientific understanding consists in only by finding out what scientific explanation is and vice versa. On the other hand, although we have no clear independent notion of scientific understanding, I think we do have some general ideas about what features such a notion should have, and we can use these general ideas to judge the adequacy of philosophical theories of explanation. At any rate, this is the method I will follow. I will argue that traditional accounts of scientific explanation result in notions of scientific understanding that have objectionable or counterintuitive features. From my discussion of traditional theories I will extract some general properties that a concept of scientific understanding ought to have. Finally, I will propose an account of scientific explanation that possesses these desirable properties.

It seems to me that the philosophical literature on explanation falls into two rough groups. Some philosophers, like Hempel and Nagel, have relatively precise proposals as to the nature of the explanation relation, but relatively little to say about the connection between their proposals and scientific understanding, i.e., about what it is about the relation they propose that gives us understanding of the world. Other philosophers, like Toulmin, Scriven, and Dray, have a lot to say about understanding, but relatively vague ideas about just what relation it is that produces this understanding. To illustrate this situation I will discuss three attempts at explicating the explanation relation that have been prominent in the literature.

1. The best known philosophical account of explanation, the D-N model, is designed primarily as a theory about the explanation of particular events, but the view that the explanation relation is basically a deductive relation applies equally to our present concern. According to the D-N model, a description of one phenomenon can explain a description of a second phenomenon only if the first description entails the second. Of course, a deductive relation between two such descriptions is not sufficient for one to be an explanation of the other, as expounders of the D-N model readily admit.[1]

The entailment requirement puts a constraint on the explanation relation, but it does not by itself tell us what it is about the explanation relation that gives us understanding of the explained phenomenon, that makes the world more intelligible. In some of their writings defenders of the D-N model give the impression that they consider such a task to lie outside the province of the philosopher of science, because concepts like 'understanding' and 'intelligibility' are psychological or pragmatic. For example, Hempel writes: "such expressions as 'realm of understanding' and 'comprehensible' do not belong to the vocabulary of logic, for they refer to psychological or pragmatic aspects of explanation" (413). He goes on to characterize the pragmatic aspects of explanation as those which vary from individual to individual. Explanation in its pragmatic aspects is "a relative notion, something can be sgnificantly said to constitute an explanation in this sense only for this or that individual" (426). The philosopher of science, according to Hempel, aims at explicating the nonpragmatic aspects of explanation, the sense of 'explanation' on which *A* explains *B simpliciter* and not *for you or for me.*

I completely agree with Hempel's contention that the philosopher of science should be interested in an objective notion of explanation, a notion that doesn't vary capriciously from individual to individual. However, the considerations Hempel advances can serve as an argument against the attempt to connect explanation and understanding only by an equivocation on 'pragmatic'. In the sense in which such concepts as 'understanding', and 'comprehensible' are clearly pragmatic, 'pragmatic' means roughly the same as 'psychological', i.e., having to do with the thoughts, beliefs, attitudes,

---

[1] Compare C. G. Hempel and P. Oppenheim. "Studies in the Logic of Explanation," in Hempel's *Aspects of Scientific Explanation* (New York: Free Press, 1965), p. 273, note 33; parenthetical page references to Hempel are to this article. The difficulty is that the conjunction of two laws entails each of its conjuncts but does not necessarily explain them.

etc. of persons. However, 'pragmatic' can also mean subjective as opposed to objective. In this sense, a pragmatic notion varies from individual to individual, and is therefore a relative notion. But a concept can be pragmatic in the first sense without being pragmatic in the second. Take the concept of rational belief, for example— presumably, if it is rational to believe a given sentence on given evidence it is so for anyone, and not merely for this or that individual. Similarly, although the notion of understanding, like knowledge and belief but unlike truth, just is a psychological notion, I don't see why it can't be a perfectly objective one. I don't see why there can't be an objective or rational sense of 'scientific understanding', a sense on which what is scientifically comprehensible is constant for a relatively large class of people. Therefore, I don't see how the philosopher of science can afford to ignore such concepts as 'understanding' and 'intelligibility' when giving a theory of the explanation relation.

Despite his reluctance, Hempel as a matter of fact does try to connect his model of explanation with the notion of understanding. He writes: "the [D-N] argument shows that, given the particular circumstances and the laws in question, the occurrence of the phenomenon *was to be expected*; and it is in this sense that the explanation enables us to *understand why* the phenomenon occurred" (327). Here, showing that a phenomenon was to be expected comes to this: if one had known "the particular circumstances and laws in question" before the explained phenomenon occurred, one would have had rational grounds for expecting the explained phenomenon to occur. The phenomenon would not have taken one by surprise.

This attempt at connecting explanation and understanding is clearly best suited to the special case of the explanation of particular events; for only particular events occur at definite times, and can thus actually be expected before their occurrence. Nevertheless, the account seems to fail even here, since understanding and rational expectation are quite distinct notions. To have grounds for rationally expecting some phenomenon is not the same as to understand it. I think that this contention is conclusively established by the well-known examples of prediction via so-called "indicator laws"—the barometer and the storm, Koplick spots and measles, etc. In these examples, one is able to predict some phenomenon on the basis of laws and initial conditions, but one has not thereby enhanced one's understanding of why the phenomenon occurred. To the best of my knowledge, Hempel himself accepts these counterexamples, and, because of them, would concede today that the D-N model provides at best necessary conditions for the explanation of particular events.

When we come to the explanation of general regularities or patterns of behavior the situation is even worse. Since general regularities do not occur at definite times, there is no question of literally expecting them. In this context, showing a phenomenon to have been expected can only mean having rational grounds for believing that the phenomenon does occur. And it is clear that having grounds for believing that a phenomenon occurs, though it may be part of understanding that phenomenon, is not sufficient for such understanding. Scientific explanations may involve the provision of grounds for believing that the explained phenomena occur, but it is not in virtue of the provision of such grounds that they give us understanding.

I conclude that the D-N model has the following advantages: It provides a clear, precise, and simple condition—entailment—that the explanation relation must satisfy, which, as a necessary condition, is not obviously mistaken. Also, it makes explanation relatively objective—what counts as an explanation does not depend on the arbitrary tastes of the scientist or the age. However, D-N theorists have not succeeded in saying what it is about the explanation relation that provides understanding of the world.

2. A second view, which has been surprisingly popular, holds that scientific explanations give us understanding of the world by relating (or reducing) unfamiliar phenomena to familiar ones. This view is inspired by such examples as the kinetic theory of gases, which, it is thought, gain their explanatory power by comparing unfamiliar phenomena, such as the Boyle-Charles law, to familiar phenomena, such as the movements of tiny billiard balls. P. W. Bridgman states this view very clearly: "I believe that examination will show that the essence of explanation consists in reducing a situation to elements with which we are so familiar that we accept them as a matter of course, so our curiosity rests." [2] Among contemporary writers, William Dray seems to favor this view:

> Why does the theory of geometrical optics explain the length of particular shadows? . . . it is surely because a ray diagram goes along with it, allowing us to think of light as travelling along ray lines, some of the lines passing over the wall and others coming to a dead halt on its surface. The shadow length is explained when . . . we think of

[2] *The Logic of Modern Physics* (New York: Macmillan, 1968), p. 37.

light as 'something travelling', i.e., when we apply to it a very
familiar and perhaps anthropomorphic way of thinking. . . . Thus
the role of theory in such explanations is really *parasitic* upon the fact
that it suggests, with the aid of postulated, unobservable, entities, a
'hat-doffing' series of happenings which we are licensed to fill in . . .
But if the travelling of observable entities along observable rails in
a similar way would not explain a similar pattern of impact on encountering a wall, and if the jostling of a tightly packed crowd
would not explain the straining and collapsing of the walls of a tent
in which they were confined, then the corresponding scientific theories
would not explain shadow lengths and the behavior of gases.[3]

The implication here is clearly that theories like the kinetic theory
of gases are able to explain phenomena only to the extent that they
relate them to more familiar processes and events.

This view, although it is initially attractive and does make an
honest attempt to relate explanation and understanding, is rather
obviously inadequate. As a matter of fact, many scientific explanations relate relatively familiar phenomena, such as the reflecting and refracting of light, to relatively unfamiliar phenomena,
such as the behavior of electromagnetic waves. If the view under
consideration were correct, most of the explanations offered by
contemporary physics, which postulate phenomena stranger and
less familiar than any that they explain, could not possibly explain.
But, on reflection, it is not hard to see why this account fails so
completely. For, being familiar, just like being expected, is not at
all the same thing as being understood. We are all familiar with
the behavior of household appliances like radios, televisions, and
refrigerators; but how many of us understand why they behave
the way they do?

Michael Scriven, although he explicitly rejects the "familiarity"
account of explanation, appears to hold a view which is similar to
it in important respects. He believes that in any given context each
person possesses a "realm of understanding"—a set of phenomena
which that person understands at a given time. The job of explanation is to relate phenomena that are not in this privileged set to
phenomena that are in it:

. . . the request for an explanation presupposes that *something* is understood, and a complete answer is one that relates the object of
inquiry to the realm of understanding in some comprehensible and
appropriate way. What this way is varies from subject matter to subject matter just as what makes something better than something else

[3] *Laws and Explanation in History* (New York: Oxford, 1964), pp. 79/80.

varies from the field of automobiles to solutions of chess problems, but the *logical function* of explanation, as of evaluation, is the same in each field.[4]

Thus, whereas on the "familiarity" view of the explanation relation the phenomenon being explained must be related to a phenomenon that is familiar, on Scriven's account the phenomenon being explained must be related to a phenomenon that is already understood. On both views the phenomenon doing the explaining must have a special epistemological status. Not just any phenomenon will do. Both views conflict with the D-N account on this point. For, according to the D-N model, any phenomenon (regardless of its familiarity or epistemological status) that bears the appropriate deductive relation to the phenomenon being explained will do.

Scriven's view seems to me to be inadequate for the same reason as the "familiarity" view is. There just are many explanations in science which relate the phenomena being explained to phenomena that are not themselves understood in the relevant sense; i.e., we do not understand why these latter phenomena occur. This is true whenever a phenomenon is explained by reducing it to some "basic" or "fundamental" processes; e.g., an explanation in terms of the behavior of the fundamental particles of physics. In such cases the phenomenon doing the explaining is not itself understood; it is simply a brute fact. But its ability to explain *other* phenomena is not thereby impaired. Thus, I think that neither Scriven nor the "familiarity" theorists have given us good reason to suppose that it is a necessary feature of the explanation relation that the phenomenon doing the explaining must itself have some special epistemological status. It does not have to be a familiar or "hat doffing" phenomenon, nor do we have to understand why *it* occurs. It merely has to explain the phenomenon to which it is related.

3. A third approach to the explanation relation can be rather un-charitably labeled the "intellectual fashion" view. Like the "familiarity" theorists, holders of this view believe that the phenomenon doing the explaining must have a special epistemological status, but, unlike the "familiarity" theorists, they think that this status varies from scientist to scientist and from historical period to historical period. At any given time certain phenomena are regarded as

[4] "Explanations, Predictions, and Laws," in H. Feigl and G. Maxwell, eds., *Minnesota Studies in the Philosophy of Science*, vol. III (Minneapolis: Univ. of Minnesota Press, 1970), p. 202.

somehow self-explanatory or natural. Such phenomena need no explanation; they represent ideals of intelligibility. Explanation, within a particular historical tradition, consists in relating other phenomena to such ideals of intelligibility. Perhaps the clearest statement of this view is that of Stephen Toulmin, who calls such self-explanatory phenomena "ideals of natural order":

> . . . the scientist's prior expectations are governed by certain rational ideas or conceptions of the regular order of Nature. Things which happen according to these ideas he finds unmysterious; the cause or explanation of an event comes in question . . . through seemingly deviating from this regular way; its classification among the different sorts of phenomena (e.g., 'anomalous refraction') is decided by contrasting it with the regular, intelligible case; and, before the scientist can be satisfied, he must find some way of applying or extending or modifying his prior ideas about Nature so as to bring the deviant event into the fold.[5]

Thus, the meaning of 'scientific understanding' varies with historical tradition, since what counts as an ideal of intelligibility does. Consequently, the very same theory may count as explanatory for one tradition but may fail to explain for another.

In all fairness, it should be pointed out that most supporters of this account do not believe that the choice of such ideals of intelligibility is completely capricious, depending only on the whims and prejudices of particular scientists. On the contrary, most believe that there can be good reasons, usually having to do with predictive power, for choosing one ideal over another. Indeed, one writer, N. R. Hanson, practically identifies predictive power with intelligibility. He argues that scientific theories typically go through three stages. When they are first proposed they are regarded as mere algorithms or "black boxes." As they begin to make more successful predictions than already existing theories, they become more respectable "grey boxes." Finally, through their ability to connect previously diverse areas of research, they become standards of intelligibility; they become what Hanson calls "glass boxes." The phenomena described by a theory in this third stage are taken as paradigms of naturalness and comprehensibility. According to Hanson, when a theory has successfully gone through these three stages "our very idea of what 'understanding' means will have grown and changed with the growth and change of the theory. So also will our idea of 'explanation'."[6]

[5] *Foresight and Understanding* (New York: Harper & Row, 1963), pp. 45–46.
[6] *The Concept of the Positron* (New York: Cambridge, 1963), p. 38.

This view clearly has a lot of historical support. There are many cases in the history of science where what seems explanatory to one scientist is a mere computational device for another; and there are cases where what is regarded as intelligible changes with tradition.[1] However, it seems to me that it would be desirable, if at all possible, to isolate a common, objective sense of explanation which remains constant throughout the history of science; a sense of 'scientific understanding' on which the theories of Newton, Maxwell, Einstein, and Bohr all produce scientific understanding. It would be desirable to find a concept of explanation according to which what counts as an explanation does not depend on what phenomena one finds particularly natural or self-explanatory. In fact, although there may be good reasons for picking one "ideal of natural order" over another, I cannot see any reason but prejudice for regarding some phenomena as somehow more natural, intelligible, or self-explanatory than others. All phenomena, from the commonest everyday event to the most abstract processes of modern physics, are equally in need of explanation—although it is impossible, of course, that they all be explained at once.

Therefore, although the "intellectual fashion" account may ultimately be the best that we can do, I don't see how it can give us what we are after: an objective and rational sense of 'understanding' according to which scientific explanations give us understanding of the world. We should try every means possible of devising an objective concept of explanation before giving in to something like the "intellectual fashion" account.

From the above discussion of three important contemporary theories of the explanation relation we can extract three desirable properties that a theory of explanation should have:

1. It should be sufficiently general—most, if not all, scientific theories that we all consider to be explanatory should come out as such according to our theory. This is where the "familiarity" theory fails, since, according to that view, theories whose basic phenomena are strange and unfamiliar—e.g., all of contemporary physics—cannot be explanatory. Although it is unreasonable to demand that a philosophical account of explanation should show that every theory that has ever been thought to be explanatory really is explanatory, it must at least square with most of the important, central cases.

[1] Examples can be found in Toulmin, Hanson, and T. Mischel, "Pragmatic Aspects of Explanation," *Philosophy of Science*, xxxiii, 1 (March 1966): 40–60.

2. It should be objective—what counts as an explanation should not depend on the idiosyncracies and changing tastes of scientists and historical periods. It should not depend on such nonrational factors as which phenomena one happens to find somehow more natural, intelligible, or self-explanatory than others. This is where the "intellectual fashion" account gives us less than we would like. If there is some objective and rational sense in which scientific theories explain, a philosophical theory of explanation should tell us what it is.

3. Our theory should somehow connect explanation and understanding—it should tell us what kind of understanding scientific explanations provide and how they provide it. This is where D-N theorists have been particularly negligent, although none of our three theories has given a satisfactory account of scientific understanding.

Thus, none of the three theories of explanation we have examined satisfies all our three conditions; none of them has succeeded in isolating a property of the explanation relation which is possessed by most of the clear, central cases of scientific explanation, which is common to the theories of scientists from various historical periods, and which has a demonstrable connection with understanding. Is there such a property?

Consider a typical scientific theory—e.g., the kinetic theory of gases. This theory explains phenomena involving the behavior of gases, such as the fact that gases approximately obey the Boyle-Charles law, by reference to the behavior of the molecules of which gases are composed. For example, we can deduce that any collection of molecules of the sort that gases are, which obeys the laws of mechanics will also approximately obey the Boyle-Charles law. How does this make us understand the behavior of gases? I submit that if this were all the kinetic theory did we would have added nothing to our understanding. We would have simply replaced one brute fact with another. But this is not all the kinetic theory does—it also permits us to derive other phenomena involving the behavior of gases, such as the fact that they obey Graham's law of diffusion and (within certain limits) that they have the specific-heat capacities that they do have, from the same laws of mechanics. The kinetic theory effects a significant *unification* in what we have to accept. Where we once had three independent brute facts—that gases approximately obey the Boyle-Charles law, that they obey Graham's law, and that they have the specific-heat capacities they do have—we now have only one—that molecules obey the laws of mechanics. Furthermore, the kinetic theory also allows us to integrate the behavior of gases with other phenomena, such as the motions of the planets and of falling bodies near the earth. This is because the laws of mechanics also permit us to derive both the fact that planets obey Kepler's laws and the fact that falling bodies obey Galileo's laws. From the fact that all bodies obey the laws of mechanics it follows that the planets behave as they do, falling bodies behave as they do, and gases behave as they do. Once again, we have reduced a multiplicity of unexplained, independent phenomena to one. I claim that this is the crucial property of scientific theories we are looking for; this is the essence of scientific explanation—science increases our understanding of the world by reducing the total number of independent phenomena that we have to accept as ultimate or given. A world with fewer independent phenomena is, other things equal, more comprehensible than one with more.

Many philosophers have of course noticed the unifying effect of scientific theories to which I have drawn attention; e.g., Hempel in one place writes: "a worthwhile scientific theory explains an empirical law by exhibiting it as one aspect of more comprehensive underlying regularities, which have a variety of other testable aspects as well, i.e., which also imply various other empirical laws. Such a theory thus provides a systematically unified account of many different empirical laws" (144). However, the only writer that I am aware of who has suggested that this unification or reduction in the number of independent phenomena is the essence of explanation in science is William Kneale:

> When we explain a given proposition we show that it follows logically from some other proposition or propositions. But this can scarcely be a complete account of the matter. . . . An explanation must in some sense simplify what we have to accept. Now the explanation of laws by showing that they follow from other laws is a simplification of what we have to accept because it reduces the number of untransparent necessitations we need to assume. . . . What we can achieve . . . is a reduction of the number of independent laws we need to assume for a complete description of nature.[8]

But does this idea really make sense? Can we give a clear meaning to 'reduce the total number of independent phenomena'? Can we make our account a little less sketchy? First of all, I will suppose that we can represent what I have been calling *phenomena*—i.e., general uniformities or patterns of behavior—by law-like *sentences*;

8 *Probability and Induction* (New York: Oxford, 1949), pp. 91–92.

and that instead of speaking of the total number of independent phenomena we can speak of the total number of (logically) independent law-like sentences. Secondly, since what is reduced is the total number of phenomena we have to accept, I will suppose that at any given time there is a set K of *accepted* law-like sentences, a set of laws accepted by the scientific community. Furthermore, I will suppose that the set K is deductively closed in the following sense: if S is a law-like sentence, and $K \vdash S$, then S is a member of K; i.e., K contains all law-like consequences of members of K. Our problem now is to say when a given law-like sentence permits a reduction of the number of independent sentences of K. For an example of what we want to characterize, let K contain the Boyle-Charles law, Graham's law, Galileo's law of free fall, and Kepler's laws, and let S be the conjunction of the laws of mechanics. Intuitively, we think S permits a reduction of the total number of independent sentences of K because we can replace a large number of independent laws by one (or at least by a smaller number); i.e., we can replace the set containing the Boyle-Charles law, Graham's law, Galileo's law, and Kepler's laws by $\{S\}$. The trouble with this intuition is that it is not at all clear what counts as *one* law and what counts as *two*. For example, why haven't we reduced the number of independent laws if we replace the set containing the Boyle-Charles law, Graham's law, etc. by the unit set of their *conjunction*? It won't do to say that this conjunction is really not one law but four since it is logically equivalent to a set of four independent laws. For *any* sentence is equivalent to a set of n sentences for any finite n —e.g., S is equivalent to $\{P, P \supset S\}$, where P is any consequence of S. I think the answer to this difficulty may be the following: although every sentence is equivalent to a set of n independent sentences, it is not the case that every sentence is equivalent to a set of n *independently acceptable* sentences—e.g., the members of the set $\{P, P \supset S\}$ may not be acceptable independently of S; for our only grounds for accepting $P \supset S$, say, might be that it is a consequence of S.

I don't have anything very illuminating to say about what it is for one sentence to be acceptable independently of another. Presumably, it means something like: there are sufficient grounds for accepting one which are not also sufficient grounds for accepting the other. If this is correct, the notion of independent acceptability satisfies the following conditions:

(1) If $S \vdash Q$, then S is not acceptable independently of Q.
(2) If S is acceptable independently of P and $Q \vdash P$, then S is acceptable independently of Q.

(assuming that sufficient grounds for accepting S are also sufficient for accepting any consequence of S).

Given such a concept of independent acceptability, the notion of 'reducing the number of independent sentences' can be made relatively precise in the following way. Let a *partition* of a sentence S be a set of sentences Γ such that Γ is logically equivalent to S and each S' in Γ is acceptable independently of S. Thus, if S is the conjunction of the Boyle-Charles law, Graham's law, Galileo's law, and Kepler's laws, the set Γ containing the conjuncts is a partition of S. I will say that a sentence S is *K-atomic* if it has no partition; i.e., if there is no pair $\{S_1, S_2\}$ such that S_1 and S_2 are acceptable independently of S and S_1 & S_2 is logically equivalent to S. Thus, the above conjunction is not *K*-atomic. Let a *K-partition* of a set of sentences Δ be a set Γ of *K*-atomic sentences which is logically equivalent to Δ (I assume that such a *K*-partition exists for every set Δ). Let the *K-cardinality* of a set of sentences Δ, *K*-card (Δ), be inf {card (Γ): Γ a *K*-partition of Δ}. Thus, if S is the above conjunction, *K*-card ($\{S\}$) is at least 4. Finally, I will say that S *reduces* the set Δ iff *K*-card ($\Delta \cup \{S\}$) < *K*-card (Δ). Thus, if S is the above conjunction and Γ is the set of its conjuncts, S does not reduce Γ.

How can we define *explanation* in terms of these ideas? If S is a candidate for explaining some S' in K, we want to know whether S permits a reduction in the number of independent sentences. I think that the relevant set we want S to reduce is the set of *independently acceptable* consequences of S ($\text{con}_K(S)$). For instance, Newton's laws are a good candidate for explaining Boyle's law, say, because Newton's laws reduce the set of their independently acceptable consequences—the set containing Boyle's law, Graham's law, etc. On the other hand, the *conjunction* of Boyle's law and Graham's law is not a good candidate, since it does not reduce the set of its independently acceptable consequences. This suggests the following definition of explanation between laws:

(D1) S_1 explains S_2 iff $S_2 \in \text{con}_K(S_1)$ and S_1 reduces $\text{con}_K(S_1)$

Actually this definition seems to me to be too strong; for if S_1 explains S_2 and S_3 is some independently acceptable law, then S_1 & S_3 will not explain S_2—since S_1 & S_3 will not reduce $\text{con}_K(S_1$ & $S_3)$. This seems undesirable—why should the conjunction of a completely irrelevant law to a good explanation destroy its explanatory power? So I will weaken (D1) to

(D1') S_1 explains S_2 iff there exists a partition Γ of S_1 and an $S_i \in \Gamma$ such that $S_2 \in \text{con}_K(S_i)$ and S_i reduces $\text{con}_K(S_i)$.

Thus, if S_1 explains S_2, then so does S_1 & S_3; for $\{S_1, S_3\}$ is a partition of S_1 & S_3, and S_1 reduces $con_K(S_1)$ by hypothesis.

Note that this definition is not vulnerable to the usual "conjunctive" trivialization of deductive theories of explanation alluded to in footnote 1 above; i.e., the conjunction of two independent laws does not explain its conjuncts. Furthermore, my account shows why such a conjunction cannot be a good explanation. It does not increase our understanding since it does not reduce its independently acceptable consequences.

On the view of explanation that I am proposing, the kind of understanding provided by science is global rather than local. Scientific explanations do not confer intelligibility on individual phenomena by showing them to be somehow natural, necessary, familiar, or inevitable. However, our over-all understanding of the world is increased; our total picture of nature is simplified via a reduction in the number of independent phenomena that we have to accept as ultimate. It seems to me that previous attempts at connecting explanation and understanding have failed through ignoring the global nature of scientific understanding. If one concentrates only on the local aspects of explanation—the phenomenon being explained, the phenomenon doing the explaining, and the relation (deductive or otherwise) between them—one ends up trying to find some special epistemological status—familiarity, naturalness, or being an "ideal of natural order"—for the phenomenon doing the explaining. For how else is understanding conferred on the phenomenon being explained? However, attention to the global aspects of explanation—the relation of the phenomena in question to the total set of accepted phenomena—allows one to dispense with any special epistemological status for the phenomenon doing the explaining. As long as a reduction in the total number of independent phenomena is achieved, the basic phenomena to which all others are reduced can be as strange, unfamiliar, and unnatural as you wish—even as strange as the basic facts of quantum mechanics.

This global view of scientific understanding also, it seems to me, provides the correct answer to the old argument that science is incapable of explaining anything because the basic phenomena to which others are reduced are themselves neither explained nor understood. According to this argument, science merely transfers our puzzlement from one phenomenon to another; it replaces one surprising phenomenon by another equally surprising phenomenon.

The standard answer to this old argument is that phenomena are explained one at a time; that a phenomenon's being itself unexplained does not prevent it from explaining other phenomena in turn. I think this reply is not quite adequate. For the critic of science may legitimately ask how our total understanding of the world is increased by replacing one puzzling phenomenon with another. The answer, as I see it, is that scientific understanding is a global affair. We don't simply replace one phenomenon with another. We replace one phenomenon with a *more comprehensive* phenomenon, and thereby effect a reduction in the total number of accepted phenomena. We thus genuinely increase our understanding of the world.

MICHAEL FRIEDMAN

Harvard University

Causal Explanation*

I. CAUSAL HISTORIES

Any particular event that we might wish to explain stands at the end of a long and complicated causal history. We might imagine a world where causal histories are short and simple; but in the world as we know it, the only question is whether they are infinite or merely enormous.

An explanandum event has its causes. These act jointly. We have the icy road, the bald tire, the drunk driver, the blind corner, the approaching car, and more. Together, these cause the crash. Jointly they suffice to make the crash inevitable, or at least highly probable, or at least much more probable than it would otherwise have been. And the crash depends on each. Without any one it would not have happened, or at least it would have been very much less probable than it was.

But these are by no means all the causes of the crash. For one thing, each of these causes in turn has its causes; and those too are causes of the crash. So in turn are their causes, and so, perhaps, *ad infinitum*. The crash is the culmination of countless distinct, converging causal chains.

* This paper is descended, distantly, from my Hägerstrom Lectures in Uppsala in 1977, and more directly from my Howison Lectures in Berkeley in 1979.

Roughly speaking, a causal history has the structure of a tree. But not quite: the chains may diverge as well as converge. The roots in child-hood of our driver's reckless disposition, for example, are part of the causal chains via his drunkenness, and also are part of other chains via his bald tire.

Further, causal chains are dense. (Not necessarily, perhaps—time might be discrete—but in the world as we mostly believe it to be.) A causal chain may go back as far as it can go and still not be complete, since it may leave out intermediate links. The blind corner and the oncoming car were not immediate causes of the crash. They caused a swerve; that and the bald tire and icy road caused a skid; that and the driver's drunkenness caused him to apply the brake, which only made matters worse .... And still we have mentioned only a few of the most salient stages in the last second of the causal history of the crash. The causal process was in fact a continuous one.

Finally, several causes may be lumped together into one big cause. Or one cause may be divisible into parts. Some of these parts may themselves be causes of the explanandum event, or of parts of it. (Indeed, some parts of the explanandum event itself may be causes of others.) The baldness of the tire consists of the baldness of the inner half plus the baldness of the outer half; the driver's drunkenness con-sists of many different disabilities, of which several may have con-tributed in different ways to the crash. There is no one right way—though there may be more or less natural ways—of carving up a causal history.

The multiplicity of causes and the complexity of causal histories are obscured when we speak, as we sometimes do, of *the* cause of some-thing. That suggests that there is only one. But in fact it is common-place to speak of "the *X*" when we know that there are many *X*'s, and even many *X*'s in our domain of discourse, as witness McCawley's sentence "the dog got in a fight with another dog." If someone says that the bald tire was the cause of the crash, another says that the driver's drunkenness was the cause, and still another says that the cause was the bad upbringing which made him so reckless, I do not think any of them disagree with me when I say that the causal history includes all three. They disagree only about which part of the causal history is most salient for the purposes of some particular inquiry. They may be looking for the most remarkable part, the most remedi-able or blameworthy part, the least obvious of the discoverable parts, ... Some parts will be salient in some contexts, others in others. Some will not be at all salient in any likely context, but they belong to the causal history all the same: the availability of petrol, the birth of the driver's paternal grandmother, the building of the fatal road, the position and velocity of the car a split second before the impact.[1]

(It is sometimes thought that only an aggregate of conditions inclusive enough to be sufficient all by itself—Mill's "whole cause"—deserves to be called "the cause." But even on this eccentric usage, we still have many deserving candidates for the title. For if we have a whole cause at one time, then also we have other whole causes at later times, and perhaps at earlier times as well.)

A causal history is a relational structure. Its *relata* are *events*: local matters of particular fact, of the sorts that may cause or be caused. I have in mind events in the most ordinary sense of the word: flashes, battles, conversations, impacts, strolls, deaths, touchdowns, falls, kisses, ... But also I mean to include events in a broader sense: a moving object's continuing to move, the retention of a trace, the presence of copper in a sample. (See my "Events," in this volume.)

These events may stand in various relations, for instance spatiotemporal relations and relations of part to whole. But it is their causal relations that make a causal history. In particular, I am concerned with relations of causal dependence. An event depends on others, which depend in turn on yet others, . . . ; and the events to which an event is thus linked, either directly or stepwise, I take to be its causes. Given the full structure of causal dependence, all other causal relations are given. Further, I take causal dependence itself to be counterfactual dependence, of a suitably non-backtracking sort, between distinct events: in Hume's words, "if the first . . . had not been, the second never had existed."[2] (See "Causation," in this volume.) But this paper is not meant to rely on my views about the analysis of causation.

---

[1] On definite descriptions that do not imply uniqueness, see "Scorekeeping in a Language Game," in my *Philosophical Papers*, Volume I; and James McCawley, "Presupposition and Discourse Structure," in *Syntax and Semantics* 11, ed. by David Dineen and Choon-kyu Oh (New York: Academic Press, 1979). On causal selection, see Morton G. White, *Foundations of Historical Knowledge* (New York: Harper & Row, 1965), Chapter IV. Peter Unger, in "The Uniqueness of Causation," *American Philosophical Quarterly* 14 (1977): 177–88, has noted that not only "the cause of" but also the verb "caused" may be used selectively. There is something odd—inconsistent, he thinks—in saying with emphasis that each of two distinct things caused something. Even "a cause of" may carry some hint of selectivity. It would be strange, though I think not false, to say in any ordinary context that the availability of petrol was a cause of the crash.

[2] *An Enquiry Concerning Human Understanding*, Section VII.

Whatever causation may be, there are still causal histories, and what I shall say about causal explanation should still apply.[3]

I include relations of probabilistic causal dependence. Those who know of the strong scientific case for saying that our world is an indeterministic one, and that most events therein are to some extent matters of chance, never seriously renounce the commonsensical view that there is plenty of causation in the world. (They may preach the "downfall of causality" in their philosophical moments. But whatever that may mean, evidently it does not imply any shortage of causation.) For instance, they would never dream of agreeing with those ignorant tribes who disbelieve that pregnancies are caused by events of sexual intercourse. The causation they believe in must be probabilistic. And if, as seems likely, our world is indeed thoroughly indeterministic and chancy, its causal histories must be largely or entirely structures of probabilistic causal dependence. I take such dependence to obtain when the objective chances of some events depend counterfactually upon other events: if the cause had not been, the effect would have been very much less probable than it actually was. (See Postscript B to "Causation," in this volume.) But again, what is said in this paper should be compatible with any analysis of probabilistic causation.

The causal history of a particular event includes that event itself, and all events which are part of it. Further, it is closed under causal dependence: anything on which an event in the history depends is itself an event in the history. (A causal history need not be closed under the converse relation. Normally plenty of omitted events will depend on included ones.) Finally, a causal history includes no more than it must to meet these conditions.

II. EXPLANATION AS INFORMATION

Here is my main thesis: *to explain an event is to provide some information about its causal history.*

In an act of explaining, someone who is in possession of some infor-

---

[3] One author who connects explanation and causation in much the same way that I do, but builds on a very different account of causation, is Wesley C. Salmon. See his "Theoretical Explanation," in *Explanation*, ed. by Stephen Körner (New Haven: Yale University Press, 1975); "A Third Dogma of Empiricism," in *Basic Problems in Methodology and Linguistics*, ed. by R. Butts and J. Hintikka (Dordrecht: Reidel, 1977); and "Why Ask 'Why?'?" *Proceedings of the American Philosophical Association* 51 (1978): 683–705.

mation about the causal history of some event—*explanatory information*, I shall call it—tries to convey it to someone else. Normally, to
someone who is thought not to possess it already, but there are exceptions: examination answers and the like. Afterward, if the recipient
understands and believes what he is told, he too will possess the information. The why-question concerning a particular event is a request
for explanatory information, and hence a request that an act of explaining be performed.

In one sense of the word, an explanation of an event is such an act of
explaining. To quote Sylvain Bromberger, "an explanation may be
something about which it makes sense to ask: How long did it take?
Was it interrupted at any point? Who gave it? When? Where? What
were the exact words used? For whose benefit was it given?"[4] But it is
not clear whether just any act of explaining counts as an explanation.
Some acts of explaining are unsatisfactory; for instance the explanatory
information provided might be incorrect, or there might not be enough
of it, or it might be stale news. If so, do we say that the performance
was no explanation at all? Or that it was an unsatisfactory explanation?
The answer, I think, is that we will gladly say either— thereby making
life hard for those who want to settle, once and for all, the necessary
and sufficient conditions for something to count as an explanation.
Fortunately that is a project we needn't undertake.

Bromberger goes on to say that an explanation "may be something
about which none of [the previous] questions makes sense, but about
which it makes sense to ask: Does anyone know it? Who thought of it
first? Is it very complicated?" An explanation in this second sense of
the word is not an act of explaining. It is a chunk of explanatory information—information that may once, or often, or never, have been conveyed in an act of explaining. (It might even be information that never
could be conveyed, for it might have no finite expression in any
language we could ever use.) It is a proposition about the causal history
of the explanandum event. Again it is unclear—and again we needn't
make it clear—what to say about an unsatisfactory chunk of explanatory information, say one that is incorrect or one that is too small to
suit us. We may call it a bad explanation, or no explanation at all.

Among the true propositions about the causal history of an event,
one is maximal in strength. It is the whole truth on the subject—the
biggest chunk of explanatory information that is free of error. We

---

[4] "An Approach to Explanation," in *Analytical Philosophy: Second Series*, ed. by R. J.
Butler (Oxford: Blackwell, 1965).

might call this the *whole* explanation of the explanandum event, or simply *the* explanation. (But "the explanation" might also denote that one out of many explanations, in either sense, that is most salient in a certain context.) It is, of course, very unlikely that so much explanatory information ever could be known, or conveyed to anyone in some tremendous act of explaining!

One who explains may provide not another, but rather himself, with explanatory information. He may think up some hypothesis about the causal history of the explanandum event, which hypothesis he then accepts. Thus Holmes has explained the clues (correctly or not, as the case may be) when he has solved the crime to his satisfaction, even if he keeps his solution to himself. His achievement in this case probably could not be called "an explanation"; though the chunk of explanatory information he has provided himself might be so called, especially if it is a satisfactory one.

Not only a person, but other sorts of things as well, may explain. A theory or a hypothesis, or more generally any collection of premises, may provide explanatory information (correct or incorrect) by implying it. That is so whether or not anyone draws the inference, whether or not anyone accepts or even thinks of the theory in question, and whether or not the theory is true. Thus we may wonder whether our theories explain more than we will ever realize, or whether other undreamt-of theories explain more than the theories we accept.

Explanatory information comes in many shapes and sizes. Most simply, an explainer might give information about the causal history of the explanandum by saying that a certain particular event is included therein. That is, he might specify one of the causes of the explanandum. Or he might specify several. And if so, they might comprise all or part of a cross-section of the causal history: several events, more or less simultaneous and causally independent of one another, that jointly cause the explanandum. Alternatively, he might trace a causal chain. He might specify a sequence of events in the history, ending with the explanandum, each of which is among the causes of the next. Or he might trace a more complicated, branching structure that is likewise embedded in the complete history.

An explainer well might be unable to specify fully any particular event in the history, but might be in a position to make existential statements. He might say, for instance, that the history includes an event of such-and-such kind. Or he might say that the history includes several events of such-and-such kinds, related to one another in such-and-such ways. In other words, he might make an existential statement to the effect that the history includes a pattern of events of a certain sort. (Such a pattern might be regarded, at least in some cases, as one complex and scattered event with smaller events as parts.) He might say that the causal history has a certain sort of cross-section, for instance, or that it includes a certain sort of causal chain.

If someone says that the causal history includes a pattern of events having such-and-such description, there are various sorts of description that he might give. A detailed structural specification might be given, listing the kinds and relations of the events that comprise the pattern. But that is not the only case. The explainer might instead say that the pattern that occupies a certain place in the causal history is some biological, as opposed to merely chemical, process. Or he might say that it has some global structural feature: it is a case of underdamped negative feedback, a dialectical triad, or a resonance phenomenon. (And he might have reason to say this even if he has no idea, for instance, what sort of thing it is that plays the role of a damper in the system in question.) Or he might say that it is a process analogous to some other, familiar process. (So in this special case, at least, there is something to the idea that we may explain by analogizing the unfamiliar to the familiar. At this point I am indebted to David Velleman.) Or he might say that the causal process, whatever it may be, is of a sort that tends in general to produce a certain kind of effect. I say "we have lungs because they keep us alive"; my point being that lungs were produced by that process, whatever it may be, that can and does produce all manner of life-sustaining organs. (In conveying that point by those words, of course I am relying on the shared presupposition that such a process exists. In explaining, as in other communication, literal meaning and background work together.) And I might say this much, whether or not I have definite opinions about what sort of process it is that produces life-sustaining organs. My statement is neutral between evolution, creation, vital forces, or what have you; it is also neutral between opinionation and agnosticism.

In short: information about what the causal history includes may range from the very specific to the very abstract. But we are still not done. There is also negative information: information about what the causal history does *not* include. "Why was the CIA man there when His Excellency dropped dead?—Just coincidence, believe it or not." Here the information given is negative, to the effect that a certain sort of pattern of events—namely, a plot—does not figure in the causal history. (At least, not in that fairly recent part where one might have been suspected. Various ancient plots doubtless figure in the causal histories of all current events, this one included.)

A final example. The patient takes opium and straightway falls asleep; the doctor explains that opium has a dormitive virtue. Doubtless the doctor's statement was not as informative as we might have wished, but observe that it is not altogether devoid of explanatory information. The test is that it suffices to rule out at least some hypotheses about the causal history of the explanandum. It rules out this one: the opium merchants know that opium is an inert substance, yet they wish to market it as a soporific. So they keep close watch; and whenever they see a patient take opium, they sneak in and administer a genuine soporific. The doctor has implied that this hypothesis, at least, is false; whatever the truth may be, at least it somehow involves distinctive intrinsic properties of the opium.

Of course I do not say that all explanatory information is of equal worth; or that all of it equally deserves the honorific name "explanation." My point is simply that we should be aware of the variety of explanatory information. We should not suppose that the only possible way to give some information about how an event was caused is to name one or more of its causes.

III. NON-CAUSAL EXPLANATION?

It seems quite safe to say that the provision of information about causal histories figures very prominently in the explaining of particular events. What is not so clear is that it is the whole story. Besides the causal explanation that I am discussing, is there also any such thing as non-causal explanation of particular events? My main thesis says there is not. I shall consider three apparent cases of it, one discussed by Hempel and two suggested to me by Peter Railton.[5]

First case. We have a block of glass of varying refractive index. A beam of light enters at point *A* and leaves at point *B*. In between, it passes through point *C*. Why? Because *C* falls on the path from *A* to *B* that takes light the least time to traverse; and according to Fermat's principle of

---

[5] Carl G. Hempel, *Aspects of Scientific Explanation and other Essays in the Philosophy of Science* (New York: Free Press, 1965), p. 353; Peter Railton, *Explaining Explanation* (Ph. D. dissertation, Princeton University, 1979). I am much indebted to Railton throughout this paper, both where he and I agree and where we do not. For his own views on explanation, see also his "A Deductive-Nomological Model of Probabilistic Explanation," *Philosophy of Science* 45 (1978): 206–26; and "Probability, Explanation, and Information," *Synthese* 48 (1981): 233–56.

least time, that is the path that any light going from A to B must follow. That seems non-causal. The light does not get to C because it looks ahead, calculates the path of least time to its destination B, and steers accordingly! The refractive index in parts of the glass that the light has not yet reached has nothing to do with causing it to get to C, but that is part of what makes it so that C is on the path of least time from A to B.

I reply that it is by no means clear that the light's passing through C has been explained. But if it has, that is because this explanation combines with information that its recipient already possesses to imply something about the causal history of the explanandum. Any likely recipient of an explanation that mentions Fermat's principle must already know a good deal about the propagation of light. He probably knows that the bending of the beam at any point depends causally on the local variation of refractive index around that point. He probably knows, or at least can guess, that Fermat's principle is somehow provable from some law describing that dependence together with some law relating refractive index to speed of light. Then he knows this: (1) the pattern of variation of the refractive index along some path from A to C is part of the causal history of the light's passing through C, and (2) the pattern is such that it, together with a pattern of variation elsewhere that is not part of the causal history, makes the path from A to C be part of a path of least time from A to B. To know this much is not to know just what the pattern that enters into the causal history looks like, but it is to know something—something relational—about that pattern. So the explanation does indeed provide a peculiar kind of information about the causal history of the explanandum, on condition that the recipient is able to supply the extra premises needed.

Second case. A star has been collapsing, but the collapse stops. Why? Because it's gone as far as it can go. Any more collapsed state would violate the Pauli Exclusion Principle. It's not that anything caused it to stop—there was no countervailing pressure, or anything like that. There was nothing to keep it out of a more collapsed state. Rather, there just was no such state for it to get into. The state-space of physical possibilities gave out. (If ordinary space had boundaries, a similar example could be given in which ordinary space gives out and something stops at the edge.)

I reply that information about the causal history of the stopping has indeed been provided, but it was information of an unexpectedly negative sort. It was the information that the stopping had no causes at all, except for all the causes of the collapse which was a precondition of the stopping. Negative information is still information. If you request information about arctic penguins, the best information I can give you is that there aren't any.

Third case. Walt is immune to smallpox. Why? Because he possesses antibodies capable of killing off any smallpox virus that might come along. But his possession of antibodies doesn't *cause* his immunity. It *is* his immunity. Immunity is a disposition, to have a disposition is to have something or other that occupies a certain causal role, and in Walt's case what occupies the role is his possession of antibodies.

I reply that it's as if we'd said it this way: Walt has some property that protects him from smallpox. Why? Because he possesses antibodies, and possession of antibodies is a property that protects him from smallpox. Schematically: Why is it that something is F? Because A is F. An existential quantification is explained by providing an instance. I agree that something has been explained, and not by providing information about its causal history. But I don't agree that any particular event has been non-causally explained. The case is outside the scope of my thesis. That which protects Walt—namely, his possession of antibodies—is indeed a particular event. It is an element of causal histories; it causes and is caused. But that was not the explanandum. We could no more explain that just by saying that Walt possesses antibodies than we could explain an event just by saying that it took place. What we did explain was something else: the fact that something or other protects Walt. The obtaining of this existential fact is not an event. It cannot be caused. Rather, events that would provide it with a truth-making instance can be caused. We explain the existential fact by identifying the truth-making instance, by providing information about the causal history thereof, or both. (For further discussion of explanation of facts involving the existence of patterns of events, see Section VIII of "Events," in this volume.)

What more we say about the case depends on our theory of dispositions.[6] I take for granted that a disposition requires a causal basis: one has the disposition iff one has a property that occupies a certain

[6] See the discussions of dispositions and their bases in D. M. Armstrong, *A Materialist Theory of the Mind* (London: Routledge & Kegan Paul, 1968), pp. 85–88; Armstrong, *Belief, Truth and Knowledge* (Cambridge: Cambridge University Press, 1973), pp. 11–16; Elizabeth W. Prior, Robert Pargetter, and Frank Jackson, "Three Theses about Dispositions," *American Philosophical Quarterly* 19 (1982): 251–57; and Elizabeth W. Prior, *Dispositions* (Aberdeen: Aberdeen University Press, 1985). See also Section VIII of "Events," in this volume. Parallel issues arise for functionalist theories of mind. See my "An Argument for the Identity Theory" and "Mad Pain and Martian Pain," in *Philosophical Papers*, Volume I; and Jackson, Pargetter, and Prior, "Functionalism and Type-Type Identity Theories," *Philosophical Studies* 42 (1982): 209–25.

causal role. (I would be inclined to require that this be an intrinsic property, but that is controversial.) Shall we then identify the disposition with its basis? That would make the disposition a cause of its manifestations, since the basis is. But the identification might vary from case to case. (It surely would, if we count the unactualized cases.) For there might be different bases in different cases. Walt might be disposed to remain healthy if exposed to virus on the basis of his possession of antibodies, but Milt might be so disposed on the basis of his possession of dormant antibody-makers. Then if the disposition is the basis, immunity is different properties in the cases of Walt and Milt. Or better: "immunity" denotes different properties in the two cases, and there is no property of immunity *simpliciter* that Walt and Milt share.

That is disagreeably odd. But Walt and Milt do at least share something: the existential property of having some basis or other. This is the property such that, necessarily, it belongs to an individual X iff X has some property that occupies the appropriate role in X's case. So perhaps we should distinguish the disposition from its various bases, and identify it rather with the existential property. That way, "immunity" could indeed name a property shared by Walt and Milt. But this alternative has a disagreeable oddity of its own. The existential property, unlike the various bases, is too disjunctive and too extrinsic to occupy any causal role. There is no event that is essentially a having of the existential property; *a fortiori*, no such event ever causes anything. (Compare the absurd double-counting of causes that would ensue if we said, for instance, that when a match struck in the evening lights, one of the causes of the lighting is an event that essentially involves the property of being struck in the evening or twirled in the morning. I say there is no such event.) So if the disposition is the existential property, then it is causally impotent. On this theory, we are mistaken whenever we ascribe effects to dispositions.

Fortunately we needn't decide between the two theories. Though they differ on the analysis of disposition-names like "immunity," they agree about what entities there are. There is one genuine event— Walt's possession of antibodies. There is a truth about Walt to the effect that he has the existential property. But there is no second event that is essentially a having of the existential property, but is not essentially a having of it in any particular way. Whatever "Walt's immunity" may denote, it does not denote such an event. And since there is no such event at all, there is no such event to be non-causally explained.

IV. GENERAL EXPLANATION

My main thesis concerns the explanation of particular events. As it stands, it says nothing about what it is to explain general kinds of events. However, it has a natural extension. All the events of a given kind have their causal histories, and these histories may to some extent be alike. Especially, the final parts of the histories may be much the same from one case to the next, however much the earlier parts may differ. Then information may be provided about what is common to all the parallel causal histories—call it *general explanatory information* about events of the given kind. To explain a kind of event is to provide some general explanatory information about events of that kind.

Thus explaining why struck matches light in general is not so very different from explaining why some particular struck match lit. In general, and in the particular case, the causal history involves friction, small hot spots, liberation of oxygen from a compound that decomposes when hot, local combustion of a heated inflammable substance facilitated by this extra oxygen, further heat produced by this combustion, and so on.

There are intermediate degrees of generality. If we are not prepared to say that every event of such-and-such kind, without exception, has a causal history with so-and-so features, we need not therefore abjure generality altogether and stick to explaining events one at a time. We may generalize modestly, without laying claim to universality, and say just that quite often an event of such-and-such kind has a causal history with so-and-so features. Or we may get a bit more ambitious and say that it is so in most cases, or at least in most cases that are likely to arise under the circumstances that prevail hereabouts. Such modest generality may be especially characteristic of history and the social sciences; but it appears also in the physical sciences of complex systems, such as meteorology and geology. We may be short of known laws to the effect that storms with feature X always do Y, or always have a certain definite probability of doing Y. Presumably there are such laws, but they are too complicated to discover either directly or by derivation from first principles. But we do have a great deal of general knowledge of the sorts of causal processes that commonly go on in storms.

The pursuit of general explanations may be very much more widespread in science than the pursuit of general laws. And not necessarily because we doubt that there are general laws to pursue. Even if the scientific community unanimously believed in the existence of powerful general laws that govern all the causal processes of nature, and whether or not those laws were yet known, meteorologists and geologists and physiologists and historians and engineers and laymen would still want general knowledge about the sorts of causal processes that go on in the systems they study.

V. EXPLAINING WELL AND BADLY

An act of explaining may be more or less satisfactory, in several different ways. It will be instructive to list them. It will *not* be instructive to fuss about whether an unsatisfactory act of explaining, or an unsatisfactory chunk of explanatory information, deserves to be so-called, and I shall leave all such questions unsettled.

1. An act of explaining may be unsatisfactory because the explanatory information provided is unsatisfactory. In particular, it might be misinformation: it might be a false proposition about the causal history of the explanandum. This defect admits of degree. False is false, but a false proposition may or may not be close to the truth.[7] If it has a natural division into conjuncts, more or fewer of them may be true. If it has some especially salient consequences, more or fewer of those may be true. The world as it is may be more or less similar to the world as it would be if the falsehood were true.

2. The explanatory information provided may be correct, but there may not be very much of it. It might be a true but weak proposition; one that excludes few (with respect to some suitable measure) of the alternative possible ways the causal history of the explanandum might be. Or the information provided might be both true and strong, but unduly disjunctive. The alternative possibilities left open might be too widely scattered, too different from one another. These defects too

---

[7] The analysis of verisimilitude has been much debated. A good survey is Ilkka Niiniluoto, "Truthlikeness: Comments on Recent Discussion," *Synthese* 38 (1978): 281–329. Some plausible analyses have failed disastrously, others conflict with one another. One conclusion that emerges is that it is probably a bad move to try to define a single virtue of verisimilitude-cum-strength. It's hard to say whether strength is a virtue in the case of false information, especially if we have no uniquely natural way of splitting the misinformation into true and false parts. Another conclusion is that even if this lumping together is avoided, verisimilitude still seems to consist of several distinguishable virtues.

admit of degree. Other things being equal, it is better if more correct explanatory information is provided, and it is better if that information is less disjunctive, up to the unattainable limit in which the whole explanation is provided and there is nothing true and relevant left to add.

3. The explanatory information provided may be correct, but not thanks to the explainer. He may have said what he did not know and had no very good reason to believe. If so, the act of explaining is not fully satisfactory, even if the information provided happens to be satisfactory.

4. The information provided, even if satisfactory in itself, may be stale news. It may add little or nothing to the information the recipient possesses already.

5. The information provided may not be of the sort the recipient most wants. He may be especially interested in certain parts of the causal history, or in certain questions about its overall structure. If so, no amount of explanatory information that addresses itself to the wrong questions will satisfy his wants, even if it is correct and strong and not already in his possession.

6. Explanatory information may be provided in such a way that the recipient has difficulty in assimilating it, or in disentangling the sort of information he wants from all the rest. He may be given more than he can handle, or he may be given it in a disorganized jumble.[8] Or he may be given it in so unconvincing a way that he doesn't believe what he's told. If he is hard to convince, just telling him may not be an effective way to provide him with information. You may have to argue for what you tell him, so that he will have reason to believe you.

7. The recipient may start out with some explanatory misinformation, and the explainer may fail to set him right.

This list covers much that philosophers have said about the merits and demerits of explanations, or about what does and what doesn't deserve the name. And yet I have not been talking specifically about explanation at all! What I have been saying applies just as well to acts of providing information about *any* large and complicated structure. It might as well have been the rail and tram network of Melbourne rather than the causal history of some explanandum event. The information provided, and the act of providing it, can be satisfactory or not in pre-

---

[8] As in the square peg example of Hilary Putnam, "Philosophy and our Mental Life," in his *Mind, Language and Reality* (Cambridge: Cambridge University Press, 1975), pp. 295–97.

cisely the same ways. There is no special subject: pragmatics of explanation.

Philosophers have proposed further *desiderata*. A good explanation ought to show that the explanandum event had to happen, given the laws and the circumstances; or at least that it was highly probable, and could therefore have been expected if we had known enough ahead of time; or at least that it was less surprising than it may have seemed. A good explanation ought to show that the causal processes at work are of familiar kinds; or that they are analogous to familiar processes; or that they are governed by simple and powerful laws; or that they are not too miscellaneous. But I say that a good explanation ought to show none of these things unless they are true. If one of these things is false in a given case, and if the recipient is interested in the question of whether it is true, or mistakenly thinks that it is true, then a good explanation ought to show that it is false. But that is nothing special: it falls under points 1, 5, and 7 of my list.

It is as if someone thought that a good explanation of any current event had to be one that revealed the sinister doings of the CIA. When the CIA really does play a part in the causal history, we would do well to tell him about it: we thereby provide correct explanatory information about the part of the causal history that interests him most. But in case the CIA had nothing to do with it, we ought not to tell him that it did. Rather we ought to tell him that it didn't. Telling him what he hopes to hear is not even a merit to be balanced off against the demerit of falsehood. In itself it has no merit at all. What does have merit is addressing the right question.

This much is true. We are, and we ought to be, biased in favor of believing hypotheses according to which what happens is probable, is governed by simple laws, and so forth. That is relevant to the credibility of explanatory information. But credibility is not a separate merit alongside truth; rather, it is what we go for in seeking truth as best we can.

Another proposed *desideratum* is that a good explanation ought to produce understanding. If understanding involves seeing the causal history of the explanandum as simple, familiar, or whatnot, I have already registered my objection. But understanding why an event took place might, I think, just mean possession of explanatory information about it—the more of that you possess, the better you understand. If so, of course a good explanation produces understanding. It produces possession of that which it provides. But this *desideratum*, so construed, is empty. It adds nothing to our understanding of explanation.

history of my visiting Melbourne; and if I had gone to one of the other places instead, presumably that would not have been part of the causal history of my going there. It would have been wrong to answer: Because I like going to places with good friends, good philosophy, cool weather, nice scenery, and plenty of trains. That liking is also part of the causal history of my visiting Melbourne, but it would equally have been part of the causal history of my visiting any of the other places, had I done so.

The same effect can be achieved by means of contrastive stress. Why did I *fly* to Brisbane when last I went there? I had my reasons for wanting to get there, but I won't mention those because they would have been part of the causal history no matter how I'd travelled. Instead I'll say that I had too little time to go by train. If I had gone by train, my having too little time could not have been part of the causal history of my so doing.

If we distinguish plain from contrastive why-questions, we can escape a dilemma about explanation under indeterminism. On the one hand, we seem quite prepared to offer explanations of chance events. Those of us who think that chance is all-pervasive (as well as those who suspend judgment) are no less willing than the staunchest determinist to explain the events that chance to happen.[10] On the other hand, we balk at the very idea of explaining why a chance event took place—for is it not the very essence of chance that one thing happens rather than another for no reason whatsoever? Are we of two minds?

No; I think we are right to explain chance events, yet we are right also to deny that we can ever explain why a chance process yields one outcome rather than another. According to what I've already said, indeed we cannot explain why one happened *rather than the other*. (That is so regardless of the respective probabilities of the two.) The actual causal history of the actual chance outcome does not differ at all

---

[10] A treatment of explanation in daily life, or in history, dare not set aside the explanation of chance events as a peculiarity arising only in quantum physics. If current scientific theory is to be trusted, chance events are far from exceptional. The misguided hope that determinism might prevail in history if not in physics well deserves Railton's mockery: "All but the most basic regularities of the universe stand forever in peril of being interrupted or upset by intrusion of the effects of random processes . . . . The success of a social revolution might appear to be explained by its overwhelming popular support, but this is to overlook the revolutionaries' luck: if all the naturally unstable nuclides on earth had commenced spontaneous nuclear fission in rapid succession, the triumph of the people would never have come to pass." ("A Deductive-Nomological Model of Probabilistic Explanation," pp. 223–24.) On the same point, see my Postscript B to "A Subjectivist's Guide to Objective Chance," in this volume.

107

VI. WHY-QUESTIONS, PLAIN AND CONTRASTIVE

A why-question, I said, is a request for explanatory information. All questions are requests for information of some or other sort.[9] But there is a distinction to be made. Every question has a maximal true answer: the whole truth about the subject matter on which information is requested, to which nothing could be added without irrelevancy or error. In some cases it is feasible to provide these maximal answers. Then we can reasonably hope for them, request them, and settle for nothing less. "Who done it?—Professor Plum." There's no more to say.

In other cases it isn't feasible to provide maximal true answers. There's just too much true information of the requested sort to know or to tell. Then we do not hope for maximal answers and do not request them, and we always settle for less. The feasible answers do not divide sharply into complete and partial. They're all partial, but some are more partial than others. There's only a fuzzy line between enough and not enough of the requested information. "What's going on here?"—No need to mention that you're digesting your dinner. "Who is Bob Hawke?"—No need to write the definitive biography. Less will be a perfectly good answer. Why-questions, of course, are among the questions that inevitably get partial answers.

When partial answers are the order of the day, questioners have their ways of indicating how much information they want, or what sort. "In a word, what food do penguins eat?" "Why, in economic terms, is there no significant American socialist party?"

One way to indicate what sort of explanatory information is wanted is through the use of contrastive why-questions. Sometimes there is an explicit "rather than. . . . " Then what is wanted is information about the causal history of the explanandum event, not including information that would also have applied to the causal histories of alternative events, of the sorts indicated, if one of them had taken place instead. In other words, information is requested about the difference between the actualized causal history of the explanandum and the unactualized causal histories of its unactualized alternatives. Why did I visit Melbourne in 1979, rather than Oxford or Uppsala or Wellington? Because Monash University invited me. That is part of the causal

---

[9] Except perhaps for questions that take imperative answers: "What do I do now, Boss?"

from the unactualized causal history that the other outcome would have had, if that outcome had happened. A contrastive why-question with "rather" requests information about the features that differentiate the actual causal history from its counterfactual alternative. There are no such features, so the question can have no positive answer. Thus we are right to call chance events inexplicable, if it is contrastive explanation that we have in mind. (Likewise, we can never explain why a chance event *had* to happen, because it didn't have to.) But take away the "rather" (and the "had") and explanation becomes possible. Even a chance event has a causal history. There is information about that causal history to be provided in answer to a plain why-question. And thus we are right to proceed as we all do in explaining what we take to be chance events.

VII. THE COVERING-LAW MODEL

The covering-law model of explanation has long been the leading approach. As developed in the work of Hempel and others, it is an elegant and powerful theory. How much of it is compatible with what I have said?

Proponents of the covering-law model do not give a central place to the thesis that we explain by providing information about causes. But neither do they say much against it. They may complain that the ordinary notion of causation has resisted precise analysis; they may say that mere mention of a cause provides less in the way of explanation than might be wished; they may insist that there are a few special cases in which we have good non-causal explanations of particular occurrences. But when they give us their intended examples of covering-law explanation, they almost always pick examples in which—as they willingly agree—the covering-law explanation does include a list of joint causes of the explanandum event, and thereby provides information about its causal history.

The foremost version of the covering-law model is Hempel's treatment of explanation in the non-probabilistic case.[11] He proposes that an explanation of a particular event consists, ideally, of a correct deductive-nomological (henceforth D–N) argument. There are law premises and particular-fact premises and no others. The conclusion

---

[11] For a full presentation of Hempel's views, see the title essay in his *Aspects of Scientific Explanation*.

says that the explanandum event took place. The argument is valid, in the sense that the premises could not all be true and the conclusion false. (We might instead define validity in syntactic terms. If so, we should be prepared to included mathematical, and perhaps definitional, truths among the premises.) No premise could be deleted without destroying the validity of the argument. The premises are all true.

Hempel also offers a treatment for the probabilistic case; but it differs significantly from his deductive-nomological model, and also it has two unwelcome consequences. (1) An improbable event cannot be explained at all. (2) One requirement for a correct explanation—"maximal specificity"—is relative to our state of knowledge; so that our ignorance can make correct an explanation that would be incorrect if we knew more. Surely what's true is rather that ignorance can make an explanation seem to be correct when really it is not. Therefore, instead of Hempel's treatment of the probabilistic case, I prefer to consider Railton's "deductive-nomological model of probabilistic explanation".[12] This closely parallels Hempel's D–N model for the non-probabilistic case, and it avoids both the difficulties just mentioned. Admittedly, Railton's treatment is available only if we are prepared to speak of chances—single-case objective probabilities. But that is no price at all if we have to pay it anyway. And we do, if we want to respect the apparent content of science. (Which is not the same as

---

[12] See Railton's paper of the same name. In what follows I shall simplify Railton's position in two respects. (1) I shall ignore his division of a D-N argument for a probabilistic conclusion into two parts, the first deriving a law of uniform chances from some broader theory and the second applying that law to the case at hand. (2) I shall pretend, until further notice, that Railton differs from Hempel only in his treatment of probabilistic explanation; in fact there are other important differences, to be noted shortly.

It is important to distinguish Railton's proposal from a different way of using single-case chances in a covering-law model of explanation, proposed in James H. Fetzer, "A Single Case Propensity Theory of Explanation," *Synthese* 28 (1974), pp. 171–98. For Fetzer, as for Railton, the covering laws are universal generalizations about single-case chances. But for Fetzer, as for Hempel, the explanatory argument, without any addendum, is the whole of the explanation; it is inductive, not deductive; and its conclusion says outright that the explanandum took place, not that it had a certain chance. This theory shares some of the merits of Railton's. However, it has one quite peculiar consequence. For Fetzer, as for Hempel, an explanation is an argument; however, a good explanation is not necessarily a good argument. Fetzer, like Railton, wants to have explanations even when the explanandum is extremely improbable. But in that case a good explanation is an extremely bad argument. It is an inductive argument whose premises not only fail to give us any good reason to believe the conclusion, but in fact give us very good reason to *dis*believe the conclusion.

respecting the positivist philosophy popular among scientists.) Frequencies—finite or limiting, actual or counterfactual—are fine things in their own right. So are degrees of rational belief. But they just do not fit our ordinary conception of objective chance, as exemplified when we say that any radon-222 atom at any moment has a 50% chance of decaying within the next 3.825 days. If chances are good enough for theorists of radioactive decay, they are good enough for philosophers of science.

Railton proposes that an explanation of a particular chance event consists, ideally, of two parts. The first part is a D–N argument, satisfying the same constraints that we would impose in the nonprobabilistic case, to a conclusion that the explanandum event had a certain specified chance of taking place. The chance can be anything: very high, middling, or even very low. The D–N argument will have probabilistic laws among its premises—preferably, laws drawn from some powerful and general theory—and these laws will take the form of universal generalizations concerning single-case chances. The second part of the explanation is an addendum—not part of the argument—which says that the event did in fact take place. The explanation is correct if both parts are correct: if the premises of the D–N argument are all true, and the addendum also is true.

Suppose we have a D–N argument, either to the explanandum event itself or to the conclusion that it has a certain chance. And suppose that each of the particular-fact premises says, of a certain particular event, that it took place. Then those events are jointly sufficient, given the laws cited, for the event or for the chance. In a sense, they are a minimal jointly sufficient set; but a proper subset might suffice given a different selection of true law premises, and also it might be possible to carve off parts of the events and get a set of the remnants that is still sufficient under the original laws. To perform an act of explaining by producing such an argument and committing oneself to its correctness is, in effect, to make two claims: (1) that certain events are jointly sufficient, under the prevailing laws, for the explanandum event or for a certain chance of it; and (2) that only certain of the laws are needed to establish that sufficiency.

It would make for reconciliation between my account and the covering-law model if we had a covering-law model of causation to go with our covering-law model of explanation. Then we could rest assured that the jointly sufficient set presented in a D–N argument was a set of causes of the explanandum event. Unfortunately, that assurance is not to be had. Often, a member of the jointly sufficient set presented in a D–N argument will indeed be one of the causes of the explanandum event. But it may not be. The counterexamples are well known; I need only list them.

1. An irrelevant non-cause might belong to a non-minimal jointly sufficient set. Requiring minimality is not an adequate remedy; we can get an artificial minimality by gratuitously citing weak laws and leaving stronger relevant laws uncited. That is the lesson of Salmon's famous example of the man who escapes pregnancy because he takes birth control pills, where the only cited law says that nobody who takes the pills becomes pregnant, and hence the premise that the man takes pills cannot be left out without spoiling the validity of the argument.[13]

2. A member of a jointly sufficient set may be something other than an event. For instance, a particular-fact premise might say that something has a highly extrinsic or disjunctive property. I claim that such a premise cannot specify a genuine event; see "Events," in this volume.

3. An effect might belong to a set jointly sufficient for its cause, as when there are laws saying that a certain kind of effect can be produced in only one way. That set might be in some appropriate sense minimal, and might be a set of events. That would not suffice to make the effect be a cause of its cause.

4. Such an effect might also belong to a set jointly sufficient for another effect, perhaps a later effect, of the same cause. Suppose that, given the laws and circumstances, the appearance of a beer ad on my television could only have been caused by a broadcast which would also cause a beer ad to appear on your television. Then the first appearance may be a member of a jointly sufficient set for the second; still, these are not cause and effect. Rather they are two effects of a common cause.

5. A preempted potential cause might belong to a set jointly sufficient for the effect it would have caused, since there might be nothing that could have stopped it from causing that effect without itself causing the same effect.

In view of these examples, we must conclude that the jointly sufficient set presented in a D–N argument may or may not be a set of causes. We do not, at least not yet, have a D–N analysis of causation. All the same, a D–N argument may present causes. If it does, or rather

---

[13] See Wesley C. Salmon et al., *Statistical Explanation and Statistical Relevance* (Pittsburgh: University of Pittsburgh Press, 1971), p. 34.

if it appears to the explainer and audience that it does, then on my view it ought to look explanatory. That is the typical case with sample D–N arguments produced by advocates of the covering-law model.

If the D–N argument does not appear to present causes, and it looks explanatory anyway, that is a problem for me. In Section III, I discussed three such problem cases; the alleged non-causal explanations there considered could readily have been cast as D–N arguments, and indeed I took them from Hempel's and Railton's writings on covering-law explanation. In some cases, I concluded that information was after all given about how the explanandum was caused, even if it happened in a more roundabout way than by straightforward presentation of causes. In other cases, I concluded that what was explained was not really a particular event. Either way, I'm in the clear.

If the D–N argument does not appear to present causes, and therefore fails to look explanatory, that is a problem for the covering-law theorist. He might just insist that it *ought* to look explanatory, and that our customary standards of explanation need reform. To the extent that he takes this high-handed line, I lose interest in trying to agree with as much of his theory as I can. But a more likely response is to impose constraints designed to disqualify the offending D–N arguments. Most simply, he might say that an explanation is a D–N argument of the sort that does present a set of causes, or that provides information in some more roundabout way about how the explanandum was caused. Or he might seek some other constraint to the same effect, thereby continuing the pursuit of a D–N analysis of causation itself. Railton is one covering-law theorist who acknowledges that not just any correct D–N argument (or probabilistic D–N argument with addendum) is explanatory; further constraints are needed to single out the ones that are. In sketching these further constraints, he does not avoid speaking in causal terms. (He has no reason to, since he is not attempting an analysis of causation itself.) For instance, he distinguishes D–N arguments that provide an "account of the mechanism" that leads up to the explanandum event; by which he means, I take it, that there ought to be some tracing of causal chains. He does not make this an inescapable requirement, however, because he thinks that not all covering-law explanation is causal.[14]

A D–N argument may explain by presenting causes, or otherwise giving information about the causal history of the explanandum; is it

---

[14] See his *Explaining Explanation*, "A Deductive-Nomological Model of Probabilistic Explanation," and "Probability, Explanation, and Information."

also true that any causal history can be characterized completely by means of the information that can be built into D–N arguments? That would be so if every cause of an event belongs to some set of causes that are jointly sufficient for it, given the laws; or, in the probabilistic case, that are jointly sufficient under the laws for some definite chance of it. Is it so that causes fall into jointly sufficient sets of one or the other sort? That does not follow, so far as I can tell, from the counter-factual analysis of causation that I favor. It may nevertheless be true, at least in a world governed by a sufficiently powerful system of (strict or probabilistic) laws; and this may be such a world. If it is true, then the whole of a causal history could in principle be mapped by means of D–N arguments (with addenda in the probabilistic case) of the explanatory sort.

In short, if explanatory information is information about causal histories, as I say it is, then one way to provide it is by means of D–N arguments. Moreover, under the hypothesis just advanced, there is no explanatory information that could not in principle be provided in that way. To that extent the covering-law model is dead right.

But even when we acknowledge the need to distinguish explanatory D–N arguments from others, perhaps by means of explicitly causal constraints, there is something else wrong. It is this. The D–N argument—correct, explanatory, and fully explicit—is represented as the ideal serving of explanatory information. It is the right shape and the right size. It is enough, anything less is not enough, and anything more is more than enough.

Nobody thinks that real-life explainers commonly serve up full D–N arguments which they hope are correct. We very seldom do. And we seldom could—it's not just that we save our breath by leaving out the obvious parts. We don't know enough. Just try it. Choose some event you think you understand pretty well, and produce a fully explicit D–N argument, one that you can be moderately sure is correct and not just almost correct, that provides some non-trivial explanatory information about it. Consult any science book you like. Usually the most we can do, given our limited knowledge, is to make existential claims.[15] We can venture to claim that there exists some (correct, etc.)

---

[15] In *Foundations of Historical Knowledge*, Chapter III, Morton White suggests that "because"-statements should be seen as existential claims. You assert the existence of an explanatory argument which includes a given premise, even though you may be unable to produce the argument. This is certainly a step in the right direction. However it seems to underestimate the variety of existential statements that might be made, and also it incorporates a suspect D-N analysis of causation.

D–N argument for the explanandum that goes more or less like this, or that includes this among its premises, or that draws its premises from this scientific theory, or that derives its conclusion from its premise with the aid of this bit of mathematics, or . . . . I would commend these existential statements as explanatory, to the extent—and only to the extent—that they do a good job of giving information about the causal history of the explanandum. But if a proper explanation is a complete and correct D–N argument (perhaps plus addendum), then these existential statements are not yet proper explanations. Just in virtue of their form, they fail to meet the standard of how much information is enough.

Hempel writes "To the extent that a statement of individual causation leaves the relevant antecedent conditions, and thus also the requisite explanatory laws, indefinite it is like a note saying that there is a treasure hidden somewhere."[16] The note will help you find the treasure provided you go on working, but so long as you have only the note you have no treasure at all; and if you find the treasure you will find it all at once. I say it is not like that. A shipwreck has spread the treasure over the bottom of the sea and you will never find it all. Every dubloon you find is one more dubloon in your pocket, and also it is a clue to where the next dubloons may be. You may or may not want to look for them, depending on how many you have so far and on how much you want to be how rich.

If you have anything less than a full D–N argument, there is more to be found out. Your explanatory information is only partial. Yes. *And so is any serving of explanatory information we will ever get,* even if it consists of ever so many perfect D–N arguments piled one upon the other. There is always more to know. A D–N argument presents only one small part—a cross section, so to speak—of the causal history. There are very many other causes of the explanandum that are left out. Those might be the ones we especially want to know about. We might want to know about causes earlier than those presented. Or we might want to know about causes intermediate between those presented and the explanandum. We might want to learn the mechanisms involved by tracing particular causal chains in some detail. (The premises of a D–N argument might tell us that the explanandum would come about through one or the other of two very different causal chains, but not tell us which one.) A D–N argument might give us far from enough explanatory information, considering what sort of information we

---

[16] *Aspects of Scientific Explanation*, p. 349.

want and what we possess already. On the other hand, it might give us too much. Or it might be the wrong shape, and give us not enough and too much at the same time; for it might give us explanatory information of a sort we do not especially want. The cross-section it presents might tell us a lot about the side of the causal history we're content to take for granted, and nothing but stale news about the side we urgently want to know more about.

Is a (correct, etc.) D–N argument in *any* sense a complete serving of explanatory information? Yes in this sense, and this sense alone: it completes a jointly sufficient set of causes. (And other servings complete seventeen-membered sets, still others complete sets going back to the nineteenth century. . . .) The completeness of the jointly sufficient set has nothing to do with the sort of enoughness that we pursue. There is nothing ideal about it, in general. Other shapes and sizes of partial servings may be very much better—and perhaps also better within our reach.

It is not that I have some different idea about what is the unit of explanation. We should not demand a unit, and that demand has distorted the subject badly. It's not that explanations are things we may or may not have one of; rather, explanation is something we may have more or less of.

One bad effect of an unsuitable standard of enoughness is that it may foster disrespect for the explanatory knowledge of our forefathers. Suppose, as may be true, that seldom or never did they get the laws quite right. Then seldom or never did they possess complete and correct D–N arguments. Did they therefore lack explanatory knowledge? Did they have only some notes, and not yet any of the treasure? Surely not! And the reason, say I, is that whatever they may not have known about the laws, they knew a lot about how things were caused.

But once again, the covering-law model needn't have the drawback of which I have been complaining; and once again it is Railton who has proposed the remedy.[17] His picture is similar to mine. Associated with each explanandum we have a vast and complicated structure; explanatory information is information about this structure; an act of explaining is an act of conveying some of this information; more or less information may be conveyed, and in general the act of explaining may be more or less satisfactory in whatever ways any act of conveying information about a large and complicated structure may be more or

---

[17] See *Explaining Explanation* and "Probability, Explanation, and Information."

less satisfactory. The only difference is that whereas for me the vast structure consists of events connected by causal dependence, for Railton it is an enormous "ideal text" consisting of D–N arguments—correct, satisfying whatever constraints need be imposed to make them explanatory, and with addenda as needed—strung together. They fit together like proofs in a mathematics text, with the conclusion of one feeding in as a premise to another, and in the end we reach arguments to the occurrence, or at least a chance, of the explanandum itself. It is unobjectionable to let the subject matter come in units of one argument each, so long as the activity of giving information about it needn't be broken artificially into corresponding units.

By now, little is left in dispute. Both sides agree that explaining is a matter of giving information, and no standard unit need be completed. The covering-law theorist has abandoned any commitment he may once have had to a D–N analysis of causation; he agrees that not just any correct D–N argument is explanatory; he goes some distance toward agreeing that the explanatory ones give information about how the explanandum is caused; and he does not claim that we normally, or even ideally, explain by producing arguments. For my part, I agree that one way to explain would be to produce explanatory D–N arguments; and further, that an explainer may have to argue for what he says in order to be believed. Explanation as argument versus explanation as information is a spurious contrast. More important, I would never deny the relevance of laws to causation, and therefore to explanation; for when we ask what would have happened in the absence of a supposed cause, a first thing to say is that the world would then have evolved lawfully. The covering-law theorist is committed, as I am not, to the thesis that all explanatory information can be incorporated into D–N arguments; however, I do not deny it, at least not for a world like ours with a powerful system of laws. I am committed, as he is not, to the thesis that all explaining of particular events gives some or other sort of information about how they are caused; but when we see how many varieties of causal information there are, and how indirect they can get, perhaps this disagreement too will seem much diminished.

One disagreement remains, central but elusive. It can be agreed that information about the prevailing laws is at least highly relevant to causal information, and *vice versa*; so that the pursuit of explanation and the investigation of laws are inseparable in practice. But still we can ask whether information about the covering laws is itself *part* of explanatory information. The covering law theorist says yes; I say no. But this looks like a question that would be impossible to settle, given that there is no practical prospect of seeking or gaining information about causes without information about laws, or information about laws without information about causes. We can ask whether the work of explaining would be done if we knew all the causes and none of the laws. We can ask; but there is little point trying to answer, since intuitive judgments about such preposterous situations needn't command respect.

III/Theoretical Explanation

Wesley C. Salmon[1]

In previous discussions of the explanation of particular events,[2] I have argued—contra Hempel and many others—that such an explanation is not 'an *argument* to the effect that the event to be explained ... *was to be expected* by reason of certain explanatory facts' (my italics).[3] Indeed, in the case of 'inductive' or 'statistical' explanation at least, I have maintained that such explanations are not *arguments* of any kind, and that consequently, they need not embody the *high* probabilities that would be required to provide reasonable grounds for expectation of the explanandum event. I have argued, instead, that a statistical explanation of a particular event consists of an assemblage of factors relevant to the occurence or non-occurrence of the event to be explained, along with the associated probability values. If

[1] The author wishes to express his gratitude to the National Science Foundation (U.S.A.) for support of research on scientific explanation and other related topics.
[2] 'Statistical Explanation' in *The Nature and Function of Scientific Theories*, Robert G. Colodny, ed. (Pittsburgh: University of Pittsburgh Press, 1970), and reprinted in Wesley C. Salmon, et al., *Statistical Explanation and Statistical Relevance* (Pittsburgh: University of Pittsburgh Press, 1971).
[3] Carl G. Hempel, 'Explanation in Science and in History', in *Frontiers of Science and Philosophy*, Robert G. Colodny, ed. (Pittsburgh: University of Pittsburgh Press, 1962), p. 10.

the probabilities are high, as they will surely be in some cases, the explanation may provide the materials from which an argument can be constructed, but the argument itself is *not* an integral part of the explanation. This model has been called the 'statistical-relevance' or '*S–R* model'.[4]

In addition, I have claimed that the so-called 'deductive-nomological' model of explanation of particular events is incorrect. It is not merely that there are explanandum events which seem explainable only inductively or statistically; Hempel and Oppenheim acknowledged such cases from the very beginning. There are also cases—such as the man who consumes his wife's birth control pills and avoids pregnancy—in which an obviously defective explanation fulfils the conditions for deductive-nomological explanation. All such examples seem to me to exhibit failures of relevance. I have suggested, therefore, that even events which appear amenable to deductive-nomological explanation should also be incorporated, as limiting cases, under the statistical-relevance model.[5]

Arguments by Greeno and others[6] have convinced me that explanations of particular events seldom, if ever, have genuine scientific import (as opposed to practical value), and that explanations which are scientifically interesting are almost always explanations of classes of events. This leads to the suggestion, elegantly elaborated by Greeno,[7] that the goodness or utility of a scientific explanation should be assessed with respect to its ability to account for entire classes of phenomena, rather than by its ability to deal with any particular event in

[4] See especially the Introduction of *Statistical Explanation and Statistical Relevance*, op. cit.

[5] This approach has been elaborated in some detail in the three essays by Richard C. Jeffrey, James G. Greeno, and myself in *Statistical Explanation and Statistical Relevance*.

[6] For example William P. Alston, 'The Place of Explanation of Particular Facts in Science', *Philosophy of Science*, XXXVIII, 1 (March 1971), pp. 13–34.

[7] James G. Greeno, 'Explanation and Information', in *Statistical Explanation and Statistical Relevance*; first published as 'Evaluation of Statistical Hypotheses Using Information Transmitted', *Philosophy of Science*, XXXVII, 2 (June 1970), pp. 279–93.

isolation. If, to use Greeno's example, a sociological explanation is offered to account for delinquent behaviour in teen-age
boys, it is to be evaluated in terms of its ability to assign correct
probability values to this occurrence among various specifiable
classes of boys, not in terms of its ability to predict whether
Johnny Jones will turn delinquent. This shift of emphasis is
important, because it removes any temptation to suppose that
we cannot explain Johnny's behaviour unless we can cite
conditions in relation to which it is highly probable. Perhaps
Johnny is a member of a class in which delinquency is very
improbable, and no more can be said in the matter. This
does not mean that the explanation of *his* delinquency—which is
just part of the explanation of delinquency in boys—is defective
or weak. As Jeffrey has argued persuasively,[8] the explanation of
a low-probability event is not necessarily any weaker than the
explanation of a high-probability event. Even if Billy Smith is a
member of a class of boys in which the delinquency rate is
very high, the explanation of his delinquency by the above-
mentioned sociological theory is no better or stronger than the
explanation of Johnny Jones's delinquency. High probability is
not the desideratum, nor is it the standard by which the quality
of explanations is to be judged; rather, a correct probability
distribution across *relevant* variables is what we should seek.

At the conclusion of my elaboration of the *S–R* model, I expressed certain reservations about it. The two most important
problems concerned the involvement of causality in scientific
explanation and the nature of theoretical explanation. These
two problems are intimately related to one another, and together
they form the subject of the present paper. I shall agree from
the outset that *causal relevance* (or causal influence) plays an
indispensable rôle in scientific explanation, and I shall attempt
to show how this relation can be explicated in terms of the

[8] Richard C. Jeffrey, 'Statistical Explanation vs. Statistical Inference', in *Statistical Explanation and Statistical Relevance*; first
published in *Essays in Honour of Carl G. Hempel*, Nicholas Rescher,
ed. (Dordrecht, Holland: D. Reidel Publishing Company, 1969), pp.
104–13.

concept of statistical relevance. I shall then argue that the demand for suitable causal relations necessitates reference to theoretical entities, and thus leads to the introduction of theoretical explanations. The theme of the paper will be the centrality of certain kinds of *statistical* relevance relations in the notions of casual explanation and theoretical explanation. The result will be an account of theoretical explanation that differs fundamentally from the received deductive-nomological model.[9]

I   THE COMMON CAUSE PRINCIPLE

When all of the lights in a room go off simultaneously, especially if quite a number were on, we infer that a switch has been opened, a fuse has blown, a power line is down, or so forth, but not that all of the bulbs burned out at once. It is, of course, possible that such a chance coincidence might occur, but so improbable that it is not seriously entertained. The principle is not very different from that by which we conclude that two (or five thousand) identical copies of the same book were produced by a common source. A similar kind of inference is involved when one observes an ordinary bridge deck arranged in perfect order, starting from the ace of spades, and concludes (knowing that cards are packed that way at the factory) that this is a newly opened unshuffled deck, rather than one which arrived at the orderly state by random shuffling. The same principle is involved when two witnesses in court give testimony which is alike in content; if collusion can be ruled out, we have strong grounds for supposing that they are truthfully reporting something they both have observed.

The principle governing these examples has been pointed out by many authors. It is deeply embedded in Russell's infamous

[9] Richard Beven Braithwaite, *Scientific Explanation* (New York and London: Cambridge University Press, 1953); Carl G. Hempel, *Aspects of Scientific Explanation* (New York: The Free Press, 1965); Ernest Nagel, *The Structure of Science* (New York: Harcourt, Brace & World, 1959) are among the major proponents of the 'received view'.

Ee

'postulates of scientific inference',[10] and Reichenbach has called it 'the principle of the common cause'.[11] It may be stated roughly as follows: When apparently unconnected events occur in conjunction more frequently than would be expected if they were independent, then assume that there is a common cause. This principle demands considerable explication, for it involves such obscure concepts as *cause* and *connection*.

Let us take our departure from the standard definition of *statistical independence*. Given two types of events A and B that occur, respectively, with probabilities $P(A)$ and $P(B)$, they are statistically independent if and only if the probability of their joint occurrence, $P(A \& B)$, is simply the product of their individual occurrences; i.e.

$$P(A \& B) = P(A) \times P(B)$$

If, contrariwise, their joint occurrence is more probable (or less probable) than the product of the probabilities of their individual occurrences, we must say that they are not statistically independent of one another, but rather, that they are *statistically relevant* to each other. Statistical independence and statistical relevance, as just defined, are clearly symmetric relations.

It seems fairly clear that events which are statistically independent of each other are completely without explanatory value with regard to one another. If, for example, recovery from neurotic symptoms after psychotherapy occurs with a frequency equal to the spontaneous remission rate, then psychotherapy has no explanatory value concerning the curing of mental illness.[12] One reason that independence is of no help whatever in providing explanations is that independent

[10] Bertrand Russell, *Human Knowledge: Its Scope and Limits* (New York: Simon and Schuster, 1948), Part VI, esp. ch. IX.
[11] Hans Reichenbach, *The Direction of Time* (Berkeley and Los Angeles: University of California Press, 1956), §19.
[12] This thesis is argued at length in my essay, 'Statistical Explanation', op. cit.

events are inferentially and practically irrelevant; knowing that an event of one type has occurred is of no help in trying to predict the occurrence or non-occurrence of an event of the other type, or in determining the odds with which to bet on it. Another reason, which will demand close attention, is that statistically independent events are causally irrelevant as well.

If events of the two types are not independent of one another, the occurrence of an event of the one type *may* (but *need* not) help to explain an event of the other type. Suppose, for instance, that the picture on my television receiver occasionally breaks up into a sort of herringbone pattern. At first I may think that this is occurring randomly, but I then discover that there is a nearby police broadcasting station that goes on the air periodically. When I find a strong statistical correlation between the operation of the police transmitter and the breakup of the picture, I conclude that the police broadcast is part of the explanation of the television malfunction. Roughly speaking, the operation of the police transmitter is the cause (or a part of the cause) of the bad TV picture. Obviously, a great deal more has to be filled in to have anything like a complete explanation, but we have identified an important part.

In other cases, however, statistical correlations do not have any such direct explanatory import. The most famous example is the barometer. The rapid dropping of the barometer does not explain the subsequent storm (though, of course, it may enable us to predict it). Likewise, the subsequent storm does not explain the behaviour of the barometer. Both are explained by a common cause, namely, the meteorological conditions that cause the storm and are indicated by the barometer. In this case, there is a statistical relevance relation between the barometer reading and the storm, but neither event is invoked to explain the other. Instead, both are explained by a common cause.[13]

The foregoing two examples, of the TV interference and the

[13] The barometer example is analysed in some detail in *Statistical Explanation and Statistical Relevance*, pp. 53–5.

barometer, illustrate respectively cases in which correlated events can and cannot play an explanatory rôle. The difference is easy to see. The instance in which the event can play an explanatory rôle is one in which it is cause (or part thereof) of the explanandum event The case in which the event cannot play an explanatory rôle is one in which it is not any part of the cause of the explanandum event.

Reichenbach's *basic* principle of explanation seems to be this: *every relation of statistical relevance must be explained by relations of causal relevance.* The various possibilities can be illustrated by a single example. An instructor who receives identical essays from Adams and Brown, two different students in the same class, inevitably infers that something other than a fortuitous coincidence is responsible for their identity. Such an event might, of course, be due to sheer chance (as in the simultaneous burning out of all light bulbs in a room), but that hypothesis is so incredibly improbable that it is not seriously entertained. The instructor may seek evidence that one student copied from the other; i.e. that Adams copied from Brown or that Brown copied from Adams. In either of these cases the identity of the papers can be explained on grounds that one is cause (or part of a cause) of the other. In either of these cases there is a direct causal relation from the one paper to the other, so a causal connection is established. It may be, however, that each student copied from a common source, such as a paper in a fraternity file. In this case, neither of the students' papers is a causal antecedent of the other, but there is a coincidence that has to be explained. The explanation is found in the common cause, the paper in the file, that is a causal antecedent to each. (What university teacher has not witnessed the consternation of two students confronted with the identity of their papers when neither had copied from the other, but both had used a common file; this nicely dramatizes both the need for an explanation and the practical certainty that a common cause exists.)

The case of the common cause, according to Reichenbach's

analysis, exhibits an interesting formal property. It is an immediate consequence of our foregoing definition of statistical independence that event A is statistically relevant to event B if and only if $P(B) \neq P(A, B)$.[14] Let us assume positive statistical relevance; then

$$P(A, B) > P(B) \text{ and } P(B, A) > P(A).$$

From this it follows that

$$P(A \& B) > P(A) \times P(B).$$

To explain this improbable coincidence, we attempt to find a common cause C such that

$$P(C, A \& B) = P(C, A) \times P(C, B),$$

which is to say that, in the presence of the common cause C, A and B are once more rendered statistically independent of one another. The statistical dependency is, so to speak, swallowed up in the relation of causal relevance of C to A and C to B. Under these circumstances C must, of course, be statistically relevant to both A and B; that is,

$$P(C, A) > P(A) \text{ and } P(C, B) > P(B).$$

These *statistical* relevance relations must be explained in terms of two causal processes in which C is *causally* relevant to A and C is *causally* relevant to B.

A further indirect causal relation between two correlated events may obtain, namely, both may serve as partial causes for a common effect. Perhaps Adams and Brown are basketball stars on a championship team which can beat its chief rival if and only if either Adams or Brown plays. Caught at plagiarism, however, both are disqualified and the team loses. As Reichenbach points out, a common effect which follows a combination of partial causes cannot be used to explain the coincidence in the absence of a common cause. In the absence of any common

[14] As is my wont, I am following Reichenbach's non-standard notation, using '$P(A, B)$' to stand for the probability *from A to B*, i.e. the probability, given A, of B.

source, and in the absence of copying one from the other, we cannot attribute the identity of the two papers to a conspiracy of events to produce the team's defeat.[15] Thus, there is no 'principle of the common effect' to parallel the principle of the common cause. This fact provides a basic temporal asymmetry of explanation which is difficult to incorporate into the standard deductive-nomological account of explanation.

2   CAUSAL EXPLANATION OF STATISTICAL RELEVANCE

To provide an explanation of a particular event we may make reference to a statistically relevant event, but the statistical relevance relation itself is a statistical generalization. I agree with the standard nomological account of explanation which demands that an explanation have at least one general statement in the explanans. As indicated in the preceding section, however, we are adopting a principle which says that relations of statistical relevance must be explained in terms of relations of causal relevance. This brings us to the problem of explanations of general relations.

Most of the time (though I *am* prepared to admit exceptions) we do not try to explain statistical independencies or irrelevancies. If the incidence of sunny days in Bristol is independent of the occurrence of multiple human births in Patagonia, no explanation seems called for.[16] Statistical dependencies often do demand explanation, however, and causal relations constitute the explanatory device. Plagiarism, unfortunately, is not a unique occurrence; identical papers turn up with a frequency that cannot be attributed to chance. In such cases it is possible

---

[15] The temporal asymmetry of explanation is discussed at length in connection with the common cause principle (and lack of a parallel common effect principle) in 'Statistical Explanation', §11–12.

[16] In a situation in which we expect to find a statistical correlation and none is found, we may demand an explanation. Why, for example, is the presence of a certain insecticide irrelevant to the survival of a given species of insect? Because of an adaptation of the species? or an unnoticed difference between that species and another that finds the substance lethal? Etc.

to trace observable chains of events from the essays back to a *causal* antecedent. In these instances nothing of a theoretical nature has to be introduced, for the explanation can be given in terms of observable events and processes.[17] In other cases, such as the breakup of the television picture, it is necessary to invoke theoretical considerations if we want to give a causal explanation of the statistical dependency. The statistical relevance between the events of the two types may help to explain the breakup of the picture, and this correlation is essentially observable—for example by telephoning the station to ask if they have just been on the air. The statistical dependency itself, however, cannot be explained without reference to such theoretical entities as electromagnetic waves.

Spatio-temporal continuity obviously makes the critical difference in the two examples just mentioned. In the instance of cheating on the essay, we can provide spatio-temporally continuous processes from the common cause to the two events whose coincidence was to be explained. Having provided the continuous *causal* connections, we have furnished the explanation. In the case of trouble with the TV picture, a statistical correlation is discovered between events which are remote from one another spatially, and this correlation itself requires explanation in terms of such processes as the propagation of electromagnetic waves in space. We invoke a theoretic process which exhibits the desired continuity requirements. When we have provided spatio-temporally continuous connections between correlated events, we have fulfilled a major part of the demand for an explanation of the correlation. We shall return in a subsequent section to a more thorough discussion of the introduction of theoretic entities into explanatory contexts.

The propagation of electromagnetic radiation is generally taken to be a continuous causal process. In characterizing it as

[17] I realize that a full and complete explanation would require references to the theoretical aspects of perception and other psychophysiological mechanisms, but for the moment the example is being taken in common-sense terms.

continuous we mean, I suppose, that given any two spatio-temporally distinct events in such a process, we can interpolate other events between them in the process.[18] But, over and above continuity, what do we mean by characterizing a process as causal? At the very least, it would seem reasonable to insist that events that are causally related exhibit statistical dependencies. This suggests that we require, as a necessary but not sufficient condition, that explanation of statistical dependencies between events that are not contiguous be given by means of statistical relevance between neighbouring or contiguous events.[19]

We have been talking about *causes* and *causal relations*; these seem to figure essentially in the concept of explanation. The principle we are considering (as enunciated by Reichenbach) is the principle of the common *cause*; Russell's treatment of scientific knowledge relies heavily and explicitly upon *causal* relations. It seems to me a serious shortcoming of the received doctrine of scientific explanation that it does not incorporate any full-blooded requirement of causality.[20] But we must not forget the lessons Hume has taught us. The question is whether we can explicate the concept of causality in terms that do not surreptitiously introduce any 'occult' concepts of 'power' or 'necessary connection'. Statistical relevance relations represent the type of constant conjunction Hume relied upon, and spatio-temporal contiguity is also consonant with his strictures. Hume's attempt to explicate causal relations in terms of

[18] For present purposes I ignore the distinction between denseness and genuine continuity in the Cantorean sense. For a detailed discussion of this distinction and its relevance to physics, see my anthology *Zeno's Paradoxes* (Indianapolis: Bobbs-Merrill Co., 1970), especially my Introduction and the selections by Adolf Grünbaum.

[19] In the present context I am ignoring the perplexities about discontinuities and causal anomalies in quantum mechanics.

[20] In Hempel's account of deductive-nomological explanation, there is some mention of nomological relations constituting causal relations, but this passing mention of causality is too superficial to capture the features of causal processes with which we are concerned, and which seems ineradicably present in our intuitive notions about explanation.

constant conjunction was admittedly inadequate because it was an oversimplification; Russell's was also inadequate for the same reason, as I shall show in the next section. Our problem is to see whether we can provide a more satisfactory account of causal processes using only such notions as statistical relevance. We shall see in a moment that processes which satisfy the conditions of continuity and mutual statistical relevance are not necessarily causal processes. We shall, however, remain true to the Humean spirit if we can show that more complicated patterns of statistical relevance relations will suffice to do the job.

3 CAUSAL PROCESSES AND PSEUDO-PROCESSES

Reichenbach tried, in various ways, to show how the concept of causal relevance could be explicated in terms of statistical relevance. He believed, essentially, that causal relevance is a special case of statistical relevance. One of his most fruitful suggestions, in my opinion, employs the concept of a *mark*.[21] Since we are not, in this context, attempting to deal with the problem of 'time's arrow', and correlatively, with the nature and existence of irreversible processes, let us assume that we have provided an adequate physical basis for identifying irreversible processes and ascertaining their temporal direction. Thus, to use one of Reichenbach's favourite examples, we can 'mark' a beam of light by placing a red filter in its path. A beam of white light, encountering such a filter, will lose all of its frequencies except those in the red range, and the red colour of the beam will thus be a mark transmitted onward from the point at which the filter is placed in its path. Such marking procedures can obviously be used for the transmission of information along causal processes.

In the context of relativity theory, it is essential to distinguish causal processes, such as the propagation of a light

[21] Although Reichenbach often discussed the 'mark method' of dealing with causal relevance, the following discussion is based chiefly upon *The Direction of Time*, op. cit., §23.

ray, from various pseudo-processes, such as the motion of a spot of light cast upon a wall by a rotating beacon. The light ray itself can be marked by the use of a filter, or it can be modulated to transmit a message. The same is not true of the spot of light. If it is made red at one place because the light beam creating it passes through a red filter, that red mark is not passed on to the successive positions of the spot. The motion of the spot is a well-defined process of some sort, but it is not a causal process. The causal processes involved are the passages of light rays from the beacon to the wall, and these can be marked to transmit a message. But the direction of message transmission is from the beacon to the wall, not across the wall. This fact has great moment for special relativity, for the light beam can travel no faster than the universal constant c, while the spot can move across the wall at arbitrarily high velocities. Causal processes can be used to synchronize clocks; pseudo-processes cannot. The arbitrarily high velocities of pseudo-processes cannot be exploited to undermine the relativity of simultaneity.[22]

Consider a car travelling along a road on a sunny day. The car moves along in a straight line at 60 m.p.h., and its shadow moves along the verge at the same speed. If the shadow encounters another car parked on the verge, it will be distorted, but will continue on unaffected thereafter. If the car collides with another car and continues on, it will bear the marks of the collision If the car passes a building tall enough to cut off the sunlight the shadow will be destroyed, but it will exist again immediately when the car passes out of the shadow of the building. If the car is totally destroyed, say by an atomic explosion, it will not automatically pop back into existence after the blast and continue on its journey as if nothing had happened.

There are many causal processes in this physical world;

[22] See Hans Reichenbach, *The Philosophy of Space and Time* (New York: Dover Publications, 1957), §23, for a discussion of 'unreal sequences', which I have chosen to call 'pseudo-processes'.

among the most important are the transmission of electromagnetic waves, the propagation of sound waves and other
deformations in various material media, and the motion of
physical objects. Such processes transpire at finite speeds no
greater than that of light; they involve the transportation of
energy from one place to another, and they can carry messages.
Assuming, as we are, that a temporal direction has been
established, we can say that the earlier members of such
causal processes are *causally relevant* to the later ones, but not
conversely.[23] Causal relevance thus becomes an asymmetric
relation, one which we might also call 'causal influence'. We can
test for the relation of causal relevance by making marks in the
processes we suspect of being causal and seeing whether the
marks are, indeed, transmitted. Radioactive 'tagging' can,
for example, be used to trace physiological causal processes.
The notion of causal relevance has been aptly characterized by
saying, 'You wiggle something over here and see if anything
wiggles over there.' This formulation suggests, of course, some
form of human intervention, but that is obviously no essential
part of the definition. It does not matter what agency is
responsible for the marking of the process. At the same time,
experimental science is built upon the idea that *we* can do the
wiggling.[24] There is an obvious similarity between this approach
and Mill's methods of difference and concomitant variation.

Just as it is necessary to distinguish causal processes from
pseudo-processes, so also is it important to distinguish the
relation of causal relevance from the relation of statistical
relevance, especially in view of the fact that pseudo-processes
exhibit striking instances of statistical relevance. Given the
moving spot of light on a wall produced by our rotating beacon,

[23] Although Reichenbach seemed to maintain in his earlier writings,
such as *The Philosophy of Space and Time*, that the mark method could
be taken as an independent criterion of temporal direction (without
any other basis for distinguishing irreversible processes), he abandoned
that view in the later work *The Direction of Time*.

[24] We must, however, resist the strong temptation to use intervention as a criterion of temporal direction; see *The Direction of
Time*, §6.

the occurrence of the spot at one point makes it highly probable that the spot will appear at a nearby point (in the well-established path) at some time very soon thereafter. This is not a certainty, of course, for the light may burn out, an opaque object may block the beam, or the beacon may stop rotating in its accustomed fashion. The same is true of causal processes. Given an occurrence at some point in the process, there is a high probability of another occurrence at a nearby point in the well-established path. Again, however, there is no certainty, for the process may be disturbed or stopped by some other agency. These considerations show that pseudo-processes may exhibit both continuity and statistical relevance among members; this establishes our earlier contention that these two properties, though perhaps necessary, are not sufficient to characterize causal processes.

Pseudo-processes exhibit the same basic characteristics as correlated events or improbable coincidences which require explanation in terms of a common cause. There is a strong correlation between the sudden drop of the barometer and the occurrence of a storm; however, fiddling with a barometer will have no effect upon the storm, and marking or modifying the storm (assuming we had power to do so) would not be transmitted to the (earlier) barometer reading. The pseudo-process is, in fact, just a fairly elaborate pattern of highly correlated events produced by a common cause (the rotating beacon). Pseudo-processes, like other cases of non-causal statistical relevance, require explanation; they do not provide it, even when they possess the sought-after property of spatio-temporal continuity.

One very basic and important principle concerning causal relevance—i.e. the transmission of marks—is, nevertheless, that it seems to be embedded in continuous processes. Marks (or information) are transmitted continuously in space and time. Spatio-temporal continuity, I shall argue, plays a vital rôle in theoretical explanation. The fact that it seems to break down in quantum mechanics—that quantum mechanics seems unavoidably to engender causal anomalies—is a source of great distress. It is far more severe, to my mind, than the discomfort we should experience on account of the apparent breakdown of determinism in that domain. The failure of determinism is one thing; the violation of causality quite another. As I understand it, determinism is the thesis that (loosely speaking) the occurrence of an event has probability zero or one in the presence of a complete set of statistically relevant conditions. Indeterminism, by contrast, obtains if there are complete sets of statistically relevant conditions (i.e. homogeneous reference classes) with respect to which the event may either happen or not—the probability of its occurrence has some intermediate value other than zero or one.[25] The breakdown of causality lies in the fact that (in the quantum domain) causal influence is not transmitted with spatio-temporal continuity. This, I take it, formulates a fundamental aspect of Bohr's principle of complementarity as well as Reichenbach's principle of anomaly.[26] Causal influence need not be deterministic to exhibit continuity; we are construing causal relevance as a species of statistical relevance. Causality, in this sense, is entirely compatible with indeterminism, but quantum mechanics goes beyond indeterminism in its admission of familiar spatio-temporal discontinuities.[27] In classical physics and relativity theory, however, we retain the principle that all causal influence is via action by contact. It is

[25] This conception of determinism, which seems to me especially suitable in the context of discussions of explanation, is elaborated in my essay 'Statistical Explanation', §4. Note also that technically it is illegitimate to identify probability one with invariable occurrence and probability zero with universal absence, but that technicality need not detain us, and I ignore it in the text of this paper.

[26] *The Direction of Time*, p. 216. See also Hans Reichenbach, *Philosophic Foundations of Quantum Mechanics* (Berkeley and Los Angeles: University of California Press, 1946).

[27] It would be completely compatible with indeterminism and causality to suppose that a 'two-slit experiment' performed with macroscopic bullets would yield a two-slit statistical distribution which is just the superposition of two one-slit patterns when large numbers of bullets are involved. At the same time, it might be the case that no trajectory of any individual bullet is precisely determined by the physical conditions. This imaginary situation differs sharply, of course, from the familiar two-slit experiment of quantum mechanics.

doubtful, to say the least, that action by contact can be maintained in quantum mechanics. Even in the macrocosm, however, pseudo-processes may display obvious discontinuities, as for example, when the spot of light from the rotating beacon must 'jump' from the edge of a wall to a cloud far in the background.

Another fundamental characteristic of causal influence is its asymmetric character; in this respect it differs from the relation of statistical relevance. It is an immediate consequence of our foregoing definition of statistical relevance that A is relevant to B if B is relevant to A.[28] This has the consequence that effects are statistically relevant to causes if (as must be the case) causes are statistically relevant to their effects. As we shall see below, Reichenbach defines the *screening-off relation* in terms of statistical relevance; it is a non-symmetric relation from which the relation of causal relevance inherits its asymmetry. The property of asymmetry is crucial, for the common cause which explains a coincidence always precedes it.

4.   THEORETICAL EXPLANATION

In our world the principle of the common cause works rather nicely. We can explain the identical essays by tracing them back to a common cause via two continuous causal processes. These causal processes are constituted, roughly speaking, of events that are in principle observable, and which were in fact observed by the two plagiarists. Many authors, including Hume very conspicuously, have explained how we may endow our world of everyday physical objects with a high degree of spatio-temporal continuity by suitably interpolating observable objects and events between observed objects and events. Russell has discussed at length the way in which similar structures grouped around a centre could be explained in terms of the propagation of continuous causal influence from the

[28] See *Statistical Explanation and Statistical Relevance*, p. 55, esp. note 53.

common centre; indeed, this principle became one of Russell's postulates of scientific inference.[29] In many of his examples, if not all, the continuous process is in principle observable at any point in its propagation from the centre to more distant points at later times.

Although we can endow our world with lots of continuity by reference to observable (though unobserved) entities, we cannot do a very complete job of it. In order to carry through the task we must introduce some entities that are unobservable, at least for ordinary human capabilities of perception. If, for example, we notice that the kitchen windows tend to get foggy on cold days when water is boiling on the stove, we connect the boiling on the stove with the fogging of the windows by hypothesizing the existence of water molecules that are too small to be seen by the naked eye, and by asserting that they travel continuous trajectories from the pan to the window. Similar considerations lead to the postulation of microbes, viruses, and genes for the explanation of such phenomena as the spread of disease and the inheritance of biological characteristics. Note, incidentally, how fundamental a role the transmission of a mark or information plays in modern molecular biology. Electromagnetic waves are invoked to fulfil the same kind of function; in the explanation of the TV picture disturbance, the propagation of electromagnetic waves provided the continuous connection. These unobservable entities are not fictions—not simple-minded fictions at any rate—for we maintain that it is possible to detect them at intermediate positions in the causal process. Hertz detected electromagnetic waves; he could have positioned his detector (or additional detectors) at intermediate places. The high correlation between a spark in the detecting loop and a discharge at the emitter had to be explained by a causal process travelling continuously in space and time. Moreover, the water molecules from the boiling pan will condense on a chilled tumbler anywhere in the kitchen. Microbes

---

[29] Russell, *Human Knowledge*, Part VI, ch. VI. It is called 'the structural postulate'.

and viruses, chromosomes and genes, all can be detected with suitable microscopes; even heavy atoms can now, it seems, be observed with the electron scanning microscope. The claim that there are continuous causal processes involving unobservable objects and events is one that we are willing to test; along with this claim goes some sort of theory about how these intermediate parts of the process can be detected. The existence of causal relevance relations is also subject to test, of course, by the use of marking processes.

Many philosophers, most especially Berkeley, have presented detailed arguments against the view that there are unobserved physical objects. Berkeley did, nevertheless, tacitly admit the common cause principle, and consequently invoked God as a theoretical entity to explain statistical correlations among observed objects. Many other philosophers, among them Mach, presented detailed arguments against the view that there are unobservable objects. Such arguments lead either to phenomenalism (as espoused, for example, by C. I. Lewis) or instrumentalism (as espoused by many early logical positivists). Reichenbach strenuously opposed both of these views, and in the course of his argument he offers a strange analogy—namely, his cubical world. [30]

Reichenbach invites us to consider an observer who is confined to the interior of a cube in which a bunch of shadows appear on the various walls. Careful observation reveals a close correspondence between the shadows on the ceiling and those on one of the walls; there is a high statistical correlation between the shadow events on the ceiling and those on the wall. For example, when the observer notices what appears to be the shadow of one bird pecking at another on the ceiling, he finds the same sort of shadow-pattern on the wall. Reichenbach argues that these correlations should be explained as shadows of the same birds cast on the ceiling and the wall; that is, birds

[30] Hans Reichenbach, *Experience and Prediction* (Chicago: University of Chicago Press, 1938), esp. §14. Contrary to popular opinion, Reichenbach was never a logical positivist, and he regarded *Experience and Prediction* as his *refutation* of positivism

outside of the cube should be postulated. It is further postulated that they are illuminated by an exterior source which makes the shadows of the same birds appear on the translucent material of both the ceiling and the wall. He stipulates that the inhabitant of the cube cannot get to the ceiling or walls to poke holes in them, or any such thing, so that it is physically impossible for the inhabitant to observe the birds directly. Nevertheless, according to Reichenbach, he should infer their existence.[31] Reichenbach is doing precisely what he advocated explicitly in his later work; he is explaining a relation of statistical relevance in terms of relations of causal relevance, invoking a common cause to explain the observed non-contiguous coincidences. The causal processes he postulates are, of course, spatio-temporally continuous.

In *Experience and Prediction* Reichenbach claims that the theory of probability enables us to infer, with a reasonable degree of probability, the existence of entities of unobservable types. This claim seems problematic, to say the least, and I was never quite clear how he thought it could be done. One could argue that all we can observe in the cubical world are constant conjunctions between patterns on the ceiling and patterns on the wall. If constant (temporal) conjunction were the whole story as far as causality is concerned, then we could say that the patterns on the ceiling cause the patterns on the wall, or vice versa. There would be no reason to postulate anything beyond the shadows, for the constant conjunctions are given observationally, and they are all we need. The fact that they are not connected to one another by continuous causal lines would be no ground for worry; there would be no reason to postulate a common cause to link the observed coincidences via *continuous* causal processes. This, a very narrow Humean might say, is the

[31] Reichenbach does not say whether there are any birds *inside* the cube, so that the inference is to entities outside the cube quite like those on the inside, or no birds on the inside to give a clue to the nature of the inferred exterior birds. If his analogy is to be interesting we must adopt the latter interpretation and demand that the observer postulate theoretical entities quite unlike those he has observed.

entire empirical content of the situation; we cannot infer even with probability that the common cause exists. Such counter-arguments might be offered by phenomenalists or instrumenta-lists.

Reichenbach is evidently invoking (though not explicitly in 1938) his principle that statistical relevance must be explained by causal relevance, where causal relevance is defined by continuity and the ability to transmit a mark. In the light of this principle, we may say that there is a certain probability $P(A)$ that a particular pattern (the shadow of one bird pecking at another) will appear on the ceiling, and a certain probability $P(B)$ that a similar pattern will appear on the wall. There is another probability $P(A \& B)$ that this pattern will appear both on the ceiling and on the wall at the same time. This latter probability seems to be much larger than it would be if the events were independent, i.e.

$$P(A \& B) \gg P(A) \times P(B).$$

Reichenbach's principle asserts that this sort of statistical dependency demands causal explanation if, as in this example, A and B are not spatio-temporally contiguous. Using this principle, Reichenbach can certainly claim that the existence of the common cause can be inferred with a probability; otherwise we would have to say that the probability of $A \& B$ is *equal to* the product of the two individual probabilities, and that we were misled into thinking that an inequality holds because the observed frequency of $A \& B$ is much larger than the actual probability. In other words, the choice is between a common cause and an exceedingly improbable coincidence. This makes the common cause the less improbable hypothesis. But the high frequency of the joint occurrence is statistically miraculous only if there are no alternatives except fortuitous coincidence or a continuous connection to a common cause. If we could have causal relevance without spatio-temporal contiguity, no explanation would be required, and hence there would be no probabilistic evidence for the existence of the common cause.

If, on the other hand, we can find an adequate basis for adopting the principle that statistical relevancies must be explained by continuous causal processes, then it seems we have sufficient ground for postulating or inferring the existence of theoretical entities.

In rejecting the notion that we have an impression of necessary connection, Hume analysed the causal relation in terms of constant conjunction. As he realized explicitly, his analysis of causation leaves open the possibility of filling the spatio-temporal gaps in the causal chain by interpolating events between observed causes and observed effects. In so doing, he maintained, we simply discover a larger number of relations of constant conjunction with higher degrees of spatio-temporal contiguity. In recognition of the fact that causal relations often serve as a basis for inference, Hume attempts to provide this basis in the 'habit' or 'custom' to which observed constant conjunction naturally gives rise.

Russell has characterized causal lines as continuous series of events in which it is possible to infer the nature of some members of the series from the characteristics of other events in the same series. This means, in our terms, that there are relations of statistical relevance among the members of such series. Although causal series have enormous epistemological significance for Russell, providing a basis for our knowledge of the physical world, his characterization of causal series is by no means subjective. It is by virtue of factual relations among the members of causal series that we are enabled to make the inferences by which causal processes are characterized.

Statistical relevance relations do provide a basis for making certain kinds of inferences, but they do not have all of the characteristics of causal relevance as defined by Reichenbach; in particular, they do not always have the ability to transmit a mark. Although Russell did not make explicit use of mark transmission in his definitions, his approach would seem hospitable to the addition of this property as a further criterion

141

of causal processes. Russell emphasizes repeatedly the idea that perception is a causal process by which structure can be transmitted. He frequently cites processes like radio transmission as physical analogues of perception, and he obviously considers such examples extremely important. The transmission of messages by the modulations of radio waves is a paradigm of a mark. In similar ways, the absorption of all frequencies but those in the green range from white light falling upon a leaf is a suggestive case of the marking of a causal process involved in human perception. The transmitted mark conveys information about the interaction that is responsible for the mark. The mark principle thus seems to me to be a desirable addition to Russell's definition of causal processes, and one that can be fruitfully incorporated into his postulates of scientific knowledge.

I do not wish to create the impression that ability to transmit a mark is any mysterious kind of necessary connection or 'power' of the sort Hume criticized in Locke. Ability to transmit a mark is simply a species of constant conjunction. We observe that certain kinds of events tend to be followed by others in certain kinds of processes. Rays of white light are series of undulatory events which are spatio-temporally distributed in well-defined patterns. Events which we would describe as passage of light through a red filter are followed by undulations with frequencies confined to the red range; undulations characterized by other frequencies do not normally follow thereupon. It is a fact about this world (at least as long as we stay out of the quantum realm) that there are many continuous causal processes that do transmit marks. This is fortunate for us, for such processes are highly informative. Russell was probably right in saying that without them we would not have anything like the kind of knowledge of the physical world we actually do have. It is not too surprising that causal processes capable of carrying information figure significantly in our notion of scientific explanation. To maintain that such processes are continuous, we must invoke theoretical entities.

Let us then turn to the motivation for the continuity requirement.

5 SPATIO-TEMPORAL CONTINUITY

Throughout this paper we have been discussing the continuity requirement on causal processes; it is now time to see why they figure so importantly in the discussion. If special relativity is correct, there are no essential spatio-temporal limitations upon relations of statistical relevance, but there are serious limitations upon relations of causal relevance. Any event A that we choose can be placed at the apex of a Minkowski light cone, and this cone establishes *a cone of causal relevance*. The backward section of the cone, containing all of the events in the absolute past of A, contains all events that bear the relation of causal relevance to A. The forward part of the light cone, which contains all events in the absolute future of A, contains all events to which A may bear the relation of causal relevance. In contrast, an event B which is in neither the backward nor the forward sections of the cone cannot bear the relation of causal relevance to A, nor can A bear that relation to B. Nevertheless, B can sustain a relation of *statistical* relevance to A. When this occurs, according to Reichenbach's principle, there must be a common cause C somewhere in the region of overlap of the backward sections of the light cones of A and B. The relation of statistical relevance is *not* explainable, as mentioned above, by a common effect in the region of overlap of the forward sections of the two light cones.[32]

If our claims are correct, any *statistical* relevance relation between two events can be explained in terms of *causal* relevance relations. Causal relevance relations are embedded in continuous causal processes. If, therefore, an event C is causally relevant to A, then we can, so to speak, mark off a boundary in

[32] These statements obviously represent factual claims about this world. We believe they are true, and if they are true they are very important. But we have no reason to think they are true in all possible worlds.

the backward part of the light cone (i.e. the causal relevance cone) and be sure that C is either within that part of the cone or else that it is connected with A by a continuous causal process which crosses that boundary. Hence, to investigate the question of what events are causally relevant to A, we have only to examine the interior and boundary of some spatial neighbourhood of A for a certain time in the immediate past of A. We can thus ascertain whether such an event lies within that neighbourhood, or whether a connecting causal process crosses the boundary. We have been assuming, let us recall, that a continuous causal process can be detected anywhere along its path. This means that we do not have to search the whole universe to find out what events bear relations of *causal* relevance to A.[33]

If we make it our task to find out what events are *statistically* relevant to A, all of the events in the universe are potential candidates. There are, in principle, no spatio-temporal limitations on statistical relevance. But, it might be objected, statistical relevance relations can serve as a basis for inductive inference, or at least for inductive behaviour (for example, betting). How are we therefore justified, if knowledge is our aim, in restricting our considerations to events which are causally relevant? The answer lies in the *screening-off* relation.[34]

If A and B are two events that are statistically relevant to one another, but neither is causally relevant to the other, then there must be a common cause C in the region of overlap of the past light cones of A and B. It is possible to demonstrate the causal relevance of C to A by showing that a mark can be transmitted along the causal process from C to A, and the causal relevance of C to B can be demonstrated in a similar fashion. There is, however, no way of transmitting a mark from B to A or from A to B. When we have that kind of situation,

[33] In this connection it is suggestive to remember Popper's distinction between falsifiable and unfalsifiable existential statements.

[34] Reichenbach, *The Direction of Time*, p. 189, and Salmon, 'Statistical Explanation', §7.

which can be unambiguously defined by the use of marking techniques, we find that the statistical relevance of B to A is absorbed in the *statistical* relevance of C to A. That is just what the screening-off relation amounts to. Given that B is statistically relevant to A, and C is statistically relevant to A, we have

$$P(B, A) > P(A) \text{ and } P(C, A) > P(A).$$

To say that C screens off B from A means that, given C, B becomes statistically irrelevant to A; i.e.

$$P(B \& C, A) = P(C, A).$$

Thus, for example, though the barometer drop indicates a storm and is statistically relevant to the occurrence of the storm, the barometer becomes statistically irrelevant to the occurrence of the storm given the meteorological conditions which led to the storm and which are indicated by the barometer reading. The claim that statistical relevance relations can always be explained in terms of causal relevance relations therefore means that causal relevance relations screen off other kinds of statistical relevance relations.

The screening-off relation can be used, moreover, to deal with questions of causal proximity. We can say in general that more remote causal relevance relations are screened off by more immediate causal relevance relations. Part of what we mean by saying that causation operates via action by contact is that the more proximate causes absorb the entire influence of more remote causes. Thus, we do not even have to search the entire backward section of the light cone to find *all* factors relevant to the occurrence of A. A *complete* set of factors statistically relevant to the occurrence of a given event can be found by examining the interior and boundary of an appropriate neighbouring section of its past light cone. Any factor outside of that portion of the cone that is, by itself, statistically relevant to the occurrence of the event in question is screened off by events within that neighbouring portion of the light cone. These are strong factual claims; if correct, they have an enormous bearing upon our conception of explanation.

6   CONCLUSIONS

In this paper I have been trying to elaborate the view of scientific explanation which is present, at least implicitly I think, in the works of Russell and Reichenbach. Such explanation is causal in a very deep and pervasive sense; yet I believe it does not contain causal notions that have been proscribed by Hume's penetrating critique. This causal treatment accounts in a natural way for the invocation of theoretical entities in scientific explanation. It is therefore, I hope, an approach to scientific explanation that fits especially well with scientific realism (as opposed to instrumentalism). Still, I do not wish to claim that this account of explanation establishes the realistic thesis regarding theoretical entities. An instrumentalist might well ask: Is the world understandable because it contains continuous causal processes, or do we make it understandable by imputing continuous causal processes? This is a difficult and far-reaching query.

It is tempting to try to argue for the realist alternative by saying that it would be a statistical miracle of overwhelming proportions if there were statistical dependencies between remote events which reflect precisely the kinds of dependencies we should expect if there were continuous causal connections between them. At the same time, the instrumentalist might retort: what makes remote statistical dependencies any more miraculous than contiguous ones? Unless one is willing to countenance (as I am not) some sort of pre-Humean concept of power or necessary connection, I do not know quite what answer to give. We may have reached a point at which a pragmatic vindication, a posit, or a postulate is called for. It may be possible to argue that scientific understanding can be achieved most efficiently (if such understanding is possible at all), by searching for spatio-temporally continuous processes

146

capable of transmitting marks. This may be the situation with which Russell attempted to cope by offering his postulates of scientific inference.[35] The preceding section was an attempt to spell out the methodological advantages we gain if the world is amenable to explanations of this sort, but I do not intend to suggest that the world is otherwise totally unintelligible. After all, we still have to cope with quantum mechanics, and that does not make scientific understanding seem hopeless.

Regardless of the merits of the foregoing account of explanation, and regardless of the stand one decides to take on the realism-instrumentalism issue, it is worthwhile, I think, to contrast this account with the standard deductive-nomological account. According to the received view, empirical laws, whether universal or statistical, are explained by deducing them from more general laws or theories. Deductive subsumption is the key to theoretical explanation. According to the present account, statistical dependencies are explained by, so to speak, filling in the causal connections in terms of spatio-temporally continuous causal processes. I do not mean to deny, of course, that there are physical laws or theories which characterize the causal processes to which we are referring—laws of mechanics which govern the motions of material bodies, laws of optics and electromagnetism that govern the propagation of electromagnetic waves, etc. The point is, rather, that explanations are not arguments on this view. Causal or theoretical explanation of a statistical correlation between distinct types of events is an exhibition of the way in which those regularities fit into the causal structure of the world—an exhibition of the causal connections between them which give rise to the statistical relevance relations.

---

[35] I have discussed Russell's views on his postulates in some detail in 'Russell on Scientific Inference *or* Will the Real Deductivist Please Stand Up', in George Nakhnikian, ed., *Bertrand Russell's Philosophy* (London: Gerald Duckworth & Co., 1974), pp. 183–208. In the same essay I have discussed aspects of Popper's methodological approach that are relevant to this context.

A DEDUCTIVE-NOMOLOGICAL MODEL OF PROBABILISTIC EXPLANATION*

PETER RAILTON[†]

Princeton University

It has been the dominant view that probabilistic explanations of particular facts must be inductive in character. I argue here that this view is mistaken, and that the aim of probabilistic explanation is not to demonstrate that the explanandum fact was nomically expectable, but to give an account of the chance mechanism(s) responsible for it. To this end, a deductive-nomological model of probabilistic explanation is developed and defended. Such a model has application only when the probabilities occurring in covering laws can be interpreted as measures of objective chance, expressing the strength of physical propensities. Unlike inductive models of probabilistic explanation, this deductive model stands in no need of troublesome requirements of maximal specificity or epistemic relativization.

What if some things happen by chance—can they nonetheless be explained? How?

Some things *do* happen by chance, according to the dominant interpretation of our present physical theory, the probabilistic interpretation of quantum mechanics. Nonetheless, they can be explained: by that theory, in virtually the same way as deterministic phenomena—deductive-nomologically. At least, that is what I hope to show in this paper.

Our universe may not be deterministic, but all is not chaos. It is governed by laws of two kinds: probabilistic (such as the laws concerning barrier penetration and certain other quantum phenomena) and non-probabilistic (such as the laws of conservation of mass-energy, charge, momentum, etc.).[1] Were the probabilism of laws of the first sort remediable by suitable elaboration of laws of the second sort,

---

*Received September, 1977.

†I would like to thank C. G. Hempel, Richard C. Jeffrey, and David Lewis for helpful criticisms of earlier drafts. I am especially indebted to David Lewis for the idea that a propensity interpretation of probability sits best with the account of probabilistic explanation given here. I have greatly benefited from discussions of related matters with Sam Scheffler and David Fair.

[1]Let us say rather loosely that a system is deterministic if, for any one instant, its state is physically compatible with only one (not necessarily different) state at each other instant. A system is indeterministic otherwise, but lawfully so if a complete description of its state at some one instant plus all true laws together entail a distribution of probabilities over possible states at later times.

Philosophy of Science, 45 (1978) pp. 206–226.
Copyright © 1978 by the Philosophy of Science Association.

the universe would be deterministic after all, and the problem of explaining chance phenomena would no longer be with us. However, indications are that physical indeterminism is irremediable, and that the universe exhibits not only chances, but lawful chances. I will argue that we come to understand chance phenomena, even when the chance involved is extremely remote, by subsuming them under these irremediably probabilistic laws.

1. Introductory Remarks on Explanation. Do I offer a deductive-nomological $(D\text{-}N)$ model of probabilistic explanation because I believe that nomic subsumption always explains?—No. There are familiar-enough kinds of non-explanatory $D\text{-}N$ arguments, for example, those that deduce the explanandum from nomically-related symptoms or after-the-fact conditions alone, citing no causes.

Yet it will not do simply to add to the $D\text{-}N$ model a requirement that the explanans contain causes whenever the explanandum is a particular fact. First, some particular facts may be explained non-causally, e.g., by subsumption under structural laws such as the Pauli exclusion principle. Second, even where causal explanation is called for, the existence of general, causal laws that cover the explanandum has not always been sufficient for explanation: the search for explanation has also taken the form of a search for mechanisms that underlie these laws. 'Mechanisms,' however, is not meant to suggest a parochial attitude toward the nomic connections—deterministic or otherwise—that tie the world together and make explanation possible.

An example may help clarify the notion of mechanism appealed to here. The following $D\text{-}N$ argument suffices to forecast *that* nasty weather lies ahead, but not to explain *why* this is so:

(S)   The glass is falling.
Whenever the glass falls the weather turns bad.

The weather will turn bad. ([5], p. 106)

Now nothing works like a barometer for predicting the weather, but nothing like a barometer works for changing it. So it is often maintained that (S) lacks explanatory efficacy because barometers lack the appropriate causal efficacy. The following inference, then, remedies the lack of the first because "it proves that the fact is a fact by citing causes and not mere symptoms" ([5], p. 107):

(C)   The glass is falling.
Whenever the glass is falling the atmospheric pressure is falling.

Whenever the atmospheric pressure is falling the weather turns bad.

The weather will turn bad. ([5], p. 106)

Yet as explanations go, (C) is also lacking: we remain in the dark as to *why* the weather will turn bad. No connection between cause and effect, no mechanism by which falling atmospheric pressure produces a change for the worse in the weather, has been revealed. I do not doubt that some account of this mechanism exists; my point is that its existence is what makes (C) superior to (S) for explanatory purposes.

(C), if moderated by boundary conditions and put less qualitatively, would supply us the capability to predict *and* control the weather (whenever, as in a laboratory simulator, we can manipulate the atmospheric pressure). While prediction and control may exhaust our practical problems in the natural world, the unsatisfactoriness of (C) shows that explanation is an activity not wholly practical in purpose. The goal of understanding the world is a theoretical goal, and if the world is a machine—a vast arrangement of nomic connections—then our theory ought to give us some insight into the structure and workings of the mechanism, above and beyond the capability of predicting and controlling its outcomes. Until supplemented with an account of the nomic links connecting changes in atmospheric pressure to changes in the weather, (C) will explain but poorly. Knowing enough to subsume an event under the right kind of laws is not, therefore, tantamount to knowing the *how* or *why* of it. As the explanatory inadequacies of successful practical disciplines remind us: explanations must be more than potentially-predictive inferences or law-invoking recipes.

Is the deductive-nomological model of explanation therefore unacceptable?—No, just incomplete. Calling for an account of the mechanism leaves open the nature of that account, and as far as I can see, the model explanations offered in scientific texts are *D-N* when complete, *D-N* sketches when not. What is being urged is that *D-N* explanations making use of true, general, causal laws may legitimately be regarded as unsatisfactory unless we can back them up with an account of the mechanism(s) at work. "An account of the mechanism(s)" is a vague notion, and one obviously admitting of degrees of thoroughness, but I will not have much to say here by way of demystification. If one sees what is lacking in (C)—a characterization, whether sketchy or blow-by-blow, of how it is that declining atmospheric pressure effects the changes we describe as "a worsening of the weather," i.e., a more or less complete filling-in of the links in the causal chains—one has the rough idea.

The *D-N* probabilistic explanations to be given below do not explain by giving a deductive argument terminating in the explanandum, for it will be a matter of chance, resisting all but *ex post facto* demonstration. Rather, these explanations subsume a fact in the sense of giving a *D-N* account of the chance mechanism responsible for it, and showing that our theory implies the existence of some physical possibility, however small, that this mechanism will produce the explanandum in the circumstances given. I hope the remarks just made about the importance of revealing mechanisms have eased the way for an account of probabilistic explanation that focuses on the indeterministic mechanisms at work, rather than the "nomic expectability" of the explanandum.

2. Hempel's Inductive-Statistical Model. For Hempel, a statistical explanation (what is called elsewhere in this paper 'a probabilistic explanation') is one that "makes essential use of at least one law or theoretical principle of statistical form" ([3], p. 380). Since Hempel distinguishes between statistical laws and mere statistical generalizations, and asserts that the former apply only where "peculiar, namely probabilistic, modes of connection" exist among the phenomena ([3], p. 377), his characterization permits statistical explanation only of genuinely indeterministic processes.[2] Were some process to have the appearance of indeterminism owing to arcane workings or uncontrolled initial conditions, then no "peculiar . . . probabilistic" modes of connection would figure essentially in explaining this "pseudo-random" process's outcomes. Not only would statistical explanation be unnecessary for such a process, it would be impossible: no probabilistic *laws* would govern it.

For example, it has been observed that 99% of all cases of infectious mononucleosis involve lymph-gland swelling. The exceptions might be due to a process that randomly misfires 1% of the time. Or, they might arise from the operation of an unknown deterministic mechanism that works to inhibit swelling whenever a patient begins in a particular initial condition, which as a mere matter of fact is typical of 1% of the population. If initial conditions could be partitioned into two mutually exclusive and jointly exhaustive classes S and $-S$, such that all Ss by law eventually develop swelling, and all $-S$s do not, the generalization '99% of all cases of infectious mononucleosis develop lymph-gland swelling' would have been shown to be no law, but

---

[2]Although there is some difficulty in reconciling all that is said in [3] with this conclusion, Hempel now accepts it (personal communication).

merely a descriptive report of observed relative frequencies. No law, it cannot support a statistical explanation. But discovering it not to be a law is just discovering that statistical explanation is uncalled for, since each case of mononucleosis will have been of type S or type $-S$ from the outset.

On the other hand, suppose that no such partition of initial conditions exists. Then the presence or absence of swelling is presumably due to a "peculiar . . . probabilistic" connection between disease and symptom, i.e., a real causal indeterminism with probability .99 in each case to produce swelling. The generalization in question would thus be nomological, creating both the possibility and the necessity of statistical explanation.

Given such genuine statistical laws, how does Hempel claim statistical explanation should proceed? He begins his account by distinguishing two sorts of statistical explanation. The first, *deductive-statistical* (D-S) explanation, involves "the deductive subsumption of a narrow statistical uniformity under more comprehensive ones" ([3], p. 380). The second, he argues, is of a qualitatively different sort:

> Ultimately . . . statistical laws are meant to be applied to particular occurrences and to establish explanatory and predictive connections among them. ([3], p. 381)

To make such laws relevant to "particular occurrences," Hempel believes we must go beyond the reach of deduction, and so he proposes an inductive model of statistical explanation.

Inductive-statistical (I-S) explanation proceeds by adducing statistical laws and associated initial conditions relative to which the explanandum is highly probable. High relative probability is required because, on Hempel's view, statistical laws become explanatorily relevant to an individual chance event only by giving us a basis upon which to inductively infer its occurrence with "practical certainty." Yet although an I-S explanation shows the explanandum to have been "nomically expectable" relative to the explanans, it does not permit detachment of a conclusion; it is less an inference than the expression of an inferential relationship: the explanandum receives a high degree of epistemic support from the explanans. If, for example, we learn that Jones has contracted infectious mononucleosis, we may infer with practical certainty that he will develop lymph-gland swelling. The same inference serves as an I-S explanation of the swelling, should it occur. Should it not occur, we would have no explanation for *this*, on Hempel's model.

However, further investigation of Jones' medical history might reveal that he suffered mononucleosis once before, and failed to develop any swelling. Let us suppose that such individuals have a much higher than normal probability of *not* showing swelling in any later bouts with mononucleosis, say .9 rather than .01. This new law and new information about Jones together permit an inference with practical certainty to the conclusion that he will *not* develop swelling, and thus support a corresponding *I-S* explanation. Relative to these new facts, however, no *I-S* explanation would be available should Jones, improbably, develop swelling. What are we to say now about the previous *I-S* explanation, which had just the opposite result? Hempel would reject it as no longer *maximally specific* relative to what we believe about Jones' case. The requirement of maximal specificity is a complicated affair,[3] but the basic idea is that we refer each case to the narrowest class of cases to which our present beliefs assign it in which the explanandum has a characteristically different probability. In Jones' case, the narrower class is clearly the class of those contracting mononucleosis for a second time who failed to develop lymph-gland swelling the first time.

If more information about Jones or new discoveries about mononucleosis turn up, we may be forced to move on to still another explanation. *I-S* explanations must be relativized to our current "epistemic situation," and are subject to change along with it. Hempel notes that this sets off *I-S* explanations from *D-N* and *D-S* explanations in a fundamental way:

> . . . the concept of statistical explanation for particular events is essentially relative to a given knowledge situation as represented by a class K of accepted statements. . . . [W]e can significantly speak of true *D-N* and *D-S* explanations: they are those potential *D-N* and *D-S* explanations whose premises (and hence also conclusions) are true—no matter whether this happens to be known or believed, and thus no matter whether the premises are included in *K*. But this idea has no significant analogue for *I-S* explanation . . . . ([3], pp. 402–403)

On Hempel's view, neither of the two contradictory explanations concerning Jones contains false premises, and the explananda in each case do indeed receive the degree of support indicated. It is just that we no longer regard the evidential relationship expressed by the first as explanatorily relevant. Were Jones to develop swelling after all, it would now have to be regarded as inexplicable.

What I take to be the two most bothersome features of *I-S* arguments as models for statistical explanation—the requirement of high proba-

---

[3]See, for example, [4].

---

bility and the explicit relativization to our present epistemic situation (bringing with it an exclusion of questions about the truth of *I-S* explanations)—derive from the inductive character of such inferences, not from the nature of statistical explanation itself. If a non-inductive model for the statistical explanation of particular facts is given, there need be no temptation to require high probability or exclude truth.

3. Jeffrey's Criticism of *I-S* Explanation. Richard C. Jeffrey has criticized Hempel's account on the grounds that statistical explanation is not a form of inference at all, except when the probability of the explanandum is "so high as to allow us to reason, in *any* decision problem, as if its probability were 1" ([5], p. 105). For such exceptional, "beautiful" cases, Jeffrey accepts *I-S* inferences as explanatory because they provide virtual "proof that the phenomenon *does* take place" ([5], p. 106).

For unbeautiful cases, there is no way of proving (in advance) that the explanandum phenomenon will occur. According to Jeffrey, the explanation *why* such unbeauties come to be is a curt "By chance." He has more to say on *how* they come about:

> . . . in the statistical case I find it strained to speak of knowledge *why* the outcome is such-and-such. I would rather speak of *understanding the process*, for the explanation is the same no matter what the outcome: it consists of a statement that the process is a stochastic one, following such-and-such a law.[4] ([5], p. 24)

Jeffrey is surely right, as against Hempel, that probable and improbable outcomes of indeterministic processes are equally explicable, and explicable in the same way. After all, why should it be explicable that a genuinely random wheel of fortune with 99 red stops and 1 black stop came to a halt on red, but inexplicable that it halted on black? Worse, on Hempel's view, halting at any *particular* stop would be inexplicable, even though the wheel must halt at some particular stop in order to yield the explicable outcome *red*.

But I fail to see how Jeffrey can defend his exemption of beautiful cases against a similar line of argument. If the burden in statistical explanation really lies with "*understanding the process* . . . no matter what the outcome," then why should it matter whether the outcome is so highly probable "as to allow us to reason, in *any* decision problem, as if its probability were 1?" The neglect Jeffrey shows here toward minute chances is appropriate for the practical task of decision-making (and perhaps explained by his generally subjectivist approach to

---

[4] A typographical error has been corrected.

probability), but we must not overlook them in the not-entirely-practical task of explaining. Virtually impossible events may occur, and they deserve and can receive the same explanation as the merely improbable or the virtually certain.

4. A *D-N* Model of Probabilistic Explanation. I will present my account of probabilistic explanation by developing an example of just such "practically negligible"—but physically real and lawful—chance: alpha-decay in long-lived radioactive elements. The mean-life of the more stable radionuclides is so long as to make the probability for any particular nucleus of such an element to decay during our lifetimes effectively zero. But our nuclear theory shows that it is *not* zero, and explains how such rarities can occur.

On the account offered here, probabilistic explanations will be either true or false independent of our epistemic situation. Moreover, to explain, they must be true. Here I am following Hempel's usage in calling an explanatory argument *true* just in case it is valid and its premises are true ([3], p. 338). Such an explanation will *not* be true if the probabilistic laws it invokes are not true; in particular, it will not be true unless the process responsible for the explanandum is genuinely indeterministic. If alpha-decay is to serve as our paradigm for probabilistic explanation, we must be correct in assuming that the probabilistic wave-mechanical account of particle transmission through the nuclear potential barrier tells us all there is to know about the cause of alpha-decay. At least, it must be true that there are no hidden variables characterizing unknown initial conditions that suffice to account for alpha-decay deterministically. However, I take it to be uncontroversial that alpha-decay is an indeterministic process, if any is.

Let us suppose that we are given an individual instance of alpha-decay to explain: a nucleus of radionuclide Uranium^{238}, call it '*u*', has emitted an alpha-particle during the time interval lasting from t_0 to $t_0 + \theta$, where θ is very small and expressed in standard units. Since the mean-life of U^{238} is 6.5×10^9 years, the probability of observing a decay by *u* during this interval is exceedingly small, but unquestionably exists (witness the decay). This probability can be given precisely by using the radioactive decay constant λ_{238} characteristic of all atoms of U^{238}. Significantly, we need not know when in the course of the history of *u* time t_0 occurs: the probability of decay is unaffected by the age of the atom. Therefore, as long as decay has not yet occurred, individual "trials"—consisting of observing a single isolated radioactive nucleus for successive intervals of the same length—are statistically independent. Using these two facts we can determine the probability of decay for individual nuclei during any time interval chosen: it will be 1 minus the probability that any such nucleus *survives* the interval intact; for u, $(1 - \exp(-\lambda_{238} \cdot \theta))$.

To obtain experimental confirmation of this value, we infer *from* the probability to decay of individual nuclei *to* statistical features of sample populations of nuclei, e.g., half-life and mean-life. These predicted statistical features are then checked against actual observed relative frequencies in large populations over long intervals. *Physical* probabilities of the sort being considered here are therefore to be contrasted with *statistical* probabilities; the former express the strength of a certain physical possibility for a given system, while the latter reduce to claims about the (limiting) relative frequencies of traits in sample populations. Much well-founded doubt has been expressed about the applicability of statistical probabilities to single cases, but physical probabilities are *located* in the features of the single case. Therefore, we can understand our nuclear theory as implying strictly universal (physical) probability-attributing laws of the form:

(1) All nuclei of radioelement E have probability $(1 - \exp(-\lambda_E \cdot t))$ to emit an alpha-particle during any time interval of length t, unless subjected to environmental radiation.

Because schema (1) is universal in form, its instances are candidates for law premises in deductive-nomological inferences concerning individual nuclei. Thus, for u:

(2) (a) All nuclei of U^{238} have probability $(1 - \exp(-\lambda_{238} \cdot \theta))$ to emit an alpha-particle during any interval of length θ, unless subjected to environmental radiation.

(b) u was a nucleus of U^{238} at time t_0, and was subjected to no environmental radiation before or during the interval $t_0 - (t_0 + \theta)$.

(c) u had probability $(1 - \exp(-\lambda_{238} \cdot \theta))$ to emit an alpha-particle during the interval $t_0 - (t_0 + \theta)$.

(2), it appears, gives a *D-N* explanation only of the fact that u had such-and-such a probability to decay during the interval in question, but we should look a bit closer. I submit that (2), when supplemented as follows, is the probabilistic explanation of u's decay:

(3) A derivation of (2a) from our theoretical account of the mechanism at work in alpha-decay.
The *D-N* inference (2).
A parenthetic addendum to the effect that u did alpha-decay during the interval $t_0 - (t_0 + \theta)$.

Am I merely making a virtue of necessity, and saying that since (3) contains all we can say about u's decay, (3) must explain it? In fact, there is a great deal more we could say about u's decay. Deliberately left out of (3) are innumerable details about the experimental apparatus (temperature, pressure, location, etc.), about the beliefs and expectations of those monitoring the experiment, and about the epistemic position of the scientific community at the time. These facts are omitted as *explanatorily irrelevant* to u's decay because they are *causally irrelevant* to the physical possibility for decay that obtained during the interval in question, and to whether or not that possibility was realized.[5] A full account of these notions of explanatory and causal relevance is not possible here, so instead I will go on to argue that what (3) comprises *is* explanatorily relevant, and explanatory.

I must begin this task with a defense of the nomological status of (2a), and of the legitimacy of treating it as a covering law for u's decay. The following criterion of nomologicality will be used: a law is a universal truth derivable from our theory without appeal to particular facts. This criterion of course lacks generality (what counts as theory if not the laws themselves?), fails to segregate natural from logical laws, picks out only so-called "universal" (as opposed to "local") laws, and is entirely too vague (how to distinguish "particular facts" from the rest?). But I trust it will do for now. The motive for excluding "particular facts" is that some true, universal statements derivable from our theory *plus* particular facts would not normally be regarded as universal laws, but would at best be "local laws," e.g., 'All *Homo neanderthalensis* live during the late Pleistocene age.'

The generalization in question here, (2a), is derived by solving the Schrödinger wave-equation for an alpha-particle of energy ≈ 4.2 MeV for the potential regions in and around the nucleus of an element with atomic number 92 and atomic weight 238, none of which are "particular facts," plus some simplifying assumptions about the structure of the nucleus and the distinctness of the alpha-particle within it prior to decay. While it is forbidden by classical physics for a low-energy particle like the ≈ 4.2 MeV alpha-particle associated with U^{238} to pass through the 24.2 MeV potential barrier surrounding

---

[5]Causal relevance is established here via the wave-equation. I do not mean to suggest that *causal* relevance is the only explanatory kind; cf. the mention of structural laws in section 1.

Some such notion of causal relevance appears to lie behind Salmon's "statistical-relevance" model of probabilistic explanation. Yet what matters is whether a factor enters into the probabilities present, not the statistics they produce.

so massive a nucleus, the quantum theory predicts that the probability amplitude for finding such an alpha-particle outside the potential barrier is nonzero. Thus a transmission coefficient for U^{238} alpha-particles is determined, which, given certain simplifying assumptions about the goings-on inside the nucleus, yields the probability that such a particle will tunnel out of the potential well "per unit time for one nucleus," namely, λ_{238} ([1], p. 175). (2a) thus neither reports a summary of past observations nor expresses a mere statistical uniformity that scattered initial conditions would lead us to anticipate. Instead, it is a law of irreducibly probabilistic form, assigning definite, physically-determined probabilities to individual systems.

It follows that the derivation of conclusions from (2a) by universal instantiation and *modus ponens* is unexceptionable. Were (2a) but a statistical generalization, properly understood as meaning "(1 − exp($-\lambda_{238}$ · θ))N of U^{238} nuclei in samples of sufficiently large size N, on average, decay during the interval $t_0 − (t_0 + \theta)$," it could not undergo universal instantiation, and would not permit detachment of a conclusion about the probability obtaining in a single case.

Further, if the wave-equation does indeed tell us all there is to know about the mechanism involved in nuclear barrier penetration, it follows that nothing more can be said to explain why the observed decay of u took place, once we have shown how (2a) is derived from our account of this mechanism, and established that (2) is valid and that (3)'s parenthetic addendum is true.

Still, does (3) explain why the decay took place? It does not explain why the decay *had to* take place, nor does it explain why the decay *could be expected to* take place. And a good thing, too: there is no *had to* or *could be expected to* about the decay to explain—it is not only a chance event, but a very improbable one. (3) does explain why the decay *improbably* took place, which is how it did. (3) accomplishes this by demonstrating that there existed at the time a small but definite physical possibility of decay, and noting that, by chance, this possibility was realized. The derivation of (2a) that begins (3) shows, by assimilating alpha-decay to the chance process of potential barrier tunneling, how this possibility comes to exist. If alpha-decays are chance phenomena of the sort described, then once our theory has achieved all that (3) involves, it has explained them to the hilt, however unsettling this may be to *a priori* intuitions. To insist upon stricter subsumption of the explanandum is not merely to demand what (alas) cannot be, but what decidedly should not be: sufficient reason that one probability rather than another be realized, that is, chances without chance.

Because of the peculiar nature of chance phenomena, it is explanatorily relevant whether the probability in question was realized, even
though there is no before-the-fact explanatory *argument*, deductive
or inductive, to this conclusion. Indeed, it is the absence of such
an argument that makes a place in probabilistic explanation for a
parenthetic addendum concerning whether the possibility became
actual in the circumstances given. These addenda may offend those
who believe that explanations must always be arguments, but at the
most general level explanations are *accounts*, not arguments. It so
happens that for deterministic phenomena inferences of a particular
kind—*D-N* arguments meeting the desiderata suggested in section
1—*are* explanatory accounts, and this for good reasons. However,
indeterministic phenomena are a different matter, and explanatory
accounts of them must be different as well. If the present model
is accepted, then almost all of the explanatory burden in probabilistic
explanation can be placed on deductive arguments—those characterizing the indeterministic mechanism and those attributing a certain
probability to the explanandum. But these arguments leave out a
crucial part of the story: did the chance fact obtain?

The parenthetic addendum fills this gap in the account, and commu-
nicates information that is relevant to the causal origin of the explanandum by telling us that it came about as the realization of a particular
physical possibility. Further, it permits us to chain probabilistic
explanations together to make more comprehensive explanations, in
which each link is able to bear the full explanatory burden for the
fact it covers, and is capable of leading us on to the next fact in
the causal sequence being explained. From (2) alone we cannot move
directly to an account of what the alpha-particle did to a nearby
photographic plate, but only to a probability (and a miserably low
one) that this account will be true. The parenthetic addendum to
(3) furnishes a non-probabilistic premise from which to begin an account
of the condition of the photographic plate: the occurrence of an
alpha-decay in the vicinity. Dropping off the addendum leaves an
explanation, but it is a *D-N* explanation of the occurrence of a particular
probability, not a probabilistic explanation of the occurrence of a
particular decay.

The scheme for probabilistic explanation of particular chance facts
by nomic subsumption that is being offered here, the *deductive-
nomological-probabilistic* (*D-N-P*) model, is this. First we display (or
truthfully claim an ability to display) a derivation from our theory
of a law of essentially probabilistic form, complete with an account
of how the law applies to the deterministic process in question. The

derived law is of the form:

(4a) $\forall t \forall x [F_{x,t} \rightarrow \text{Prob}(G)_{x,t} = p]$

"At any time, anything that is F has probability p to be G."

Next, we adduce the relevant fact(s) about the case at hand, e:

(4b) F_{e,t_0}

"e is F at time t_0,"

and draw the obvious conclusion:

(4c) $\text{Prob}(G)_{e,t_0} = p$

"e has probability p to be G at time t_0."

To which we add parenthetically, and according to how things turn out:

(4d) $(G_{e,t_0} / - G_{e,t_0})$

"(e did/did not become G at t_0)."

Whether a *D-N-P* explanation is true will depend solely upon the truth-values of its premises and addendum, and the validity of its logic. I leave open what becomes of a *D-N-P* explanation that contains true laws, initial facts, and addendum, but botches the theoretical account of the laws invoked. Let us simply say that the more botched, the less satisfactory the explanation.

The law premise (4a) will be true if all things at all times satisfy the conditional '$F_{x,t} \rightarrow \text{Prob}(G)_{x,t} = p$', using whatever reading of '$\rightarrow$' we decide upon for the analysis of natural laws in general. It will be false if there exists a partition of the Fs into those with *physical* probability r to be G and those with *physical* probability s to be G, where $s \neq r \neq p$. Such a partition might exist according to some *other* interpretation of probability, but this would not affect the truth of (4a). For example, suppose that a coin toss meeting certain specifications is an indeterministic event with probability $\frac{1}{2}$ of yielding heads. We now perform the experiment of repeating such a toss a great many times. Curiously, all and only even numbered tosses yield heads. This result supplies certain frequentists with grounds for saying that Prob(heads, even-numbered toss) = 1, while Prob(heads, odd-numbered toss) = 0.[6] But because all tosses met the specification laid down, the probability of heads was the same, $\frac{1}{2}$, on each toss, despite the curious behavior. Such behavior may make us suspicious of our original claims about the indeterminacy of the process or about the physical probability it has of producing heads, but is no proof against them. Indeed, the original probability attribution requires us to assign a definite physical probability to just such an untoward sequence of outcomes, the occurrence of which therefore hardly contradicts this attribution.

The particular fact premise (4b) will be true iff e is an F during the time in question, and not either an F^* (with probability $r \neq p$ to be G) or an F^{**} (with probability $q = p$ to be G, but unlike an F in other respects). Using the (let us say) true law that all F^{**}s have probability $q = p$ to be G, and the falsehood that e is an F^{**}, we could derive a true conclusion, indistinguishable from (4c). Hence the requirement that the *premises* be true if the argument is to explain; and if we reason logically from true premises, the conclusion will take care of itself.

5. Epistemic Relativity and Maximal Specificity Disowned. Have I kept my promise to give an account of probabilistic explanation free from relativization to our present epistemic situation?

Let us return to explanation (3), and admit that it is not the whole story: 23% of the alpha-particles emitted by U^{238} have kinetic energy 4.13 MeV, while the remaining 77% have 4.18 MeV. Therefore there are two different decay constants, $\lambda_{238}^{4.13}$ and $\lambda_{238}^{4.18}$; both are distinct from λ_{238}, used in (3). Hence we must be quite careful in stating what exactly (3) explains. It does *not* explain the particular *event* observed, for this was either a 4.13 or a 4.18 MeV decay, neither of which has probability λ_{238} in unit time. Instead, (3) explains the particular *fact about* the event observed that we set out to explain, namely, that an alpha-decay with unspecified energy (or direction, or angular momentum, etc.) took place at nucleus u during the time interval in question. This fact *does* have probability λ_{238} of obtaining in unit time, representing the sum of the two energy-correlated probabilities with which such a decay might occur.

---

[6]Cf. the discussion of place selections and homogeneity in [6], sections 4 and 7. Salmon's criterion, which requires formal randomness, would here fail to distinguish a *randomly-produced* regular sequence from a *deterministically-produced* one. Notwithstanding formal similarities, only the latter is appropriately explained non-probabilistically.

If we should learn that the decay of u was of a 4.18 MeV alpha-particle, an explanation of *this* fact would have to be referred to the more specific class of decays with probability $\lambda_{238}^{4.18}$ in unit time. Is the maximal specificity requirement thereby resurrected? There is no need for it. (3) is not an unspecific explanation of this more specific fact, but a fallacious one. It would be logically corrupt to conclude from law (2a) that an individual U^{238} nucleus has probability $(1 - \exp(-\lambda_{238} \cdot \theta))$ to decay *with energy 4.18 MeV* during any interval of length θ, since (2a) says nothing whatsoever about decay energies. The only relevant conclusion to draw from (2a) is (2c), which remains true in the face of our more detailed knowledge about the event in question. Nor is law (2a) falsified by the discovery of a 23:77 proportional distribution of decay energies, and the associated difference in decay rates. For according to our nuclear theory, there is no difference in initial condition between a nucleus about to emit a 4.13 MeV alpha-particle and one about to emit a 4.18 MeV alpha-particle. It remains true that *all* U^{238} nuclei have probability λ_{238} to decay in unit time, but it is further true that all have probability $\lambda_{238}^{4.13}$ to decay one way, and probability $\lambda_{238}^{4.18}$ to decay another.

It must next be determined whether the existence of a difference in probability *due to* a difference in initial condition can be handled by the D-N-P model without appeal to a maximal specificity requirement. To permit consideration of possible epistemological complications, it will be assumed that neither the difference in probability nor the partition of initial conditions is known at the start.

Imagine that, although we do not know it, in virtue of certain permanent structural features 23% of all naturally-occurring U^{238} nuclei fall into a class P, and the remaining 77% into a class $-P$, such that only those in P have any probability of emitting a 4.13 MeV alpha-particle, and only those in class $-P$ have any probability of emitting a 4.18 MeV alpha-particle. Suppose further that these two laws have been derived:

    (5) (a) All U^{238} nuclei of type P have probability $(1 - \exp(-\lambda_{238}^{4.13} \cdot t))$ to emit a 4.13 MeV alpha-particle during any time interval of length t, unless subjected to environmental radiation.

        (b) All U^{238} nuclei of type $-P$ have probability $(1 - \exp(-\lambda_{238}^{4.18} \cdot t))$ to emit a 4.18 MeV alpha-particle during any time interval of length t, unless subjected to environmental radiation.

Note that, by our assumptions, the specification of the kinetic energy of the particle (possibly) emitted may be dropped from (5a) and (5b) without altering the truth of either.

Until the structural differences between types P and $-P$ are discovered and understood, (3) will stand as the accepted explanation of u's decay. However, once (5a) and (5b) have become known, it will be clear from the fact that u's alpha-emission had kinetic energy 4.18 MeV that u must have been of type $-P$ prior to decay. Thus a more specific account of u's decay will be available to scientists, who, already familiar with the theoretical derivation of law (5b), offer the following truncated D-N-P version of this account:

(6) (a) All nuclei of U^{238} of type $-P$ have probability $(1 - \exp(-\lambda_{238}^{4.18} \cdot \theta))$ to emit an alpha-particle during any time interval of length θ, unless subjected to environmental radiation.

   (b) u was a nucleus of U^{238} of type $-P$ at t_0, and was subjected to no environmental radiation before or during the interval $t_0 - (t_0 + \theta)$.

   (c) u had probability $(1 - \exp(-\lambda_{238}^{4.18} \cdot \theta))$ to emit an alpha-particle during the interval $t_0 - (t_0 + \theta)$.

   (d) (And it did.)

On the Hempelian model (modified so as to permit I-S explanations of improbable phenomena), there is no problem in accounting for the previous acceptability of the I-S counterpart of (3), or for its present unacceptability. (3) had been maximally specific relative to our previous beliefs about alpha-decay in U^{238}, but no longer is, and so is superseded by the more specific (relatively speaking) I-S counterpart of (6).

On the D-N-P model, too, there is no problem in accounting for the acceptability of (3) prior to the discovery of class $-P$ and law (5b): (3)'s premises (and, of course, addendum) were taken to be true. The question is whether, in light of current beliefs, (3) can be ruled out—and (6) ruled in—without invocation of Hempelian constraints. Resolution of the problem (3) and (6) pose through epistemic relativization and maximal specificity requirements seems to me unacceptable. If we were to attribute to nucleus u two unequal probabilities to alpha-decay in a specified way during a single time interval, adding, "Let's pick the most specifically defined value for explanatory purposes," we'd be showing an unseemly tolerance for contradiction in our nuclear theory—and why stop there? Better to face up to the confrontation over truth between (3) and (6), and replace complex and unappealingly relativistic maximal specificity requirements with the simple requirement of truth. The D-N-P model does this. The current unacceptability of (3) is located not in premises insufficiently specific, but in premises insufficiently true, i.e., false. Contrary to (3)'s purported covering law (2a), not all nuclei of U^{238} have probability λ_{238} to decay in unit time if unperturbed by radiation—in fact, none do. In spite of giving accurate expectation values for decay rates in large samples of U^{238}, (2a) is false, and so explanation (3) is ruled out as unsound. Explanation (6), on the other hand, meets the simple requirement of truth, and rules itself in.[7]

Problems about incomplete, misleading, or false beliefs do not bear on whether D-N-P explanations have unrelativized truth-values, but concern rather difficulties in *establishing* the truth-values they unrelativistically have. Relativization to our current epistemic situation comes into play only when we begin to discuss whether a given D-N-P explanation *seems* true. Whether it *is* true is another matter.

6. Objections to the D-N-P Model. I cannot pretend to have said enough about deductive-nomological-probabilistic explanation to have characterized this model adequately. Such reservations as were expressed in section 1 about taking nomic subsumption under a causal law as sufficient for explanation are still in force, and little has been done—except by way of example—to show how the account offered here might accommodate them.

That the probabilistic laws invoked in D-N-P explanations are even (in some relevant sense) *causal* cannot be defended until a plausible account of physical probability has been worked out, a task well beyond the scope of this paper. Under a *propensity interpretation*, probability has the characteristics sought: a probability is the expression of the strength of a physical tendency in an individual chance system to produce a particular outcome; it is therefore straightforwardly applicable to single cases; and it is (in a relevant sense) causally responsible for that outcome whenever it is realized. However, propensities are notoriously unclear. For now I can at best assume that clarification is possible, point to a promising start in the attempt to do so—R. N. Giere, "Objective, Single-Case Probabilities and the Foundations of Statistics" ([2])—, and admit that the D-N-P model is viable only if sense can be made of propensities, or of objective, physical, lawful, single-case probabilities by any other name.

As for the requirement that explanations elucidate mechanisms, I can only repeat that an essential role is played in D-N-P explanations

---

[7]Explanation (6) is true, however, only under the contrary-to-fact assumption—made for the sake of the example—of the existence of a class $-P$.

by the theoretical deduction of the probabilistic law(s) covering the explanandum.

In lieu of further exposition, I offer the beginnings of a defense, hoping thereby to sketch out the account a bit more fully in those areas most likely to be controversial.

Because it applies only to genuinely indeterministic processes, of which there are few (if any), D-N-P explanation is too restricted in scope. It is widely believed that the probabilities associated with standard gambling devices, classical thermodynamics, actuarial tables, weather forecasting, etc., arise not from any underlying physical indeterminism, but from an unknown or uncontrolled scatter of initial conditions. If this is right, then *D-N-P* explanation would be inapplicable to these phenomena even though they are among the most familiar objects of probabilistic explanation. I do not, however, find this troublesome: if something does not happen by chance, it cannot be explained by chance. The use of epistemic or statistical probabilities in connection with such phenomena unquestionably has instrumental value, and should not be given up. What must be given up is the idea that *explanations* can be based on probabilities that have no role in bringing the world's explananda about, but serve only to describe deterministic phenomena.[8] Whether there *are* any probabilites that enter into the mechanisms of nature is still debated, but the successes of the quantum-mechanical formalism, and the existence of "no hidden variable" results for it, place the burden of proof on those who would insist that physical chance is an illusion.

It could be objected more justly that *D-N-P* explanation is too broad, not too narrow, in scope. Once restrictions have been lifted from the value a chance may have in probabilistic explanation, virtually all explanations of particular facts must become probabilistic. All but the most basic regularities of the universe stand forever in peril of being interrupted or upset by intrusion of the effects of random processes. It might seem a fine explanation for a light's going out that we opened the only circuit connecting it with an electrical power source, but an element of chance was involved: had enough atoms in the vicinity of the light undergone spontaneous beta-decay at the right moment, the electrons emitted could have kept it glowing. The success of a social revolution might appear to be explained by its overwhelming popular support, but this is to overlook the revolu-

[8]Of course, we might speak of statistical or epistemic probabilities as causes of, e.g., beliefs. But if belief formation is not *physically* probabilistic, then probabilistic explanation of it would be impossible, in spite of this sort of causal involvement on the part of statistical or epistemic probabilities.

tionaries' luck: if all the naturally unstable nuclides on earth had commenced spontaneous nuclear fission in rapid succession, the triumph of the people would never have come to pass.

No doubt this proliferation of probabilistic explanations is counter-intuitive, but contemporary science will not let us get away with any other sort of explanation in these cases—it simply cannot supply the requisite non-probabilistic laws. Because they figure in the way things *work*, tiny probabilities appropriately figure in explanations of the way things *are*, even though they scarcely ever show up in the way things turn out.

The *D-N-P* model breaks the link between prediction and explanation. Hempel has justified a "qualified thesis of the structural identity of explanation and prediction" with this principle:

> Any rationally acceptable answer to the question 'Why did X occur?' must offer information which shows that X was to be expected—if not definitely, then at least with reasonable certainty. ([3], pp. 367–368)

Abundantly many $D-N-P$ explanations—all those covering less than highly probable facts—violate this condition.

However, to abide by this condition and renounce explanations with meager probabilities I take to be worse. Why forgo the explanations of improbable phenomena offered by our theories, when these explanations provide as much of an account of why (and how) their explananda occur as do the explanations of "reasonably certain" phenomena that Hempel's condition sanctions?

Too restrictive as it stands, Hempel's condition may be taken in a way not incompatible with $D-N-P$ explanation. A $D-N-P$ explanation does yield one prediction that is perfectly strict, to the effect that a certain physical probability exists in the circumstances given. If this probability fails to obtain, or to have the value attributed to it, the explanation must be false. It is a complaint against the world, not against the $D-N-P$ model, that a direct, non-statistical test for the presence or value of this probability may prove impossible. Remarkably, the mechanisms of the world leave room for spontaneous nuclear disintegrations. Equally remarkably, our physical theory gives us insight into how they come about, and assigns determinate probabilities to them. These probabilities are connected to the rest of our theory by laws that permit both prediction and (where means exist) control: if undisturbed, nucleus a will have probability p to alpha-decay (so we should expect a's decay with *epistemic* probability p); and if we wish to alter p, our theory tells us how a must be disturbed.

It has been objected to the view of probability taken in this paper that unless probability attributions are interpreted as predictions about how relative frequencies will *actually* come out in the long run, probabilistic laws lack empirical content. Thus, if the relative frequency of decayed atoms in a large sample of some radioelement were, over a great length of time, to diverge significantly from the probability theoretically attributed to decay, that attribution would not be "borne out," i.e., would be falsified. Otherwise, it is argued, probabilistic laws are compatible with all frequencies, and empirically vacuous.

But it is impossible for a world to "bear out" all of its probability-attributing laws in this sense. For these laws imply, among other things, that it is extremely unlikely that *all* actual long-run sequences will show a relative frequency near to the single-case probability. Therefore, the demand that all long-run decay rates nearly match all corresponding decay constants comes to a demand that nothing improbable show up in the long run, which is itself an improbability showing up in the long run. Intended to clear things up on the epistemological front, this proposal cannot even get out of its own way.

By splitting apart probabilistic explanation and induction, the *D-N-P* model has lost the point of probabilistic explanation. Behind this objection lies the view that probabilistic (or statistical) explanation is an activity fundamentally unlike *D-N* explanation. A probabilistic explanation is seen as a piece of detective work. Unable to give a causal demonstration of the explanandum from evidence thus far assembled, we develop hypotheses, which are judged by how probable they are on the evidence, and whether they make the explanandum sufficiently probable. In the end, we put forward the most convincing inductive argument yet found—the one making the explanandum most antecedently probable, given what else we know about events leading up to it.

This view of probabilistic explanation confuses epistemic with objective probability, and induction with explanation. Perhaps responsible for this confusion is the similarity of the tasks of explaining a phenomenon, gathering support for such an explanation, and gathering before- or after-the-fact evidence for a phenomenon's occurrence. This confusion is abetted by misleading ways of talking about "strong" or "good" explanations. We should distinguish the following. (i) *A strong (good) explanation* is one that has great theoretical power, regardless of how well-confirmed it is or how probable it holds the explanandum to be. (ii) *A strong (good) candidate for explanation* is a proffered explanation with well-confirmed *premises*, regardless of how probable it holds the explanandum to be and irrespective of how theoretically powerful it happens to be. (iii) *A strong (good) reason for believing that the explanandum fact will obtain* is furnished by before-the-fact evidence that leads, via one's theory, to an expectation of the explanandum with high epistemic probability. (iv) *A strong (good) reason for believing that the explanandum fact obtained* is given by any evidence that lends high epistemic probability to the proposition that the explanandum fact is a fact. Strong after-the-fact evidence, even for very improbable events, may be easy to come by. Reasons of types (iii) and (iv) need have nothing to do with explanation, and may be based on symptoms (Will it rain today?— Harry's rheumatism is acting up) or even less causally relevant information (Was Sue upset?—Her brother is certain she would have been).

Although the link between probabilistic explanation and induction is looser on the *D-N-P* model than on the *I-S* model, this is no fault: on Hempel's account it was entirely too close. Measuring the strength or "acceptability" of an explanation by the magnitude of the probability it confers on the explanandum blurs the distinctions just made. Keeping (i)–(iv) distinct, the *D-N-P* model enables us to state quite simply the object of induction in explanation: given a particular fact, to find, and gather evidence for, an explanans that subsumes it; given a generalization, to find, and gather evidence for, a higher-level explanans that subsumes it; in all cases, then, to discover and establish a true and relevant explanans. The issue of showing the explanandum to have high (relative or absolute) probability is a red herring, distracting attention from the real issue: the truth or falsity, and applicability, of the laws and facts adduced in explanatory accounts.

REFERENCES

[1] Evans, R. D. *The Atomic Nucleus.* New York: McGraw Hill, 1965.
[2] Giere, R. N. "Objective Single-Case Probabilities and the Foundations of Statistics." In *Logic, Methodology and Philosophy of Science, Vol. IV.* Edited by P. Suppes, *et al.* Amsterdam: North-Holland, 1973.
[3] Hempel, C. G. "Aspects of Scientific Explanation." In *Aspects of Scientific Explanation and Other Essays.* New York: Free Press, 1965.
[4] Hempel, C. G. "Maximal Specificity and Lawlikeness in Probabilistic Explanation." *Philosophy of Science* 35 (1968): 116–133.
[5] Jeffrey, R. C. "Statistical Explanation vs. Statistical Inference." In *Essays in Honor of C. G. Hempel.* Edited by N. Rescher, *et al.* Dordrecht: D. Reidel, 1970.
[6] Salmon, W. C. "Statistical Explanation." In *Statistical Explanation and Statistical Relevance.* Edited by W. Salmon. Pittsburgh: University of Pittsburgh Press, 1971.

2 | Statistical Explanation
and Its Models

THE PHILOSOPHICAL THEORY of scientific explanation first entered the twentieth century in 1962, for that was the year of publication of the earliest bona fide attempt to provide a systematic account of statistical explanation in science.[1] Although the need for some sort of inductive or statistical form of explanation had been acknowledged earlier, Hempel's essay "Deductive-Nomological vs. Statistical Explanation" (1962) contained the first sustained and detailed effort to provide a precise account of this mode of scientific explanation. Given the pervasiveness of statistics in virtually every branch of contemporary science, the late arrival of statistical explanation in philosophy of science is remarkable. Hempel's initial treatment of statistical explanation had various defects, some of which he attempted to rectify in his comprehensive essay "Aspects of Scientific Explanation" (1965a). Nevertheless, the earlier article did show unmistakably that the construction of an adequate model for statistical explanation involves many complications and subtleties that may have been largely unanticipated. Hempel never held the view—expressed by some of the more avid devotees of the D-N model—that *all* adequate scientific explanations must conform to the deductive-nomological pattern. The 1948 Hempel-Oppenheim paper explicitly notes the need for an inductive or statistical model of scientific explanation in order to account for some types of legitimate explanation that actually occur in the various sciences (Hempel, 1965, pp. 250–251). The task of carrying out the construction was, however, left for another occasion. Similar passing remarks about the need for inductive or statistical accounts were made by other authors as well, but the project was not undertaken in earnest until 1962—a striking delay of fourteen years after the 1948 essay.

One can easily form the impression that philosophers had genuine feelings of ambivalence about statistical explanation. A vivid example can be

[1] Ilkka Niiniluoto (1981, p. 444) suggests that "Peirce should be regarded as the true founder of the theory of inductive-probabilistic explanation" on account of this statement, "The statistical syllogism may be conveniently termed the explanatory syllogism" (Peirce, 1932, 2:716). I am inclined to disagree, for one isolated and unelaborated statement of that sort can hardly be considered even the beginnings of any geniune theory.

found in Carnap's *Philosophical Foundations of Physics* (1966), which was based upon a seminar he offered at UCLA in 1958.[2] Early in the first chapter, he says:

> The general schema involved in *all explanation* can be expressed symbolically as follows:
>
> 1. $(x) (Px \supset Qx)$
> 2. Pa
> 3. Qa
>
> The first statement is the universal law that applies to any object x. The second statement asserts that a particular object a has the property P. These two statements taken together enable us to derive logically the third statement: object a has the property Q. (1966, pp. 7–8, italics added)

After a single intervening paragraph, he continues:

> At times, in giving an explanation, the only *known* laws that apply are statistical rather than universal. In such cases, we must be content with a statistical explanation. (1966, p. 8, italics added)

Farther down on the same page, he assures us that "these are genuine explanations," and on the next page he points out that "In quantum theory . . . we meet with statistical laws that may not be the result of ignorance; they may express the basic structure of the world." I must confess to a reaction of astonishment at being told that all explanations are deductive-nomological, but that some are not, because they are statistical. This lapse was removed from the subsequent paperback edition (Carnap, 1974), which appeared under a new title.

Why did it take philosophers so long to get around to providing a serious treatment of statistical explanation? It certainly was not due to any absence of statistical explanations in science. In antiquity, Lucretius (1951, pp. 66–68) had based his entire cosmology upon explanations involving spontaneous swerving of atoms, and some of his explanations of more restricted phenomena can readily be interpreted as statistical. He asks, for example, why it is that Roman housewives frequently become pregnant after sexual intercourse, while Roman prostitutes to a large extent avoid doing so. Conception occurs, he explains, as a result of a collision between a male seed and a female seed. During intercourse the prostitutes wiggle their

---

[2] As Carnap reports in the preface, the seminar proceedings were recorded and transcribed by his wife. Martin Gardner edited—it would probably be more accurate to say "wrote up"—the proceedings and submitted them to Carnap, who rewrote them extensively. There is little doubt that Carnap saw and approved the passages I have quoted.

hips a great deal, but wives tend to remain passive; as everyone knows, it is much harder to hit a moving target (1951, p. 170).[3] In the medieval period, St. Thomas Aquinas asserted:

> The majority of men follow their passions, which are movements of the sensitive appetite, in which movements of heavenly bodies can cooperate: but few are wise enough to resist these passions. Consequently astrologers are able to foretell the truth in the majority of cases, especially in a general way. But not in particular cases; for nothing prevents man resisting his passions by his free will. (1947, 1:Qu. 115, a. 4, *ad* Obj. 3)

Astrological explanations are, therefore, of the statistical variety. Leibniz, who like Lucretius and Aquinas was concerned about human free will, spoke of causes that incline but do not necessitate (1951, p. 515; 1965, p. 136).

When, in the latter half of the nineteenth century, the kinetic-molecular theory of gases emerged, giving rise to classical statistical mechanics, statistical explanations became firmly entrenched in physics. In this context, it turns out, many phenomena that *for all practical purposes* appear amenable to strict D-N explanation—such as the melting of an ice cube placed in tepid water—must be admitted *strictly speaking* to be explained statistically in terms of probabilities almost indistinguishable from unity. On a smaller scale, Brownian motion involves probabilities that are, both theoretically and practically, definitely less than one. Moreover, two areas of nineteenth-century biology, Darwinian evolution and Mendelian genetics, provide explanations that are basically statistical. In addition, nineteenth-century social scientists approached such topics as suicide, crime, and intelligence by means of "moral statistics" (Hilts, 1973).

In the present century, statistical techniques are used in virtually every branch of science, and we may well suppose that most of these disciplines, if not all, offer statistical explanations of some of the phenomena they treat. The most dramatic example is the statistical interpretation of the equations of quantum mechanics, provided by Max Born and Wolfgang Pauli in 1926–1927; with the aid of this interpretation, quantum theory explains an impressive range of physical facts.[4] What is even more im-

---

[3] Lucretius writes: "A woman makes conception more difficult by offering a mock resistance and accepting Venus with a wriggling body. She diverts the furrow from the straight course of the ploughshare and makes the seed fall wide of the plot. These tricks are employed by prostitutes for their own ends, so that they may not conceive *too frequently* and be laid up by pregnancy" (1951, p. 170, italics added).

[4] See (Wessels, 1982), for an illuminating discussion of the history of the statistical interpretation of quantum mechanics.

portant is that this interpretation brings in statistical considerations at the most basic level. In nineteenth-century science, the use of probability reflected limitations of human knowledge; in quantum mechanics, it looks as if probability may be an ineluctable feature of the physical world. The Nobel laureate physicist Leon Cooper expresses the idea in graphic terms: "Like a mountain range that divides a continent, feeding water to one side or the other, the probability concept is the divide that separates quantum theory from all of physics that preceded it" (1968, p. 492). Yet it was not until 1962 that any philosopher published a serious attempt at characterizing a statistical pattern of scientific explanation.

INDUCTIVE-STATISTICAL EXPLANATION

When it became respectable for empirically minded philosophers to admit that science not only describes and predicts, but also explains, it was natural enough that primary attention should have been directed to classic and beautiful examples of deductive explanation. Once the D-N model had been elaborated, either of two opposing attitudes might have been taken toward inductive or statistical explanation by those who recognized the legitimacy of explanations of this general sort. It might have been felt, on the one hand, that the construction of such a model would be a simple exercise in setting out an analogue to the D-N model or in relaxing the stringent requirements for D-N explanation in some straightforward way. It might have been felt, on the other hand, that the problems in constructing an appropriate inductive or statistical model were so formidable that one simply did not want to undertake the task. Some philosophers may unreflectingly have adopted the former attitude; the latter, it turns out, is closer to the mark.

We should have suspected as much. If D-N explanations are deductive arguments, inductive or statistical explanations are, presumably, inductive arguments. This is precisely the tack Hempel took in constructing his inductive-statistical or I-S model. In providing a D-N explanation of the fact that this penny conducts electricity, one offers an explanans consisting of two premises: the particular premise that this penny is composed of copper, and the universal law-statement that all copper conducts electricity. The explanandum-statement follows deductively. To provide an I-S explanation of the fact that I was tired when I arrived in Melbourne for a visit in 1978, it could be pointed out that I had been traveling by air for more than twenty-four hours (including stopovers at airports), and almost everyone who travels by air for twenty-four hours or more becomes fatigued. The explanandum gets strong inductive support from those premises; the event-to-be-explained is thus subsumed under a statistical generalization.

It has long been known that there are deep and striking disanalogies between inductive and deductive logic.[5] Deductive entailment is transitive; strong inductive support is not. Contraposition is valid for deductive entailments; it does not hold for high probabilities. These are *not* relations that hold in some approximate way if the probabilities involved are high enough; once we abandon strict logical entailment, and turn to probability or inductive support, they break down entirely. But much more crucially, as Hempel brought out clearly in his 1962 essay, the deductive principle that permits the addition of an arbitrary term to the antecedent of an entailment does not carry over at all into inductive logic. If A entails B, then $A.C$ entails B, whatever C may happen to stand for. However, no matter how high the probability of B given A, there is no constraint whatever upon the probability of B given both A and C. To take an extreme case, the probability of a prime number being odd is one, but the probability that a prime number smaller than 3 is odd has the value zero. For those who feel uneasy about applying probability to cases of this arithmetical sort, we can readily supply empirical examples. A thirty-year-old Australian with an advanced case of lung cancer has a low probability of surviving for five more years, even though the probability of surviving to age thirty-five for thirty-year-old Australians in general is quite high. It is *this* basic disanalogy between deductive and inductive (or probabilistic) relations that gives rise to what Hempel called *the ambiguity of inductive-statistical explanation*—a phenomenon that, as he emphasized, has no counterpart in D-N explanation. His *requirement of maximal specificity* was designed expressly to cope with the problem of this ambiguity.

Hempel illustrates the ambiguity of I-S explanation, and the need for the requirement of maximal specificity, by means of the following example (1965, pp. 394–396). John Jones recovers quickly from a streptococcus infection, and when we ask why, we are told that he was given penicillin, and that almost all strep infections clear up quickly after penicillin is administered. The recovery is thus rendered probable relative to these explanatory facts. There are, however, certain strains of streptococcus bacteria that are resistant to penicillin. If, in addition to the above facts, we were told that the infection is of the penicillin-resistant type, then we would have to say that the prompt recovery is rendered *improbable* relative to the available information. It would clearly be scientifically unacceptable to ignore such relevant evidence as the penicillin-resistant character of the

---

[5] These are spelled out in detail in (Salmon, 1965a). See (Salmon, 1967, pp. 109–111) for a discussion of the 'almost-deduction' conception of inductive inference.

infection; the requirement of maximal specificity is designed to block statistical explanations that thus omit relevant facts. It says, in effect, that when the class to which the individual case is referred for explanatory purposes—in this instance, the class of strep infections treated by penicillin—is chosen, we must not know how to divide it into subsets in which the probability of the fact to be explained differs from its probability in the entire class. If it has been ascertained that this particular case involved the penicillin-resistant strain, then the original explanation of the rapid recovery would violate the requirement of maximal specificity, and for that reason would be judged unsatisfactory.[6]

Hempel conceived of D-N explanations as valid deductive arguments satisfying certain additional conditions. Explanations that conform to his inductive-statistical or I-S model are correct inductive arguments also satisfying certain additional restrictions. Explanations of both sorts can be characterized in terms of the following four conditions:

1. The explanation is an argument with correct (deductive or inductive) logical form,
2. At least one of the premises must be a (universal or statistical) law,
3. The premises must be true, and
4. The explanation must satisfy the requirement of maximal specificity.

This fourth condition is automatically satisfied by D-N explanations by virtue of the fact that their explanatory laws are universal generalizations. If all A are B, then obviously there is no subset of A in which the probability of B is other than one. This condition has crucial importance with respect to explanations of the I-S variety. In general, according to Hempel (1962a, p. 10), an explanation is an argument (satisfying these four conditions) to the effect that the event-to-be-explained was to be expected by virtue of certain explanatory facts. In the case of I-S explanations, this means that the premise must lend high inductive probability to the conclusion—that is, the explanandum must be highly probable with respect to the explanans.

Explanations of the D-N and I-S varieties can therefore be schematized as follows (Hempel, 1965, pp. 336, 382):

$$C_1, C_2, \ldots, C_j \quad \text{(particular explanatory conditions)}$$

(D-N)

$$\frac{L_1, L_2, \ldots, L_k}{E} \quad \begin{array}{l} \text{(general laws)} \\ \text{(fact-to-be-explained)} \end{array}$$

---

[6] We shall see in chapter 3 that the requirement of maximal specificity, as formulated by Hempel in his (1965) and revised in his (1968), does not actually do the job. Nevertheless, this was clearly its intent.

The single line separating the premises from the conclusion signifies that the argument is deductively valid.

$$C_1, C_2, \ldots, C_j \quad \text{(particular explanatory conditions)}$$

(I-S)
$$L_1, L_2, \ldots, L_k \quad \text{(general laws, at least one statistical)}$$
$$\overline{\overline{}} \; [r]$$
$$E \qquad\qquad \text{(fact-to-be-explained)}$$

The double lines separating the premises from the conclusion signifies that the argument is inductively correct, and the number r expresses the degree of inductive probability with which the premises support the conclusion. It is presumed that r is fairly close to one.[7]

The high-probability requirement, which seems such a natural analogue of the deductive entailment relation, leads to difficulties in two ways. First, there are arguments that fulfill all of the requirements imposed by the I-S model, but that patently do not constitute satisfactory scientific explanations. One can maintain, for example, that people who have colds will probably get over them within a fortnight if they take vitamin C, but the use of vitamin C may not explain the recovery, since almost all colds clear up within two weeks regardless. In arguing for the use of vitamin C in the prevention and treatment of colds, Linus Pauling (1970) does not base his claims upon the high probability of avoidance or quick recovery; instead, he urges that massive doses of vitamin C have a bearing upon the probability of avoidance or recovery—that is, the use of vitamin C is relevant to the occurrence, duration, and severity of colds. A *high* probability of recovery, given use of vitamin C, does not confer explanatory value upon the use of this drug with respect to recovery. An *enhanced* probability value does indicate that the use of vitamin C may have some explanatory force. This example, along with a host of others which, like it, fulfill all of Hempel's requirements for a correct I-S explanation, shows that fulfilling these requirements does not constitute a sufficient condition for an adequate statistical explanation.

At first blush, it might seem that the type of relevance problem illustrated by the foregoing example was peculiar to the I-S model, but Henry Kyburg (1965) showed that examples can be found which demonstrate that the

---

[7] It should be mentioned in passing that Hempel (1965, pp. 380–381) offers still another model of scientific explanation that he characterizes as deductive-statistical (D-S). In an explanation of this type, a statistical regularity is explained by deducing it from other statistical laws. There is no real need, however, to treat such explanations as a distinct type, for they fall under the D-N schema, just given, provided we allow that at least one of the laws may be statistical. In the present context, we are concerned only with statistical explanations of nondeductive sorts.

D-N model is infected with precisely the same difficulty. Consider a sample of table salt that dissolves upon being placed in water. We ask why it dissolves. Suppose, Kyburg suggests, that someone has cast a dissolving spell upon it—that is, someone wearing a funny hat waves a wand over it and says, "I hereby cast a dissolving spell upon you." We can then 'explain' the phenomenon by mentioning the dissolving spell—without for a moment believing that any actual magic has been accomplished—and by invoking the true universal generalization that all samples of table salt that have been hexed in this manner dissolve when placed in water. Again, an argument that satisfies all of the requirements of Hempel's model patently fails to qualify as a satisfactory scientific explanation because of a failure of relevance. Given Hempel's characterizations of his D-N and I-S models of explanation, it is easy to construct any number of 'explanations' of either type that invoke some irrelevancy as a purported explanatory fact.[8] This result casts serious doubt upon the entire epistemic conception of scientific explanation, as outlined in the previous chapter, insofar as it takes all explanations to be arguments of one sort or another.

The diagnosis of the difficulty can be stated very simply. Hempel's requirement of maximal specificity (RMS) guarantees that *all* known relevant facts must be included in an adequate scientific explanation, but there is no requirement to insure that *only* relevant facts will be included. The foregoing examples bear witness to the need for some requirement of this latter sort. To the best of my knowledge, the advocates of the 'received view' have not, until recently, put forth any such additional condition, nor have they come to terms with counterexamples of these types in any other way.[9] James Fetzer's *requirement of strict maximal specificity*, which rules out the use in explanations of laws that mention nomically irrelevant properties (Fetzer, 1981, pp. 125–126), seems to do the job. In fact, in (Salmon, 1979a, pp. 691–694), I showed how Reichenbach's theory of nomological statements could be used to accomplish the same end.

The second problem that arises out of the high-probability requirement is illustrated by an example furnished by Michael Scriven (1959, p. 480). If someone contracts paresis, the straightforward explanation is that he

---

[8] Many examples are presented and analyzed in (Salmon et al., 1971, pp. 33–40). Nancy Cartwright (1983, pp. 26–27) errs when she attributes to Hempel the requirement that a statistical explanation increase the probability of the explanandum; this thesis, which I first advanced in (Salmon, 1965), was never advocated by Hempel. Shortly thereafter (1983, 28–29), she provides a correct characterization of the relationships among the views of Hempel, Suppes, and me.

[9] In Hempel's most recent discussion of statistical explanation, he appears to maintain the astonishing view that although such examples have *psychologically* misleading features, they do qualify as *logically* satisfactory explanations. (1977, pp. 107–111).

was infected with syphilis, which had progressed through the primary, secondary, and latent stages without treatment with penicillin. Paresis is one form of tertiary syphilis, and it never occurs except in syphilitics. Yet far less than half of those victims of untreated latent syphilis ever develop paresis. Untreated latent syphilis is the explanation of paresis, but it does not provide any basis on which to say that the explanandum-event was to be expected by virtue of these explanatory facts. Given a victim of latent untreated syphilis, the odds are that he will *not* develop paresis. Many other examples can be found to illustrate the same point. As I understand it, mushroom poisoning may afflict only a small percentage of individuals who eat a particular type of mushroom (Smith, 1958, Introduction), but the eating of the mushroom would unhesitatingly be offered as the explanation in instances of the illness in question. The point is illustrated by remarks on certain species in a guide for mushroom hunters (Smith, 1958, pp. 34, 185):

> *Helvella infula*, "Poisonous to some, but edible for most people. Not recommended."
> *Chlorophyllum molybdites*, "Poisonous to some but not to others. Those who are not made ill by it consider it a fine mushroom. The others suffer acutely."

These examples show that high probability does not constitute a necessary condition for legitimate statistical explanations. Taking them together with the vitamin C example, we must conclude—provisionally, at least—that a high probability of the explanandum relative to the explanans is neither necessary nor sufficient for correct statistical explanations, even if all of Hempel's other conditions are fulfilled. Much more remains to be said about the high-probability requirement, for it raises a host of fundamental philosophical problems, but I shall postpone further discussion of it until chapter 4.

Given the problematic status of the high-probability requirement, it was natural to attempt to construct an alternative treatment of statistical explanation that rests upon different principles. As I argued in (Salmon, 1965), statistical relevance, rather than high probability, seems to be the key explanatory relationship. This starting point leads to a conception of scientific explanation that differs fundamentally and radically from Hempel's I-S account. In the first place, if we are to make use of statistical relevance relations, our explanations will have to make reference to at least two probabilities, for statistical relevance involves a difference between two probabilities. More precisely, a factor C is statistically relevant to the occurrence of B under circumstances A if and only if

$$P(B|A.C) \neq P(B|A) \tag{1}$$

or

$$P(B|A.C) \neq P(B|A.\bar{C}). \tag{2}$$

Conditions (1) and (2) are equivalent to one another, provided that C occurs with a nonvanishing probability within A; since we shall not be concerned with the relevance of factors whose probabilities are zero, we may use either (1) or (2) as our definition of statistical relevance. We say that C is positively relevant to B if the probability of B is greater in the presence of C; it is negatively relevant if the probability of B is smaller in the presence of C. For instance, heavy cigarette smoking is positively relevant to the occurrence of lung cancer, at some later time, in a thirty-year-old Australian male; it is negatively relevant to survival to the age of seventy for such a person.

In order to construct a satisfactory statistical explanation, it seems to me, we need a *prior probability* of the occurrence to be explained, as well as one or more *posterior probabilities*. A crucial feature of the explanation will be the comparison between the prior and posterior probabilities. In Hempel's case of the streptococcus infection, for instance, we might begin with the probability, in the entire class of people with streptococcus infections, of a quick recovery. We realize, however, that the administration of penicillin is statistically relevant to quick recovery, so we compare the probability of quick recovery among those who have received penicillin with the probability of quick recovery among those who have not received penicillin. Hempel warns, however, that there is another relevant factor, namely, the existence of the penicillin-resistant strain of bacteria. We must, therefore, take that factor into account as well. Our original reference class has been divided into four parts: (1) infection by non-penicillin-resistant bacteria, penicillin given; (2) infection by non-penicillin-resistant bacteria, no penicillin given; (3) infection by penicillin-resistant bacteria, penicillin given; (4) infection by penicillin-resistant bacteria, no penicillin given. Since the administration of penicillin is irrelevant to quick recovery in case of penicillin-resistant infections, the subclasses (3) and (4) of the original reference class should be merged to yield (3') infection by penicillin-resistant bacteria. If John Jones is a member of (1), we have an explanation of his quick recovery, according to the S-R approach, not because the probability is high, but, rather, because it differs significantly from the probability in the original reference class. We shall see later what must be done if John Jones happens to fall into class (3').

By way of contrast, Hempel's earlier high-probability requirement demands only that the posterior probability be sufficiently large—whatever that might mean—but makes no reference at all to any prior probability. According to Hempel's abstract model, we ask, "Why is individual x a member of B?" The answer consists of an inductive argument having the following form:

$$P(B|A) = r$$
$$x \text{ is an } A$$
$$\overline{\qquad\qquad} \; [r]$$
$$x \text{ is a } B$$

As we have seen, even if the first premise is a statistical law, r is high, the premises are true, and the requirement of maximal specificity has been fulfilled, our 'explanation' may be patently inadequate, due to failure of relevancy.

In (Salmon, 1970, pp. 220–221), I advocated what came to be called the statistical-relevance or S-R model of scientific explanation. At that time, I thought that anything that satisfied the conditions that define that model would qualify as a legitimate scientific explanation. I no longer hold that view. It now seems to me that the statistical relationships specified in the S-R model constitute the *statistical basis* for a bona fide scientific explanation, but that this basis must be supplemented by certain *causal factors* in order to constitute a satisfactory scientific explanation. In chapters 5–9 I shall discuss the causal aspects of explanation. In this chapter, however, I shall confine attention to the statistical basis, as articulated in terms of the S-R model. Indeed, from here on I shall speak, not of the S-R model, but, rather, of the *S-R basis*.[10]

Adopting the S-R approach, we begin with an explanatory question in a form somewhat different from that given by Hempel. Instead of asking, for instance, "Why did x get well within a fortnight?" we ask, "Why did this person with a cold get well within a fortnight?" Instead of asking, "Why is x a B?" we ask, "Why is x, which is an A, also a B?" The answer—at least for preliminary purposes—is that x is also a C, where C

---

[10] I am extremely sympathetic to the thesis, expounded in (Humphreys, 1983), that probabilities—including those appearing in the S-R basis—are important tools in the construction of scientific explanations, but that they do not constitute any part of a scientific explanation per se. This thesis allows him to relax considerably the kinds of maximal specificity or homogeneity requirements that must be satisfied by statistical or probabilistic explanations. A factor that is statistically relevant may be causally irrelevant because, for example, it does not convert any contributing causes to counteracting causes or vice versa. This kind of relaxation is attractive in a theory of scientific explanation, for factors having small statistical relevance often seem otiose. Humphreys' approach does not show, however, that such relevance relations can be omitted from the S-R basis; on the contrary, the S-R basis must include such factors in order that we may ascertain whether they can be omitted from the causal explanation or not. I shall return to Humphreys' concept of aleatory explanation in chapter 9.

is *relevant* to *B* within *A*. Thus we have a prior probability $P(B|A)$—in this case, the probability that a person with a cold (*A*) gets well within a fortnight (*B*). Then we let *C* stand for the taking of vitamin C. We are interested in the posterior probability $P(B|A.C)$ that a person with a cold who takes vitamin C recovers within a fortnight. If the prior and posterior probabilities are equal to one another, the taking of vitamin C can play no role in explaining why this person recovered from the cold within the specified period of time. If the posterior probability is not equal to the prior probability, then *C* may, under certain circumstances, furnish part or all of the desired explanation. A large part of the purpose of the present book is to investigate the way in which considerations that are statistically relevant to a given occurrence have or lack explanatory import.

We cannot, of course, expect that every request for a scientific explanation will be phrased in canonical form. Someone might ask, for example, "Why did Mary Jones get well in no more than a fortnight's time?" It might be clear from the context that she was suffering from a cold, so that the question could be reformulated as, "Why did this person who was suffering from a cold get well within a fortnight?" In some cases, it might be necessary to seek additional clarification from the person requesting the explanation, but presumably it will be possible to discover what explanation is being called for. This point about the form of the explanation-seeking question has fundamental importance. We can easily imagine circumstances in which an altogether different explanation is sought by means of the same initial question. Perhaps Mary had exhibited symptoms strongly suggesting that she had mononucleosis; in this case, the fact that it was only an ordinary cold might constitute the explanation of her quick recovery. A given why-question, construed in one way, may elicit an explanation, while otherwise construed, it asks for an explanation that cannot be given. "Why did the Mayor contract paresis?" might mean, "Why did this adult human develop paresis?" or, "Why did this syphilitic develop paresis?" On the first construal, the question has a suitable answer, which we have already discussed. On the second construal, it has no answer—at any rate, we cannot give an answer—for we do not know of any fact in addition to syphilis that is relevant to the occurrence of paresis. Some philosophers have argued, because of these considerations, that scientific explanation has an unavoidably pragmatic aspect (e.g., van Fraassen, 1977, 1980). If this means simply that there are cases in which people ask for explanations in unclear or ambiguous terms, so that we cannot tell what explanation is being requested without further clarification, then so be it. No one would deny that we cannot be expected to supply explanations unless we know what it is we are being asked to explain. To this extent, scientific explanation surely has pragmatic or contextual components. Dealing with these considerations is, I believe, tantamount to choosing a suitable reference class with respect to which the prior probabilities are to be taken and specifying an appropriate sample space for purposes of a particular explanation. More will be said about these two items in the next section. In chapter 4—in an extended discussion of van Fraassen's theory—we shall return to this issue of pragmatic aspects of explanation, and we shall consider the question of whether there are any others.

THE STATISTICAL-RELEVANCE APPROACH

Let us now turn to the task of giving a detailed elaboration of the S-R basis. For purposes of initial presentation, let us construe the terms A, B, C, . . . (with or without subscripts) as referring to classes, and let us construe our probabilities in some sense as relative frequencies. This *does not mean* that the statistical-relevance approach is tied in any crucial way to a frequency theory of probability. I am simply adopting the heuristic device of picking examples involving frequencies because they are easily grasped. Those who prefer propensities, for example, can easily make the appropriate terminological adjustments, by speaking of chance setups and outcomes of trials where I refer to reference classes and attributes. With this understanding in mind, let us consider the steps involved in constructing an S-R basis for a scientific explanation:

1. We begin by selecting an appropriate reference class A with respect to which the prior probabilities $P(B_i|A)$ of the B_is are to be taken.
2. We impose an *explanandum-partition* upon the initial reference class A in terms of an exclusive and exhaustive set of attributes B_1, . . . , B_m; this defines a sample space for purposes of the explanation under consideration. (This partition was not required in earlier presentations of the S-R model.)
3. Invoking a set of statistically relevant factors C_1, . . . , C_s, we partition the initial reference class A into a set of mutually exclusive and exhaustive cells $A.C_1$, . . . , $A.C_s$. The properties C_1, . . . , C_s furnish the *explanans-partition*.
4. We ascertain the associated probability relations:
   prior probabilities

   $$P(B_i|A) = p_i$$
   for all i ($1 \leq i \leq m$)

   posterior probabilities

   $$P(B_i|A.C_j) = p_{ij}$$
   for all i and j ($1 \leq i \leq m$) and ($1 \leq j \leq s$)

5. We require that each of the cells $A.C_j$ be homogeneous with respect to the explanandum-partition $\{B_i\}$; that is, none of the cells in the partition can be further subdivided in any manner relevant to the occurrence of any B_i. (This requirement is somewhat analogous to Hempel's requirement of maximal specificity, but as we shall see, it is a much stronger condition.)

6. We ascertain the relative sizes of the cells in our explanans-partition in terms of the following marginal probabilities:

$$P(C_j|A) = q_j$$

(These probabilities were not included in earlier versions of the S-R model; the reasons for requiring them now will be discussed later in this chapter.)

7. We require that the explanans-partition be a maximal homogeneous partition, that is—with an important exception to be noted later—for $i \neq k$ we require that $p_{ji} \neq p_{jk}$. (This requirement assures us that our partition in terms of $C_1, \ldots, C_m$ does not introduce any irrelevant subdivision in the initial reference class A.)

8. We determine which cell $A.C_j$ contains the individual x whose possession of the attribute B_i was to be explained. The probability of B_i within the cell is given in the list under 4.

Consider in a rather rough and informal manner the way in which the foregoing pattern of explanation might be applied in a concrete situation; an example of this sort was offered by James Greeno (1971a, pp. 89–90). Suppose that Albert has committed a delinquent act—say, stealing a car, a major crime—and we ask for an explanation of that fact. We ascertain from the context that he is an American teen-ager, and so we ask, "Why did this American teen-ager commit a serious delinquent act?" The prior probabilities, which we take as our point of departure, so to speak, are simply the probabilities of the various degrees of juvenile delinquency (B_i) among American teen-agers (A)—that is, $P(B_i|A)$. We will need a suitable explanandum-partition; Greeno suggests B_1 = no criminal convictions, B_2 = conviction for minor infractions only, B_3 = conviction for a major offense. Our sociological theories tell us that such factors as sex, religious background, marital status of parents, type of residential community, socioeconomic status, and several others are relevant to delinquent behavior. We therefore take the initial reference class of American teen-agers and divide it into males and females; Jewish, Protestant, Roman Catholic, no religion; parents married, parents divorced, parents never married; urban, suburban, rural place of residence; upper, middle, lower class; and so forth. Taking such considerations into account, we arrive at a large number

183

s of cells in our partition. We assign probabilities of the various degrees of delinquent behavior to each of the cells in accordance with 4, and we ascertain the probability of a randomly selected American teen-ager belonging to each of the cells in accordance with 6. We find the cell to which Albert belongs—for example, male, from a Protestant background, parents divorced, living in a suburban area, belonging to the middle class. If we have taken into account all of the relevant factors, and if we have correctly ascertained the probabilities associated with the various cells of our partitions, then we have an S-R basis for the explanation of Albert's delinquency that conforms to the foregoing schema. If it should turn out (contrary to what I believe actually to be the case) that the probabilities of the various types of delinquency are the same for males and for females, then we would not use sex in partitioning our original reference class. By condition 5 we must employ *every* relevant factor; by condition 7 we must employ *only* relevant factors. In many concrete situations, including the present examples, we know that we have not found all relevant considerations; however, as Noretta Koertge rightly emphasized (1975), that is an ideal for which we may aim. Our philosophical analysis is designed to capture the notion of a fully satisfactory explanation.

Nothing has been said, so far, concerning the rationale for conditions 2 and 6, which are here added to the S-R basis for the first time. We must see why these requirements are needed. Condition 2 is quite straightforward; it amounts to a requirement that the sample space for the problem at hand be specified. As we shall see when we discuss Greeno's information-theoretic approach in chapter 4, both the explanans-partition and the explanandum-partition are needed to measure the information transmitted in any explanatory scheme. This is a useful measure of the explanatory value of a theory. In addition, as we shall see when we discuss van Fraassen's treatment of why-questions in chapter 4, his contrast class, which is the same as our explanandum-partition, is needed in some cases to specify precisely what explanation is being sought. In dealing with the question "Why did Albert steal a car?" we used Greeno's suggested explanandum-partition. If, however, we had used different partitions (contrast classes), other explanations might have been called forth. Suppose that the contrast class included: Albert steals a car, Bill steals a car, Charlie steals a car, and so forth. Then the answer might have involved no sociology whatever; the explanation might have been that, among the members of his gang, Albert is most adept at hot-wiring. Suppose, instead, that the contrast class had included: Albert steals a car, Albert steals a diamond ring, Albert steals a bottle of whiskey, and so forth. In that case, the answer might have been that he wanted to go joyriding.

The need for the marginal probabilities mentioned in 6 arises in the following way. In many cases, such as the foregoing delinquency example, the terms C_j that furnish the explanans-partition of the initial reference class are conjunctive. A given cell is determined by several distinct factors: sex *and* religious background *and* marital status of parents *and* type of residential community *and* socioeconomic status *and* . . . which may be designated D_k, E_n, F_r, ... These factors will be the probabilistic contributing causes and counteracting causes that tend, respectively, to produce or prevent delinquency. In attempting to understand the phenomenon in question, it is important to know how each factor is relevant—whether positively or negatively, and how strongly—both in the population at large and in various subgroups of the population. Consider, for example, the matter of sex. It may be that within the entire class of American teen-agers (A) the probability of serious delinquency (B_3) is greater among males than it is among females. If so, we would want to know by how much the probability among males exceeds the probability among females and by how much it exceeds the probability in the entire population. We also want to know whether being male is always positively relevant to serious delinquency, or whether in combination with certain other factors, it may be negatively relevant or irrelevant. Given two groups of teen-agers—one consisting entirely of boys and the other entirely of girls, but alike with respect to all of the other factors—we want to know how the probabilities associated with delinquency in each of the two groups are related to one another. It might be that in each case of two cells in the explanandum-partition that differ from one another only on the basis of gender, the probability of serious delinquency in the male group is greater than it is in the female group. It might turn out, however, that sometimes the two probabilities are equal, or that in some cases the probability is higher in the female group than it is in the corresponding male group. Relationships of all of these kinds are logically possible.

It is a rather obvious fact that each of two circumstances can individually be positively relevant to a given outcome, but their conjunction can be negatively relevant or irrelevant. Each of two drugs can be positively relevant to good health, but taken together, the combination may be detrimental—for example, certain antidepressive medications taken in conjunction with various remedies for the common cold can greatly increase the chance of dangerously high blood pressure (Goodwin and Guze, 1979). A factor that is a contributing cause in some circumstances can be a counteracting cause in other cases. Problems of this sort have been discussed, sometimes under the heading of "Simpson's paradox," by Nancy Cartwright (1983, essay 1) and Bas van Fraassen (1980, pp. 108, 148–151). In (Salmon, 1975c), I have spelled out in detail the complexities that arise in connection with statistical relevance relations. The moral is

185

that we need to know not only how the various factors D_k, E_n, F_r, ..., are relevant to the outcome, B_i, but how the relevance of each of them is affected by the presence or absence of the others. Thus, for instance, it is possible that being female might in general be negatively relevant to delinquency, but it might be positively relevant among the very poor.

Even if all of the prior probabilities $P(B_i|A)$ and all of the posterior probabilities $P(B_i|A.C_j)$ furnished under condition 4 are known, it is not possible to deduce the conditional probabilities of the B_i's with respect to the individual conjuncts that make up the C_j's or with respect to combinations of them. Without these conditional probabilities, we will not be in a position to ascertain all of the statistical relevance relations that are required. We therefore need to build in a way to extract that information. This is the function of the marginal probabilities $P(C_j|A)$ required by condition 6. If these are known, such conditional probabilities as $P(B_i|A.D_k)$, $P(B_i|A.E_n)$, and $P(B_i|A.D_k.E_n)$ can be derived.[11] When 2 and 6 are added to the earlier characterization of the S-R model (Salmon et al., 1971), then, I believe, we have gone as far as possible in characterizing scientific explanations at the level of statistical relevance relations.

[11] Suppose, for example, that we wish to compute $P(B_i|A.D_k)$, where $D_k = C_{j_1} \vee \ldots \vee C_{j_q}$, the cells C_{j_r} being mutually exclusive. This can be done as follows. We are given $P(C_j|A)$ and $P(B_i|A.C_j)$. By the multiplication theorem,

$$P(D_k.B_i|A) = P(D_k|A) \times P(B_i|A.D_k)$$

Assuming $P(D_k|A) \neq 0$, we have,

$$P(B_i|A.D_k) = P(D_k.B_i|A)/P(D_k|A) \qquad (*)$$

By the addition theorem

$$P(D_k \mid A) = \sum_{r=1}^{q} P(C_{j_r} \mid A)$$

$$P(D_k.B_i \mid A) = \sum_{r=1}^{q} P(C_{j_r}.B_i \mid A)$$

By the multiplication theorem

$$P(D_k.B_i \mid A) = \sum_{r=1}^{q} P(C_{j_r} \mid A) \times P(B_i \mid A.C_{j_r})$$

Substitution in (*) yields the desired relation:

$$P(B_i \mid A.D_k) = \frac{\sum_{r=1}^{q} P(C_{j_r} \mid A) \times P(B_i \mid A.C_{j_r})}{\sum_{r=1}^{q} P(C_{j_r} \mid A)}$$

Several features of the new version of the S-R basis deserve explicit mention. It should be noted, in the first place, that conditions 2 and 3 demand that the entire initial reference class A be partitioned, while conditions 4 and 6 require that *all* of the associated probability values be given. This is one of several respects in which it differs from Hempel's I-S model. Hempel requires only that the individual mentioned in the explanandum be placed within an appropriate class, satisfying his requirement of maximal specificity, but he does not ask for information about any class in either the explanandum-partition or the explanans-partition to which that individual does not belong. Thus he might go along in requiring that Bill Smith be referred to the class of American male teen-agers coming from a Protestant background, whose parents are divorced, and who is a middle-class suburban dweller, and in asking us to furnish the probability of his degree of delinquency within that class. But why, it may surely be asked, should we be concerned with the probability of delinquency in a lower-class, urban-American, female teen-ager from a Roman Catholic background whose parents are still married? What bearing do such facts have on Bill Smith's delinquency? The answer, I think, involves serious issues concerning scientific generality. If we ask why this American teen-ager becomes a delinquent, then, it seems to me, we are concerned with *all* of the factors that are relevant to the occurrence of delinquency, and with the ways in which these factors are relevant to that phenomenon (cf. Koertge, 1975). To have a satisfactory scientific answer to the question of why this A is a B_i—to achieve full scientific understanding—we need to know the factors that are relevant to the occurrence of the various B_i s for *any* randomly chosen or otherwise unspecified member of A. It was mainly to make good on this desideratum that requirement 6 was added. Moreover, as Greeno and I argued in *Statistical Explanation and Statistical Relevance*, a good measure of the value of an S-R basis is the gain in information furnished by the complete partitions and the associated probabilities. This measure cannot be applied to the individual cells one at a time.

A fundamental philosophical difference between our S-R basis and Hempel's I-S model lies in the interpretation of the concept of homogeneity that appears in condition 5. Hempel's requirement of maximal specificity, which is designed to achieve a certain kind of homogeneity in the reference classes employed in I-S explanations, is *epistemically relativized*. This means, in effect, that we must not *know* of any way to make a relevant partition, but it certainly does not demand that no possibility of a relevant partition can exist unbeknown to us. As I view the S-R basis, in contrast, condition 5 demands that the cells of our explanans-partition be *objectively* homogeneous; for this model, homogeneity is not epistemically relativized. Since this issue of epistemic relativization versus objective homogeneity is discussed at length in chapter 3, it is sufficient for now merely to call attention to this complex problem.[12]

Condition 7 has been the source of considerable criticism. One such objection rests on the fact that the initial reference class A, to which the S-R basis is referred, may not be maximal. Regarding Kyburg's hexed salt example, mentioned previously, it has been pointed out that the class of samples of table salt is not a maximal homogeneous class with respect to solubility, for there are many other chemical substances that have the same probability—namely, unity—of dissolving when placed in water. Baking soda, potassium chloride, various sugars, and many other compounds have this property. Consequently, if we take the maximality condition seriously, it has been argued, we should not ask, "Why does this sample of table salt dissolve in water?" but, rather, "Why does this sample of matter in the solid state dissolve when placed in water?" And indeed, one can argue, as Koertge has done persuasively (1975), that to follow such a policy often leads to significant scientific progress. Without denying her important point, I would nevertheless suggest, for purposes of elaborating the formal schema, that we take the initial reference class A as given by the explanation-seeking why-question, and look for relevant partitions within it. A significantly different explanation, which often undeniably represents scientific progress, may result if a different why-question, embodying a broader initial reference class, is posed. If the original question is not presented in a form that unambiguously determines a reference class A, we can reasonably discuss the advantages of choosing a wider or a narrower class in the case at hand.

Another difficulty with condition 7 arises if 'accidentally'—so to speak—two different cells in the partition, $A.C_i$ and $A.C_j$, happen to have equal associated probabilities p_{ki} and p_{kj} for all cells B_k in the explanandum-

---

[12] Cartwright (1983, p. 27) asserts that on Hempel's account, "what counts as a good explanation is an objective, person-independent matter," and she applauds him for holding that view. I find it difficult to reconcile her characterization with Hempel's repeated emphatic assertion (prior to 1977) of the doctrine of essential epistemic relativization of inductive-statistical explanation. In addition, she complains that my way of dealing with problems concerning the proper formulation of the explanation-seeking why-question—that is, problems concerning the choice of an appropriate initial reference class—"makes explanation a subjective matter" (ibid., p. 29). "What explains what," she continues, "depends on the laws and facts true in our world, and cannot be adjusted by shifting our interest or our focus" (ibid.). This criticism seems to me to be mistaken. Clarification of the question is often required to determine what it is that is to be explained, and this may have pragmatic dimensions. However, once the explanandum has been unambiguously specified, on my view, the identification of the appropriate explanans is fully objective. I am in complete agreement with Cartwright concerning the desirability of such objectivity; moreover, my extensive concern with objective homogeneity is based directly upon the desire to eliminate from the theory of statistical explanation such subjective features as epistemic relativization.

partition. Such a circumstance might arise if the cells are determined conjunctively by a number of relevant factors, and if the differences between the two cells cancel one another out. It might happen, for example, that the probabilities of the various degrees of delinquency—major offense, minor offense, no offense—for an upper-class, urban, Jewish girl would be equal to those for a middle-class, rural, Protestant boy. In this case, we might want to relax condition 7, allowing $A.C_i$ and $A.C_j$ to stand as separate cells, provided they differ with respect to at least two of the terms in the conjunction, so that we are faced with a fortuitous canceling of relevant factors. If, however, $A.C_i$ and $A.C_j$ differed with respect to only one conjunct, they would have to be merged into a single cell. Such would be the case if, for example, among upper-class, urban-dwelling, American teen-agers whose religious background is atheistic and whose parents are divorced, the probability of delinquent behavior were the same for boys as for girls. Indeed, we have already encountered this situation in connection with Hempel's example of the streptococcus infection. If the infection is of the penicillin-resistant variety, the probability of recovery in a given period of time is the same whether penicillin is administered or not. In such cases, we want to say, there is no relevant difference between the two classes—not that relevant factors were canceling one another out. I bring this problem up for consideration at this point, but I shall not make a consequent modification in the formal characterization of the S-R basis, for I believe that problems of this sort are best handled in the light of causal relevance relations. Indeed, it seems advisable to postpone detailed consideration of the whole matter of regarding the cells $A.C_j$ as being determined conjunctively until causation has been explicitly introduced into the discussion. As we shall see in chapter 9 (Humphreys, 1981, 1983) and (Rogers, 1981) provide useful suggestions for handling just this issue.

Perhaps the most serious objection to the S-R model of scientific explanation—as it was originally presented—is based upon the principle that *mere* statistical correlations explain nothing. A rapidly falling barometric reading is a sign of an imminent storm, and it is *highly correlated* with the onset of storms, but it certainly does not *explain* the occurrence of a storm. The S-R approach does, however, have a way of dealing with examples of this sort. A factor C, which is relevant to the occurrence of B in the presence of A, may be screened off in the presence of some additional factor D; the screening-off relation is defined by equations (3) and (4), which follow. To illustrate, given a series of days (A) in some particular locale, the probability of a storm occurring (B) is in general quite different from the probability of a storm if there has been a recent sharp drop in the barometric reading (C). Thus C is statistically relevant to B within A. If, however, we take into consideration the further fact that there is an actual drop in atmospheric pressure (*D*) in the region, then it is irrelevant whether that drop is registered on a barometer. In the presence of *D* and *A*, *C* becomes irrelevant to *B*; we say that *D* screens off *C* from *B*—in symbols,

$$P(B|A.C.D) = P(B|A.D). \tag{3}$$

However, *C* does not screen off *D* from *B*, that is,

$$P(B|A.C.D) \neq P(B|A.C), \tag{4}$$

for barometers sometimes malfunction, and it is the atmospheric pressure, not the reading on the barometer per se, that is directly relevant to the occurrence of the storm. A factor that has been screened off is irrelevant, and according to the definition of the S-R basis (condition 7), it is not to be included in the explanation. The falling barometer does not explain the storm.

Screening off is frequent enough and important enough to deserve further illustration. A study, reported in the news media a few years ago, revealed a positive correlation between coffee drinking and heart disease, but further investigation showed that this correlation results from a correlation between coffee drinking and cigarette smoking. It turned out that cigarette smoking screened off coffee drinking from heart disease, thus rendering coffee drinking statistically (as well as causally and explanatorily) irrelevant to heart disease. Returning to a previous example for another illustration, one could reasonably suppose that there is some correlation between low socioeconomic status and paresis, for there may be a higher degree of sexual promiscuity, a higher incidence of venereal disease, and less likelihood of adequate medical attention if the disease occurs. But the contraction of syphilis screens off such factors as degree of promiscuity, and the fact of syphilis going untreated screens off any tendency to fail to get adequate medical care. Thus when an individual has latent untreated syphilis, all other such circumstances as low socioeconomic status are screened off from the development of paresis.

As the foregoing examples show, there are situations in which one circumstance or occurrence is correlated with another because of an indirect causal relationship. In such cases, it often happens that the more proximate causal factors screen off those that are more remote. Thus 'mere correlations' are replaced in explanatory contexts with correlations that are intuitively recognized to have explanatory force. In *Statistical Explanation and Statistical Relevance*, where the S-R model of statistical explanation was first explicitly named and articulated, I held out some hope (but did not try to defend the thesis) that all of the causal factors that play any role in scientific explanation could be explicated in terms of statistical relevance relations—with the screening-off relation playing a crucial role. As I shall explain in chapter 6, I no longer believe this is possible. A large part of the material in the present book is devoted to an attempt to analyze the nature of the causal relations that enter into scientific explanations, and the manner in which they function in explanatory contexts. After characterizing the S-R model, I wrote:

> One might ask on what grounds we can claim to have characterized explanation. The answer is this. When an explanation (as herein explicated) has been provided, we know exactly how to regard any A with respect to the property B. We know which ones to bet on, which to bet against, and at what odds. We know precisely what degree of expectation is rational. We know how to face uncertainty about an A's being a B in the most reasonable, practical, and efficient way. We know every factor that is relevant to an A having the property B. We know exactly what weight should have been attached to the prediction that this A will be a B. We know all of the regularities (universal and statistical) that are relevant to our original question. What more could one ask of an explanation? (Salmon et al., 1971, p. 78)

The answer, of course, is that we need to know something about the causal relationships as well.

In acknowledging this egregious shortcoming of the S-R model of scientific explanation, I am not abandoning it completely. The attempt, rather, is to supplement it in suitable ways. While recognizing its incompleteness, I still think it constitutes a sound basis upon which to erect a more adequate account. And at a fundamental level, I still think it provides important insights into the nature of scientific explanation.

In the introduction to *Statistical Explanation and Statistical Relevance*, I offered the following succinct comparison between Hempel's I-S model and the S-R model:

I-S model: an explanation is an *argument* that renders the explanandum *highly probable*.

S-R model: an explanation is an *assembly of facts statistically relevant* to the explanandum, *regardless of the degree of probability that results*.

It was Richard Jeffrey (1969) who first explicitly formulated the thesis that (at least some) statistical explanations are not arguments; it is beautifully expressed in his brief paper, "Statistical Explanation vs. Statistical Inference," which was reprinted in *Statistical Explanation and Statistical Relevance*. In (Salmon, 1965, pp. 145–146), I had urged that positive relevance rather than high probability is the desideratum in statistical explanation.

In (Salmon, 1970), I expressed the view, which many philosophers found weird and counter-intuitive (e.g., L. J. Cohen, 1975), that statistical explanations may even embody *negative* relevance—that is, an explanation of an event may, in some cases, show that the event to be explained is less probable than we had initially realized. I still do not regard that thesis as absurd. In an illuminating discussion of the explanatory force of positively and negatively relevant factors, Paul Humphreys (1981) has introduced some felicitous terminology for dealing with such cases, and he has pointed to an important constraint. Consider a simple example. Smith is stricken with a heart attack, and the doctor says, "*Despite* the fact that Smith exercised regularly and had given up smoking several years ago, he contracted heart disease *because* he was seriously overweight." The "because" clause mentions those factors that are positively relevant and the "despite" clause cites those that are negatively relevant. Humphreys refers to them as *contributing causes* and *counteracting causes*, respectively. When we discuss causal explanation in later chapters, we will want to say that a complete explanation of an event must make mention of the causal factors that tend to prevent its occurrence as well as those that tend to bring it about. Thus it is *not* inappropriate for the S-R basis to include factors that are negatively relevant to the explanandum-event. As Humphreys points out, however, we would hardly consider as appropriate a putative explanation that had only negative items in the "despite" clause and no positive items in the "because" category. "Despite the fact that Jones never smoked, exercised regularly, was not overweight, and did not have elevated levels of triglycerides and cholesterol, he died of a heart attack," would hardly be considered an acceptable *explanation* of his fatal illness.

Before concluding this chapter on models of statistical explanation, we should take a brief look at the deductive-homological-probabilistic (D-N-P) model of scientific explanation offered by Peter Railton (1978). By employing well-established statistical laws, such as that covering the spontaneous radioactive decay of unstable nuclei, it is possible to deduce the fact that a decay-event for a particular isotope has a certain probability of occurring within a given time interval. For an atom of carbon 14 (which is used in radiocarbon dating in archaeology, for example), the probability of a decay in 5,730 years is 1/2. The explanation of *the probability of the decay-event* conforms to Hempel's deductive-nomological pattern. Such an explanation does not, however, explain the actual occurrence of a decay, for, given the probability of such an event—however high or low—the event in question may not even occur. Thus the explanation does not qualify as an argument to the effect that the event-to-be-explained was to be expected with deductive certainty, given the explanans. Railton is, of course, clearly aware of the fact. He goes on to point out, nevertheless, that if we simply attach an "addendum" to the deductive argument stating that the event-to-be-explained did, in fact, occur in the case at hand, we can claim to have a probabilistic *account*—which is not a deductive or inductive argument—of the occurrence of the event. In this respect, Railton is in rather close agreement with Jeffrey (1969) that some explanations are not arguments. He also agrees with Jeffrey in emphasizing the importance of exhibiting the physical mechanisms that lead up to the probabilistic occurrence that is to be explained. Railton's theory—like that of Jeffrey— has some deep affinities to the S-R model. In including a reference to physical mechanisms as an essential part of his D-N-P model, however, Railton goes beyond the view that statistical relevance relations, in and of themselves, have explanatory import. His theory of scientific explanation can be more appropriately characterized as causal or mechanistic. It is closely related to the two-tiered causal-statistical account that I am at-tempting to elaborate as an improvement upon the S-R model.

Although, with Kyburg's help, I have offered what seem to be damaging counterexamples to the D-N model—for instance, the one about the man who explains his own avoidance of pregnancy on the basis of his regular consumption of his wife's birth control pills (Salmon et al., 1971, p. 34)— the major emphasis has been upon statistical explanation, and that continues to be the case in what follows. Aside from the fact that contemporary science obviously provides many statistical explanations of many types of phenomena, and that any philosophical theory of statistical explanation has only lately come forth, there is a further reason for focusing upon statistical explanation. As I maintained in chapter 1, we can identify three distinct approaches to scientific explanation that do not seem to differ from one another in any important way as long as we confine our attention to contexts in which all of the explanatory laws are universal generalizations. I shall argue in chapter 4, however, that these three general conceptions of sci-entific explanation can be seen to differ radically from one another when we move on to situations in which statistical explanations are in principle the best we can achieve. Close consideration of statistical explanations, with sufficient attention to their causal ingredients, provides important insight into the underlying philosophical questions relating to our scientific understanding of the world.

WHAT IS A LAW OF NATURE?

THERE is a sense in which we know well enough what is ordinarily meant by a law of nature. We can give examples. Thus it is, or is believed to be, a law of nature that the orbit of a planet around the sun is an ellipse, or that arsenic is poisonous, or that the intensity of a sensation is proportionate to the logarithm of the stimulus, or that there are 303,000,000,000,000,000,000,000 molecules in one gram of hydrogen. It is not a law of nature, though it is necessarily true, that the sum of the angles of a Euclidean triangle is 180 degrees, or that all the presidents of the third French Republic were male, though this is a legal fact in its way, or that all the cigarettes which I now have in my cigarette case are made of Virginian tobacco, though this again is true and, given my tastes, not wholly accidental. But while there are many such cases in which we find no difficulty in telling whether some proposition, which we take to be true, is or is not a law of nature, there are cases where we may be in doubt. For instance, I suppose that most people take the laws of nature to include the first law of thermodynamics, the proposition that in any closed physical system the sum of energy is constant : but there are those who maintain that this principle is a convention, that it is interpreted in such a way that there is no logical possibility of its being falsified, and for this reason they may deny that it is a law of nature at all. There are two questions at issue in a case of this sort : first, whether the principle under discussion is in fact a convention, and secondly whether its being a convention, if it is one, would disqualify it from being a law of nature. In the same way, there may be a dispute whether statistical generalizations are to count as laws of nature, as distinct from the dispute whether certain generalizations, which have been taken to be laws of nature, are in fact statistical. And even if we were always able to tell, in the case of any given proposition, whether or not it had the form of a law of nature, there would still remain the problem of making clear what this implied.

The use of the word 'law', as it occurs in the expression 'laws of nature', is now fairly sharply differentiated from its use in legal and moral contexts : we do not conceive of the laws of nature as imperatives. But this was not always so. For instance, Hobbes in his *Leviathan* lists fifteen 'laws of nature' of which two of the most important are that men 'seek peace, and follow it' and 'that men perform their covenants made' : but he does not think that these laws are necessarily respected. On the contrary, he holds that the state of nature is a state of war, and that covenants will not in fact be kept unless there is some power to enforce them. His laws of nature are like civil laws except that they are not the commands of any civil authority. In one place he speaks of them as 'dictates of Reason' and adds that men improperly call them by the name of laws : 'for they are but conclusions or theorems concerning what conduceth to the conservation and defence of themselves : whereas Law, properly, is the word of him, that by right hath command over others'. 'But yet,' he continues, 'if you consider the same Theorems, as delivered in the word of God, that by right commandeth all things ; then they are properly called Laws.[1]

It might be thought that this usage of Hobbes was so far removed from our own that there was little point in mentioning it, except as a historical curiosity ; but I believe

[1] *Leviathan*, Part I, chapter xv.

that the difference is smaller than it appears to be. I think that our present use of the expression 'laws of nature' carries traces of the conception of Nature as subject to command. Whether these commands are conceived to be those of a personal deity or, as by the Greeks, of an impersonal fate, makes no difference here. The point, in either case, is that the sovereign is thought to be so powerful that its dictates are bound to be obeyed. It is not as in Hobbes's usage a question of moral duty or of prudence, where the subject has freedom to err. On the view which I am now considering, the commands which are issued to Nature are delivered with such authority that it is impossible that she should disobey them. I do not claim that this view is still prevalent; at least not that it is explicitly held. But it may well have contributed to the persistence of the feeling that there is some form of necessity attaching to the laws of nature, a necessity which, as we shall see, it is extremely difficult to pin down.

In case anyone is still inclined to think that the laws of nature can be identified with the commands of a superior being, it is worth pointing out that this analysis cannot be correct. It is already an objection to it that it burdens our science with all the uncertainty of our metaphysics, or our theology. If it should turn out that we had no good reason to believe in the existence of such a superior being, or no good reason to believe that he issued any commands, it would follow, on this analysis, that we should not be entitled to believe that there were any laws of nature. But the main argument against this view is independent of any doubt that one may have about the existence of a superior being. Even if we knew that such a one existed, and that he regulated nature, we still could not identify the laws of nature with his commands. For it is only by discovering what were the laws of nature that we could know what form these commands had taken. But this implies that we have some independent criteria for deciding what the laws of nature are. The assumption that they are imposed by a superior being is therefore idle, in the same way as the assumption of providence is idle. It is only if there are independent means of finding out what is going to happen that one is able to say what providence has in store. The same objection applies to the rather more fashionable view that moral laws are the commands of a superior being : but this does not concern us here.

There is, in any case, something strange about the notion of a command which it is impossible to disobey. We may be sure that some command will never in fact be disobeyed. But what is meant by saying that it cannot be ? That the sanctions which sustain it are too strong ? But might not one be so rash or so foolish as to defy them ? I am inclined to say that it is in the nature of commands that it should be possible to disobey them. The necessity which is ascribed to these supposedly irresistible commands belongs in fact to something different : it belongs to the laws of logic. Not that the laws of logic cannot be disregarded ; one can make mistakes in deductive reasoning, as in anything else. There is, however, a sense in which it is impossible for anything that happens to contravene the laws of logic. The restriction lies not upon the events themselves but on our method of describing them. If we break the rules according to which our method of description functions, we are not using it to describe anything. This might suggest that the events themselves really were disobeying the laws of logic, only we could not say so. But this would be an error. What is describable as an event obeys the laws of logic : and what is not describable as an event is not an event at all. The chains which logic puts upon nature are purely formal : being formal they weigh nothing, but for the same reason they are indissoluble.

From thinking of the laws of nature as the commands

of a superior being, it is therefore only a short step to crediting them with the necessity that belongs to the laws of logic. And this is in fact a view which many philosophers have held. They have taken it for granted that a proposition could express a law of nature only if it stated that events, or properties, of certain kinds were necessarily connected ; and they have interpreted this necessary connection as being identical with, or closely analogous to, the necessity with which the conclusion follows from the premisses of a deductive argument ; as being, in short, a logical relation. And this has enabled them to reach the strange conclusion that the laws of nature can, at least in principle, be established independently of experience : for if they are purely logical truths, they must be discoverable by reason alone.

The refutation of this view is very simple. It was decisively set out by Hume. 'To convince us', he says, 'that all the laws of nature and all the operations of bodies, without exception, are known only by experience, the following reflections may, perhaps, suffice. Were any object presented to us, and were we required to pronounce concerning the effect, which will result from it, without consulting past observation : after what manner, I beseech you, must the mind proceed in this operation ? It must invent or imagine some event, which it ascribes to the object as its effect : and it is plain that this invention must be entirely arbitrary. The mind can never find the effect in the supposed cause, by the most accurate scrutiny and examination. For the effect is totally different from the cause, and consequently can never be discovered in it.' [1]

Hume's argument is, indeed, so simple that its purport has often been misunderstood. He is represented as maintaining that the inherence of an effect in its cause is something which is not discoverable in nature ; that as a matter of fact our observations fail to reveal the existence of any

[1] *An Enquiry concerning Human Understanding*, iv. 1.25.

such relation : which would allow for the possibility that our observations might be at fault. But the point of Hume's argument is not that the relation of necessary connection which is supposed to conjoin distinct events is not in fact observable : it is that there could not be any such relation, not as a matter of fact but as a matter of logic. What Hume is pointing out is that if two events are distinct, they are distinct : from a statement which does no more than assert the existence of one of them it is impossible to deduce anything concerning the existence of the other. This is, indeed, a plain tautology. Its importance lies in the fact that Hume's opponents denied it. They wished to maintain both that the events which were coupled by the laws of nature were logically distinct from one another, and that they were united by a logical relation. But this is a manifest contradiction. Philosophers who hold this view are apt to express it in a form which leaves the contradiction latent : it was Hume's achievement to have brought it clearly to light.

In certain passages Hume makes his point by saying that the contradictory of any law of nature is at least conceivable ; he intends thereby to show that the truth of the statement which expresses such a law is an empirical matter of fact and not an *a priori* certainty. But to this it has been objected that the fact that the contradictory of a proposition is conceivable is not a decisive proof that the proposition is not necessary. It may happen, in doing logic or pure mathematics, that one formulates a statement which one is unable either to prove or disprove. Surely in that case both the alternatives of its truth and falsehood are conceivable. Professor W. C. Kneale, who relies on this objection, cites the example of Goldbach's conjecture that every even number greater than two is the sum of two primes. Though this conjecture has been confirmed so far as it has been

Probability and Induction, pp. 79 ff.

tested, no one yet knows for certain whether it is true or false: no proof has been discovered either way. All the same, if it is true, it is necessarily true, and if it is false, it is necessarily false. Suppose that it should turn out to be false. We surely should not be prepared to say that what Goldbach had conjectured to be true was actually inconceivable. Yet we should have found it to be the contradictory of a necessary proposition. If we insist that this does prove it to be inconceivable, we find ourselves in the strange position of having to hold that one of two alternatives is inconceivable, without our knowing which.

I think that Professor Kneale makes his case: but I do not think that it is an answer to Hume. For Hume is not primarily concerned with showing that a given set of propositions, which have been taken to be necessary, are not so really. This is only a possible consequence of his fundamental point that 'there is no object which implies the existence of any other if we consider these objects in themselves, and never look beyond the idea which we form of them',[1] in short, that to say that events are distinct is incompatible with saying that they are logically related. And against this Professor Kneale's objection has no force at all. The most that it could prove is that, in the case of the particular examples that he gives, Hume might be mistaken in supposing that the events in question really were distinct: in spite of the appearances to the contrary, an expression which he interpreted as referring to only one of them might really be used in such a way that it included a reference to the other.

But is it not possible that Hume was always so mistaken; that the events, or properties, which are coupled by the laws of nature never are distinct? This question is complicated by the fact that once a generalization is accepted as a law of nature it tends to change its status. The meanings

[1] *A Treatise of Human Nature*, i, iii, vi.

which we attach to our expressions are not completely constant : if we are firmly convinced that every object of a kind which is designated by a certain term has some property which the term does not originally cover, we tend to include the property in the designation ; we extend the definition of the object, with or without altering the words which refer to it. Thus, it was an empirical discovery that loadstones attract iron and steel : for someone who uses the word 'loadstone' only to refer to an object which has a certain physical appearance and constitution, the fact that it behaves in this way is not formally deducible. But, as the word is now generally used, the proposition that load-stones attract iron and steel is analytically true : an object which did not do this would not properly be called a load-stone. In the same way, it may have become a necessary truth that water has the chemical composition H_2O. But what then of heavy water which has the composition D_2O ? Is it not really water ? Clearly this question is quite trivial. If it suits us to regard heavy water as a species of water, then we must not make it necessary that water consists of H_2O. Otherwise, we may. We are free to settle the matter whichever way we please.

Not all questions of this sort are so trivial as this. What, for example, is the status in Newtonian physics of the principle that the acceleration of a body is equal to the force which is acting on it divided by its mass ? If we go by the text-books in which 'force' is defined as the product of mass and acceleration, we shall conclude that the prin-ciple is evidently analytic. But are there not other ways of defining force which allow this principle to be empirical ? In fact there are, but as Henri Poincaré has shown,[1] we may then find ourselves obliged to treat some other New-tonian principle as a convention. It would appear that in a system of this kind there is likely to be a conventional

---

[1] Cf. *La Science et l'hypothèse*, pp. 119-29.

element, but that, within limits, we can situate it where we choose. What is put to the test of experience is the system as a whole.

This is to concede that some of the propositions which pass for laws of nature are logically necessary, while implying that it is not true of all of them. But one might go much further. It is at any rate conceivable that at a certain stage the science of physics should become so unified that it could be wholly axiomatized : it would attain the status of a geometry in which all the generalizations were regarded as necessarily true. It is harder to envisage any such development in the science of biology, let alone the social sciences, but it is not theoretically impossible that it should come about there too. It would be characteristic of such systems that no experience could falsify them, but their security might be sterile. What would take the place of their being falsified would be the discovery that they had no empirical application.

The important point to notice is that, whatever may be the practical or aesthetic advantages of turning scientific laws into logically necessary truths, it does not advance our knowledge, or in any way add to the security of our beliefs. For what we gain in one way, we lose in another. If we make it a matter of definition that there are just so many million molecules in every gram of hydrogen, then we can indeed be certain that every gram of hydrogen will contain that number of molecules : but we must become correspondingly more doubtful, in any given case, whether what we take to be a gram of hydrogen really is so. The more we put into our definitions, the more uncertain it becomes whether anything satisfies them : this is the price that we pay for diminishing the risk of our laws being falsified. And if it ever came to the point where all the 'laws' were made completely secure by being treated as logically necessary, the whole weight of doubt would fall upon the statement

P

that our system had application.   Having deprived ourselves of the power of expressing empirical generalizations,
we should have to make our existential statements do the
work instead.

If such a stage were reached, I am inclined to say that
we should no longer have a use for the expression 'laws of
nature', as it is now understood.   In a sense, the tenure of
such laws would still be asserted : they would be smuggled
into the existential propositions.   But there would be
nothing in the system that would count as a law of nature :
for I take it to be characteristic of a law of nature that the
proposition which expresses it is not logically true.   In this
respect, however, our usage is not entirely clear-cut.   In a
case where a sentence has originally expressed an empirical
generalization, which we reckon to be a law of nature, we are
inclined to say that it still expresses a law of nature, even
when its meaning has been so modified that it has come to
express an analytic truth.   And we are encouraged in this
by the fact that it is often very difficult to tell whether this
modification has taken place or not.   Also, in the case where
some of the propositions in a scientific system play the rôle
of definitions, but we have some freedom in deciding which
they are to be, we tend to apply the expression 'laws of
nature' to any of the constituent propositions of the system,
whether or not they are analytically true.   But here it is
essential that the system as a whole should be empirical.
If we allow the analytic propositions to count as laws of
nature, it is because they are carried by the rest.

Thus to object to Hume that he may be wrong in
assuming that the events between which his causal relations
hold are 'distinct existences' is merely to make the point
that it is possible for a science to develop in such a way
that axiomatic systems take the place of natural laws.   But
this was not true of the propositions with which Hume was
concerned, nor is it true, in the main, of the sciences of

to-day. And in any case Hume is right in saying that we cannot have the best of both worlds; if we want our generalizations to have empirical content, they cannot be logically secure; if we make them logically secure, we rob them of their empirical content. The relations which hold between things, or events, or properties, cannot be both factual and logical. Hume himself spoke only of causal relations, but his argument applies to any of the relations that science establishes, indeed to any relations whatsoever.

It should perhaps be remarked that those philosophers who still wish to hold that the laws of nature are 'principles of necessitation' [1] would not agree that this came down to saying that the propositions which expressed them were analytic. They would maintain that we are dealing here with relations of objective necessity, which are not to be identified with logical entailments, though the two are in certain respects akin. But what are these relations of objective necessity supposed to be? No explanation is given except that they are just the relations that hold between events, or properties, when they are connected by some natural law. But this is simply to restate the problem; not even to attempt to solve it. It is not as if this talk of objective necessity enabled us to detect any laws of nature. On the contrary it is only *ex post facto*, when the existence of some connection has been empirically tested, that philosophers claim to see that it has this mysterious property of being necessary. And very often what they do 'see' to be necessary is shown by further observation to be false. This does not itself prove that the events which are brought together by a law of nature do not stand in some unique relation. If all attempts at its analysis fail, we may be reduced to saying that it is *sui generis*. But why then describe it in a way which leads to its confusion with the relation of logical necessity?

[1] Cf. Kneale, *op. cit.*

A further attempt to link natural with logical necessity is to be found in the suggestion that two events E and I are to be regarded as necessarily connected when there is some well-established universal statement U, from which, in conjunction with the proposition i, affirming the existence of I, a proposition e, affirming the existence of E, is formally deducible.[1] This suggestion has the merit of bringing out the fact that any necessity that there may be in the connection of two distinct events comes only through a law. The proposition which describes 'the initial conditions' does not by itself entail the proposition which describes the 'effect' : it does so only when it is combined with a causal law. But this does not allow us to say that the law itself is necessary. We can give a similar meaning to saying that the law is necessary by stipulating that it follows, either directly or with the help of certain further premisses, from some more general principle. But then what is the status of these more general principles ? The question what constitutes a law of nature remains, on this view, without an answer.

II

Once we are rid of the confusion between logical and factual relations, what seems the obvious course is to hold that a proposition expresses a law of nature when it states what invariably happens. Thus, to say that unsupported bodies fall, assuming this to be a law of nature, is to say that there is not, never has been, and never will be a body that being unsupported does not fall. The 'necessity' of a law consists, on this view, simply in the fact that there are no exceptions to it.

It will be seen that this interpretation can also be extended to statistical laws. For they too may be represented

[1] Cf. K. Popper, 'What Can Logic Do For Philosophy ?', *Supplementary Proceedings of the Aristotelian Society*, vol. xxii : and papers in the same volume by W. C. Kneale and myself.

as stating the existence of certain constancies in nature : only, in their case, what is held to be constant is the proportion of instances in which one property is conjoined with another or, to put it in a different way, the proportion of the members of one class that are also members of another. Thus it is a statistical law that when there are two genes determining a hereditary property, say the colour of a certain type of flower, the proportion of individuals in the second generation that display the dominant attribute, say the colour white as opposed to the colour red, is three quarters. There is, however, the difficulty that one does not expect the proportion to be maintained in every sample. As Professor R. B. Braithwaite has pointed out, 'when we say that the proportion (in a non-literal sense) of the male births among births is 51 per cent, we are not saying of any particular class of births that 51 per cent are births of males, for the actual proportion might differ very widely from 51 per cent in a particular class of births, or in a number of particular classes of births, without our wishing to reject the proposition that the proportion (in the non-literal sense) is 51 per cent.' [1] All the same the 'non-literal' use of the word 'proportion' is very close to the literal use. If the law holds, the proportion must remain in the neighbourhood of 51 per cent, for any sufficiently large class of cases : and the deviations from it which are found in selected sub-classes must be such as the application of the calculus of probability would lead one to expect. Admittedly, the question what constitutes a sufficiently large class of cases is hard to answer. It would seem that the class must be finite, but the choice of any particular finite number for it would seem also to be arbitrary. I shall not, however, attempt to pursue this question here. The only point that I here wish to make is that a statistical law is no less 'lawlike' than a causal law. Indeed, if the propositions

---

[1] *Scientific Explanation*, pp. 118-19.

which express causal laws are simply statements of what invariably happens, they can themselves be taken as expressing statistical laws, with ratios of 100 per cent. Since a 100 per cent ratio, if it really holds, must hold in every sample, these 'limiting cases' of statistical laws escape the difficulty which we have just remarked on. If henceforth we confine our attention to them, it is because the analysis of 'normal' statistical laws brings in complications which are foreign to our purpose. They do not affect the question of what makes a proposition lawlike : and it is in this that we are mainly interested.

On the view which we have now to consider, all that is required for there to be laws in nature is the existence of *de facto* constancies. In the most straightforward case, the constancy consists in the fact that events, or properties, or processes of different types are invariably conjoined with one another. The attraction of this view lies in its simplicity : but it may be too simple. There are objections to it which are not easily met.

In the first place, we have to avoid saddling ourselves with vacuous laws. If we interpret statements of the form 'All S is P' as being equivalent, in Russell's notation, to general implications of the form '$(x)\Phi x \supset \Psi x$', we face the difficulty that such implications are considered to be true in all cases in which their antecedent is false. Thus we shall have to take it as a universal truth both that all winged horses are spirited and that all winged horses are tame ; for assuming, as I think we may, that there never have been or will be any winged horses, it is true both that there never have been or will be any that are not spirited, and that there never have been or will be any that are not tame. And the same will hold for any other property that we care to choose. But surely we do not wish to regard the ascription of any property whatsoever to winged horses as the expression of a law of nature.

The obvious way out of this difficulty is to stipulate that the class to which we are referring should not be empty. If statements of the form 'All S is P' are used to express laws of nature, they must be construed as entailing that there are S's. They are to be treated as the equivalent, in Russell's notation, of the conjunction of the propositions '$(x) \Phi x \supset \Psi x$ and $(\exists x) \Phi x$'. But this condition may be too strong. For there are certain cases in which we do wish to take general implications as expressing laws of nature, even though their antecedents are not satisfied. Consider, for example, the Newtonian law that a body on which no forces are acting continues at rest or in uniform motion along a straight line. It might be argued that this proposition was vacuously true, on the ground that there are in fact no bodies on which no forces are acting ; but it is not for this reason that it is taken as expressing a law. It is not interpreted as being vacuous. But how then does it fit into the scheme ? How can it be held to be descriptive of what actually happens ?

What we want to say is that if there *were* any bodies on which no forces were acting then they *would* behave in the way that Newton's law prescribes. But we have not made any provision for such hypothetical cases : according to the view which we are now examining, statements of law cover only what is actual, not what is merely possible. There is, however, a way in which we can still fit in such 'non-instantial' laws. As Professor C. D. Broad has suggested,[1] we can treat them as referring not to hypothetical objects, or events, but only to the hypothetical consequences of instantial laws. Our Newtonian law can then be construed as implying that there are instantial laws, in this case laws about the behaviour of bodies on which forces are acting, which are such that when combined with the

[1] 'Mechanical and Teleological Causation' *Supplementary Proceedings of the Aristotelian Society*, vol. xiv, pp. 98 ff.

proposition that there are bodies on which no forces are acting, they entail the conclusion that these bodies continue at rest, or in uniform motion along a straight line. The proposition that there are such bodies is false, and so, if it is interpreted existentially, is the conclusion, but that does not matter. As Broad puts it, 'what we are concerned to assert is that this false conclusion is a necessary consequence of the conjunction of a certain false instantial supposition with certain true instantial laws of nature'.

This solution of the present difficulty is commendably ingenious, though I am not sure that it would always be possible to find the instantial laws which it requires. But even if we accept it, our troubles are not over. For, as Broad himself points out, there is one important class of cases in which it does not help us. These cases are those in which one measurable quantity is said to depend upon another, cases like that of the law connecting the volume and temperature of a gas under a given pressure, in which there is a mathematical function which enables one to calculate the numerical value of either quantity from the value of the other. Such laws have the form '$x = Fy$', where the range of the variable y covers all possible values of the quantity in question. But now it is not to be supposed that all these values are actually to be found in nature. Even if the number of different temperatures which specimens of gases have or will acquire is infinite, there still must be an infinite number missing. How then are we to interpret such a law? As being the compendious assertion of all its actual instances? But the formulation of the law in no way indicates which the actual instances are. It would be absurd to construe a general formula about the functional dependence of one quantity on another as committing us to the assertion that just these values of the quantity are actually realized. As asserting that for a value n of y, which is in fact not realized, the proposition that it is realized, in conjunction with the set of propositions describing all the actual cases, entails the proposition that there is a corresponding value m of x? But this is open to the same objection, with the further drawback that the entailment would not hold. As asserting with regard to any given value n of y that either n is not realized or that there is a corresponding value m of x? This is the most plausible alternative, but it makes the law trivial for all the values of y which happen not to be realized. It is hard to escape the conclusion that what we really mean to assert when we formulate such a law is that there is a corresponding value of x to every *possible* value of y.

Another reason for bringing in possibilities is that there seems to be no other way of accounting for the difference between generalizations of law and generalizations of fact. To revert to our earlier examples, it is a generalization of fact that all the Presidents of the Third French Republic are male, or that all the cigarettes that are now in my cigarette case are made of Virginian tobacco. It is a generalization of law that the planets of our solar system move in elliptical orbits, but a generalization of fact that, counting the earth as Terra, they all have Latin names. Some philosophers refer to these generalizations of fact as 'accidental generalizations', but this use of the word 'accidental' may be misleading. It is not suggested that these generalizations are true by accident, in the sense that there is no causal explanation of their truth, but only that they are not themselves the expression of natural laws.

But how is this distinction to be made? The formula '$(x) \Phi x \supset \Psi x$' holds equally in both cases. Whether the generalization be one of fact or of law, it will state at least that there is nothing which has the property Φ but lacks the property Ψ. In this sense, the generality is perfect in both cases, so long as the statements are true. Yet there seems to be a sense in which the generality of what we are

calling generalizations of fact is less complete. They seem
to be restricted in a way that generalizations of law are not.
Either they involve some spatio-temporal restriction, as in
the example of the cigarettes *now* in my cigarette case, or
they refer to particular individuals, as in the example of the
presidents of France. When I say that all the planets have
Latin names, I am referring definitely to a certain set of
individuals, Jupiter, Venus, Mercury, and so on, but when
I say that the planets move in elliptical orbits I am referring
indefinitely to anything that has the properties that constitute being a planet in this solar system. But it will not
do to say that generalizations of fact are simply conjunctions
of particular statements, which definitely refer to individuals ; for in asserting that the planets have Latin names,
I do not individually identify them : I may know that they
have Latin names without being able to list them all.
Neither can we mark off generalizations of law by insisting
that their expression is not to include any reference to
specific places or times. For with a little ingenuity, generalizations of fact can always be made to satisfy this condition.
Instead of referring to the cigarettes that are now in my
cigarette case, I can find out some general property which
only these cigarettes happen to possess, say the property of
being contained in a cigarette case with such and such
markings which is owned at such and such a period of his
life by a person of such and such a sort, where the descriptions are so chosen that the description of the person is in
fact satisfied only by me and the description of the cigarette
case, if I possess more than one of them, only by the one in
question. In certain instances these descriptions might
have to be rather complicated, but usually they would not :
and anyhow the question of complexity is not here at issue.
But this means that, with the help of these 'individuating'
predicates, generalizations of fact can be expressed in just
as universal a form as generalizations of law. And con-

versely, as Professor Nelson Goodman has pointed out, generalizations of law can themselves be expressed in such a way that they contain a reference to particular individuals, or to specific places and times. For, as he remarks, 'even the hypothesis "All grass is green" has as an equivalent "All grass in London or elsewhere is green"'.[1] Admittedly, this assimilation of the two types of statement looks like a dodge ; but the fact that the dodge works shows that we cannot found the distinction on a difference in the ways in which the statement can be expressed. Again, what we want to say is that whereas generalizations of fact cover only actual instances, generalizations of law cover possible instances as well. But this notion of possible, as opposed to actual, instances has not yet been made clear.

If generalizations of law do cover possible as well as actual instances, their range must be infinite ; for while the number of objects which do throughout the course of time possess a certain property may be finite, there can be no limit to the number of objects which might possibly possess it : for once we enter the realm of possibility we are not confined even to such objects as actually exist. And this shows how far removed these generalizations are from being conjunctions : not simply because their range is infinite, which might be true even if it were confined to actual instances, but because there is something absurd about trying to list all the possible instances. One can imagine an angel's undertaking the task of naming or describing all the men that there ever have been or will be, even if their number were infinite, but how would he set about naming, or describing, all the possible men ? This point is developed by F. P. Ramsey who remarks that the variable hypothetical '$(x)\,\Phi x$' resembles a conjunction (*a*) in that it contains all lesser, *i.e.* here all finite conjunctions, and appears as a sort of infinite product. (*b*) When we ask

---

[1] *Fact, Fiction and Forecast*, p. 78.

what would make it true, we inevitably answer that it is true if and only if every x has Φ ; *i.e.* when we regard it as a proposition capable of the two cases truth and falsity, we are forced to make it a conjunction which we cannot express for lack of symbolic power'.[1] But, he goes on, 'what we can't say we can't say, and we can't whistle it either', and he concludes that the variable hypothetical is not a conjunction and that 'if it is not a conjunction, it is not a proposition at all'. Similarly, Professor Ryle, without explicitly denying that generalizations of law are propositions, describes them as 'seasonal inference warrants',[2] on the analogy of season railway-tickets, which implies that they are not so much propositions as rules. Professor Schlick also held that they were rules, arguing that they could not be propositions because they were not conclusively verifiable ; but this is a poor argument, since it is doubtful if any propositions are conclusively verifiable, except possibly those that describe the subject's immediate experiences.

Now to say that generalizations of law are not propositions does have the merit of bringing out their peculiarity. It is one way of emphasizing the difference between them and generalizations of fact. But I think that it emphasizes it too strongly. After all, as Ramsey himself acknowledges, we do want to say that generalizations of law are either true or false. And they are tested in the way that other propositions are, by the examination of actual instances. A contrary instance refutes a generalization of law in the same way as it refutes a generalization of fact. A positive instance confirms them both. Admittedly, there is the difference that if all the actual instances are favourable, their conjunction entails the generalization of fact, whereas it does not entail the generalization of law : but still there

---

[1] *Foundations of Mathematics*, p. 238.
[2] '"If", "So", and "Because"', *Philosophical Analysis* (Essays edited by Max Black), p. 332.

is no better way of confirming a generalization of law than by finding favourable instances. To say that lawlike statements function as seasonal inference warrants is indeed illuminating, but what it comes to is that the inferences in question are warranted by the facts. There would be no point in issuing season tickets if the trains did not actually run.

To say that generalizations of law cover possible as well as actual cases is to say that they entail subjunctive conditionals. If it is a law of nature that the planets move in elliptical orbits, then it must not only be true that the actual planets move in elliptical orbits ; it must also be true that if anything were a planet it would move in an elliptical orbit : and here 'being a planet' must be construed as a matter of having certain properties, not just as being identical with one of the planets that there are. It is not indeed a peculiarity of statements which one takes as expressing laws of nature that they entail subjunctive conditionals : for the same will be true of any statement that contains a dispositional predicate. To say, for example, that this rubber band is elastic is to say not merely that it will resume its normal size when it has been stretched, but that it would do so if ever it were stretched : an object may be elastic without ever in fact being stretched at all. Even the statement that this is a white piece of paper may be taken as implying not only how the piece of paper does look but also how it would look under certain conditions, which may or may not be fulfilled. Thus one cannot say that generalizations of fact do not entail subjunctive conditionals, for they may very well contain dispositional predicates : indeed they are more likely to do so than not : but they will not entail the subjunctive conditionals which are entailed by the corresponding statements of law. To say that all the planets have Latin names may be to make a dispositional statement, in the sense that it implies not so much that people do always call them by such names but that they would so call them if they were speaking correctly. It does not, however, imply with regard to anything whatsoever that if it were a planet it would be called by a Latin name. And for this reason it is not a generalization of law, but only a generalization of fact.

There are many philosophers who are content to leave the matter there. They explain the 'necessity' of natural laws as consisting in the fact that they hold for all possible, as well as actual, instances : and they distinguish generalizations of law from generalizations of fact by bringing out the differences in their entailment of subjunctive conditionals. But while this is correct so far as it goes, I doubt if it goes far enough. Neither the notion of possible, as opposed to actual, instances nor that of the subjunctive conditional is so pellucid that these references to them can be regarded as bringing all our difficulties to an end. It will be well to try to take our analysis a little further if we can.

The theory which I am going to sketch will not avoid all talk of dispositions ; but it will confine it to people's attitudes. My suggestion is that the difference between our two types of generalization lies not so much on the side of the facts which make them true or false, as in the attitude of those who put them forward. The factual information which is expressed by a statement of the form 'for all x, if x has Φ then x has Ψ', is the same whichever way it is interpreted. For if the two interpretations differ only with respect to the possible, as opposed to the actual values of x, they do not differ with respect to anything that actually happens. Now I do not wish to say that a difference in regard to mere possibilities is not a genuine difference, or that it is to be equated with a difference in the attitude of those who do the interpreting. But I do think that it can best be elucidated by referring to such differences of attitude. In short I propose to explain the distinction between generalizations of law and generalizations of fact, and thereby to give some account of what a law of nature is, by the indirect method of analysing the distinction between treating a generalization as a statement of law and treating it as a statement of fact.

If someone accepts a statement of the form '$(x) \, \Phi x \supset \Psi x$' as a true generalization of fact, he will not in fact believe that anything which has the property Φ has any other property that leads to its not having Ψ. For since he believes that everything that has Φ has Ψ, he must believe that whatever other properties a given value of x may have they are not such as to prevent its having Ψ. It may be even that he knows this to be so. But now let us suppose that he believes such a generalization to be true, without knowing it for certain. In that case there will be various properties X, X_1 . . such that if he were to learn, with respect to any value of a of x, that a had one or more of these properties as well as Φ, it would destroy, or seriously weaken his belief that a had Ψ. Thus I believe that all the cigarettes in my case are made of Virginian tobacco, but this belief would be destroyed if I were informed that I had absent-mindedly just filled my case from a box in which I keep only Turkish cigarettes. On the other hand, if I took it to be a law of nature that all the cigarettes in this case were made of Virginian tobacco, say on the ground that the case had some curious physical property which had the effect of changing any other tobacco that was put into it into Virginian, then my belief would not be weakened in this way.

Now if our laws of nature were causally independent of each other, and if, as Mill thought, the propositions which expressed them were always put forward as being unconditionally true, the analysis could proceed quite simply. We could then say that a person A was treating a statement of the form 'for all x, if Φx then Ψx' as expressing a law of nature, if and only if there was no property X which was such that the information that a value a of x had X as well as Φ would weaken his belief that a had Ψ. And here we should have to admit the proviso that X did not logically entail not-Ψ, and also, I suppose, that its presence was not regarded as a manifestation of not-Ψ; for we do not wish to make it incompatible with treating a statement as the expression of a law that one should acknowledge a negative instance if it arises. But the actual position is not so simple. For one may believe that a statement of the form 'for all x, if Φx then Ψx' expresses a law of nature while also believing, because of one's belief in other laws, that if something were to have the property X as well as Φ it would not have Ψ. Thus one's belief in the proposition that an object which one took to be a loadstone attracted iron might be weakened or destroyed by the information that the physical composition of the supposed loadstone was very different from what one had thought it to be. I think, however, that in all such cases, the information which would impair one's belief that the object in question had the property Ψ would also be such that, independently of other considerations, it would seriously weaken one's belief that the object ever had the property Φ. And if this is so, we can meet the difficulty by stipulating that the range of properties which someone who treats 'for all x, if Φx then Ψx' as a law must be willing to conjoin with Φ, without his belief in the consequent being weakened, must not include those the knowledge of whose presence would in itself seriously weaken his belief in the presence of Φ.

There remains the further difficulty that we do not normally regard the propositions which we take to express laws of nature as being unconditionally true. In stating them we imply the presence of certain conditions which we do not actually specify. Perhaps we could specify them if we chose, though we might find it difficult to make the list exhaustive. In this sense a generalization of law may be weaker than a generalization of fact, since it may admit exceptions to the generalization as it is stated. This does not mean, however, that the law allows for exceptions : if the exception is acknowledged to be genuine, the law is held to be refuted. What happens in the other cases is that the exception is regarded as having been tacitly provided for. We lay down a law about the boiling point of water, without bothering to mention that it does not hold for high altitudes. When this is pointed out to us, we say that this qualification was meant to be understood. And so in other instances. The statement that if anything has Φ it has Ψ was a loose formulation of the law : what we really meant was that if anything has Φ but not X, it has Ψ. Even in the case where the existence of the exception was not previously known, we often regard it as qualifying rather than refuting the law. We say, not that the generalization has been falsified, but that it was inexactly stated. Thus, it must be allowed that someone whose belief in the presence of Ψ, in a given instance, is destroyed by the belief that Φ is accompanied by X may still be treating '$(x) \Phi \supset \Psi x$' as expressing a law of nature if he is prepared to accept '$(x) \Phi x \cdot \sim X x \supset \Psi x$' as a more exact statement of the law.

Accordingly I suggest that for someone to treat a statement of the form 'if anything has Φ it has Ψ' as expressing a law of nature, it is sufficient (i) that subject to a willingness to explain away exceptions he believes that in a non-trivial sense everything which in fact has Φ has Ψ (ii) that his belief that something which has Φ has Ψ is not liable to be weakened by the discovery that the object in question also has some other property X, provided (a) that X does not logically entail not-Ψ (b) that X is not a manifestation of not-Ψ (c) that the discovery that something had X would not in itself seriously weaken his belief that it had Φ (d) that he does not regard the statement 'if anything has Φ and

Q

not-X it has Ψ' as a more exact statement of the generalization that he was intending to express.

I do not suggest that these conditions are necessary, both because I think it possible that they could be simplified and because they do not cover the whole field. For instance, no provision has been made for functional laws, where the reference to possible instances does not at present seem to me eliminable. Neither am I offering a definition of natural law. I do not claim that to say that some proposition expresses a law of nature entails saying that someone has a certain attitude towards it; for clearly it makes sense to say that there are laws of nature which remain unknown. But this is consistent with holding that the notion is to be explained in terms of people's attitudes. My explanation is indeed sketchy, but I think that the distinctions which I have tried to bring out are relevant and important : and I hope that I have done something towards making them clear.

LAW STATEMENTS AND COUNTERFACTUAL INFERENCE[1]

By Roderick M. Chisholm

THE problems I have been invited to discuss arise from the fact that there are two types of true synthetic universal statement : statements of the one type, in the context of our general knowledge, seem to warrant counterfactual inference and statements of the other type do not. I shall call statements of the first type " law statements " and statements of the second type " non-law statements ". Both law and nonlaw statements may be expressed in the general form, " For every x, if x is S, x is P ". Law statements, unlike nonlaw statements, seem to warrant inference to statements of the form, "If *a*, which is not S, *were* S, *a* would be P " and " For every x, if x *were* S, x would be P ". I shall discuss (I) this distinction between law and non-law statements and (II) the related problem of interpreting counterfactual statements.[2]

I

Let us consider the following as examples of law statements :

L1. Everyone who drinks from this bottle is poisoned.

L2. All gold is malleable.

And let us consider the following as examples of nonlaw statements :

N1. Everyone who drinks from —— bottle wears a necktie.

N2. Every Canadian parent of quintuplets in the first half of the twentieth century is named ' Dionne '.

Let us suppose that L1 and N1 are concerned with the same bottle (perhaps it is one of short duration and has contained

---

[1] Read at the Western Division meeting of the American Philosophical Association, University of Illinois, May 7, 1954.

[2] Detailed formulations of this problem are to be found in the following works : W. E. Johnson, *Logic*, Vol. III, chapter I; C. H. Langford, review of W. B. Gallie's "An Interpretation of Causal Laws ", *Journal of Symbolic Logic*, Vol. VI (1941), p. 67 ; C. I. Lewis, *An Analysis of Knowledge and Valuation*, Part II; Roderick M. Chisholm, " The Contrary-to-fact Conditional ", *Mind*, Vol. 55 (1946), pp. 289–307 (reprinted in H. Feigl and W. S. Sellars, *Readings in Philosophical Analysis*); Nelson Goodman, " The Problem of Counterfactual Conditionals ", *Journal of Philosophy*, Vol. 44 (1947), pp. 113–128 (reprinted in L. Linsky, *Semantics and the Philosophy of Language*); F. L. Will, " The Contrary-to-fact Conditional ", *Mind*, Vol. 56 (1947), pp. 236–249; and William Kneale, " Natural Laws and Contrary to Fact Conditionals ", ANALYSIS, Vol. 10 (1950), pp. 121–125. See further references below and in Erna Schneider, "Recent Discussion of Subjunctive Conditionals ", *Review of Metaphysics*, Vol. VI (1953), pp. 623–647. My paper, referred to above, contains some serious errors.

only arsenic.) Let us suppose, further, that the blank in N1 is replaced by property terms which happen to characterize the bottle uniquely (perhaps they describe patterns of fingerprints). I shall discuss certain philosophical questions which arise when we make the following " preanalytic " assumptions. From L1 we can infer.

    L1.1   If Jones had drunk from this bottle, he would have been poisoned.

and from L2 we can infer

    L2.1   If that metal were gold, it would be malleable.

But from N1 we cannot infer

    N1.1   If Jones had drunk from —— bottle, he would have worn a necktie.

and from N2 we cannot infer

    N2.1   If Jones, who is Canadian, had been parent of quintuplets during the first half of the twentieth century, he would have been named ' Dionne '.

I shall not defend these assumptions beyond noting that, in respects to be discussed, they correspond to assumptions which practically everyone does make.

There are two preliminary points to be made concerning the interpretation of counterfactual statements. (1) We are concerned with those counterfactuals whose antecedents, " if a were S," may be interpreted as meaning the same as " if a had property S ". There is, however, another possible interpretation : " if a were S " could be interpreted as meaning the same as " if a were identical with something which in fact does have property S ".[1] Given the above assumptions, N2.1 is false according to the first interpretation, which is the interpretation with which we are concerned, but it is true according to the second (for if Jones were identical with one of the Dionnes, he would be named ' Dionne '). On the other hand, the statement

    N2.2   If Jones, who is Canadian, had been parent of quintuplets during the first half of the twentieth century, there would have been at least two sets of Canadian quintuplets.

is true according to the first interpretation and false according to the second. (2) It should be noted, secondly, that there is a respect—to be discussed at greater length below—in which our counterfactual statements may be thought of as being elliptical. If we assert L1.1, we might, nevertheless, accept the following qualification : " Of course, if Jones had emptied the bottle,

---

[1] Compare K. R. Popper, "A Note on Natural Laws and so-called ' Contrary-to-fact conditionals ' ", *Mind*, vol. 58 (1949), pp. 62–66.

cleaned it out, filled it with water, and *then* drunk from it, he might not have been poisoned." And, with respect to L2.1, we might accept this qualification : " If that metal were gold it would be malleable—provided, of course, that what we are supposing to be contrary-to-fact is that statement ' That metal is not gold ' and *not* the statement 'All gold is malleable '."

Can the relevant difference between law and non-law statements be described in familiar terminology without reference to counterfactuals, without use of modal terms such as " causal necessity ", " necessary condition ", " physical possibility ", and the like, and without use of metaphysical terms such as " real connections between matters of fact " ? I believe no one has shown that the relevant difference *can* be so described. I shall mention three recent discussions.

(1) It has been suggested that the distinction between law statements and nonlaw statements may be made with respect to the universality of the nonlogical terms which appear in the statements. A term may be thought of as being universal, it has been suggested, if its meaning can be conveyed without explicit reference to any particular object ; it is then said that law statements, unlike nonlaw statements, contain no nonlogical terms which are not universal.[1] (These points can be formulated more precisely.) This suggestion does not help, however, if applied to what we have been calling " law statements " and " nonlaw statements ", for L1 is a law statement containing the *non*universal nonlogical term " this bottle " and N1 (we have supposed) is a nonlaw statement all of whose nonlogical terms *are* universal. It may be that, with respect to ordinary usage, it is incorrect to call L1 a " law statement " ; this point does not affect our problem, however, since we are assuming that L1, whether or not it would ordinarily be called a " law statement ", does, in the context of our general knowledge, warrant the inference to L1.1

(2) It has been suggested that the two types of statement might be distinguished epistemologically. P. F. Strawson, in his *Introduction to Logical Theory*, suggests that in order to *know*, or to have good evidence or good reason for believing, that a

---

[1] Compare C. G. Hempel and Paul Oppenheim, " Studies in the Logic of Explanation ", *Philosophy of Science*, Vol. 15, 1948, pp. 135–175 (reprinted in H. Feigl and M. Brodbeck, *Readings in the Philosophy of Science*). It should be noted that these authors (i) attempt to characterize laws with respect only to formalized languages, (ii) concede that " the problem of an adequate definition of purely qualitative (universal) predicates remains open ", and (iii) propose a distinction between " derived " and " fundamental " laws. The latter distinction is similar to a distinction of Braithwaite, discussed below. See also Elizabeth Lane Beardsely, ", Non-Accidental and Counterfactual Sentences ", *Journal of Philosophy*, Vol. 46 (1949), pp. 573–591 ; review of the latter by Roderick M. Chisholm, *Journal of Symbolic Logic*, Vol. XVI (1951), pp. 63–64.

given nonlaw statement is true, it is necessary to know that all of its instances have in fact been observed ; but in order to know, or to have good evidence or good reason for believing, that a given law statement is true, it is *not* necessary to know that all of its instances have been examined. (We need not consider the problem of defining " instance " in this use.) "An essential part of our grounds for accepting " a nonlaw statement must be " evidence that there will be no more " instances and " that there never were more than the limited number of which observations have been recorded " (p. 199). Possibly this suggestion is true, but it leaves us with our problem. For the suggestion itself requires use of a modal term ; it refers to what a man *needs* to know, or what it is *essential* that he know, in order to know that a law statement is true. But if we thus allow ourselves the use of modal terms, we could have said at the outset merely that a law statement describes what is " physically necessary ", etc., and that a nonlaw statement does not.

(3) R. B. Braithwaite, in *Scientific Explanation*, suggests that a law statement, as distinguished from a nonlaw statement is one which " appears as a deduction from higher-level hypotheses which have been established independently of the statement (p. 303). " To consider whether or not a scientific hypothesis would, if true, be a law of nature is to consider the way in which it could enter into an established scientific deductive system " (Ibid). In other words, the question whether a statement is law-like may be answered by considering certain logical, or epistemological, relations which the statement bears to certain *other* statements. Our nonlaw statement N2, however, is deducible from the following two statements : (i) " Newspapers which are generally reliable report that all parents of quintuplets during the first half of the twentieth century are named ' Dionne ', " and (ii) " If newspapers which are generally reliable report that all parents of quintuplets during the first half of the twentieth century are named ' Dionne ', then such parents are named ' Dionne '." Statements (i) and (ii) may be considered as " higher level " parts of a " hypothetical-deductive system " from which the nonlaw statement N2 can be deduced ; indeed (i) and (ii) undoubtedly express the grounds upon which most people accept N2. It is not enough, therefore, to describe a nonlaw statement as a statement which " appears as a deduction from higher level hypotheses which have been established ndependently ". (I suggest, incidentally, that it is only at an dvanced stage of inquiry that one regards a synthetic universal tatement as being a *non*law statement.)

II

Even if we allow ourselves the distinction between law statements and nonlaw statements and characterize the distinction philosophically, by reference, say, to physical possibility (e.g. "All S is P" is a law statement provided it is not physically possible that anything be both S and not P, etc.), we find that contrary-to-fact conditionals still present certain difficulties of interpretation.[1] Assuming that the distinction between law statement and nonlaw statement is available to us, I shall now make some informal remarks which I hope will throw light upon the ordinary use of these conditionals.

Henry Hiz has suggested that a contrary-to-fact conditional might be interpreted as a metalinguistic statement, telling us something about what can be inferred in a given system of statements. " It says that, if something is accepted in this system to be true, then something else can be accepted in this system to be true."[2] This suggestion, I believe, can be applied to the ordinary use of contrary-to-fact conditionals, but it is necessary to make some qualifying remarks concerning the relevant " systems of statements ".

Let us consider one way of justifying the assertion of a contrary-to-fact conditional, " If *a* were S, *a* would be P ". The antecedent of the counterfactual is taken, its indicative form, as a *supposition* or *assumption*.[3] One says, in effect, " Let us *suppose* that *a* is S ", even though one may believe that *a* is not S. The indicative form of the consequent of the counterfactual— viz., " *a* is P "—is then shown to follow logically from the antecedent taken with certain other statements already accepted. This demonstration is then taken to justify the counterfactual. The point of asserting the counterfactual may be that of *calling attention to, emphasizing,* or *conveying,* one or more of the premises which, taken with the antecedent, logically imply the consequent.

In simple cases, where singular counterfactuals are asserted, we may thus think of the speaker : (i) as having deduced the consequences of a singular supposition, viz., the indicative form of the counterfactual antecedent, taken with a statement he

---

[1] Modal analyses of law statements are suggested by Hans Reichenbach, *Elements of Symbolic Logic*, Ch. VIII, and Arthur Burks, " The Logic of Causal Propositions ", *Mind*, Vol. LX (1951), pp. 363–382.

[2] Henry Hiz, " On the Inferential Sense of Contrary-to-fact Conditionals ", *Journal of Philosophy*, Vol. 48 (1949), pp, 586–587.

[3] Compare S. Jaskowski, " On the Rules of Suppositions in Formal Logic ", *Studia Logica*, No. 1 (Warsaw, 1934), and A. Meinong, *Über Annahmen*, concerning this use of " assumption."

interprets as a law statement; and (ii) as being concerned in part to call attention to, emphasize, or convey, the statement interpreted as a law statement. We can usually tell, from the context of a man's utterance, what the supposition is and what the other statements are with which he is concerned. He may say, " If that were gold, it would be malleable " ; it is likely, in this case, that the statement interpreted as a law statement is L2, "All gold is malleable " ; it is also likely that this is the statement he is concerned to emphasize.

F. H. Bradley suggested, in his *Principles of Logic*, that when a man asserts a singular counterfactual " the real judgment is concerned with the individual's *qualities*, and asserts no more than a connection of adjectives."[1] Bradley's suggestion, as I interpret it, is that the *whole* point of asserting a singular counterfactual, normally, is to call attention to, emphasize, or convey the statement interpreted as a law statement. It might be misleading, however, to say that the man is *affirming* or *asserting* what he takes to be a law statement, or statement describing a " connection of adjectives ", for he has not formulated it explicitly. It would also be misleading to say, as Bradley did (p. 89), that the man is merely *supposing* the law statement to be true, for the law statement is something he *believes*, and not merely supposes, to be true. If he were merely supposing "All gold is malleable," along with " That is gold ", then it is likely he would include this supposition in the antecedent of his counterfactual and say " If that were gold and if all gold were malleable, then that would be malleable ". Let us say he is *presupposing* the law statement.

We are suggesting, then, that a man in asserting a counterfactual is telling us something about what can be deduced from some " system of statements " when the indicative version of the antecedent is added to this system as a *supposition*. We are referring to the statements of this system (other than the indicative version of the antecedent) as the *presuppositions* of his assertion. And we are suggesting that, normally, at least part of the point of asserting a counterfactual is to *call attention to, emphasize,* or *convey,* one or more of these presuppositions.

The statements a man presupposes when he asserts a counterfactual will, presumably, be statements he accepts or believes. But they will not include the denial of the antecedent of his counterfactual (even if he believes this denial to be true) and

---

[1] Op. cit., p. 90. Compare D. J. O'Connor, " The Analysis of Conditional Sentences ", *Mind*, Vol. LX (1951), p. 360 ; Robert Brown and John Watling, " Counterfactual Conditionals ", *Mind*, Vol. LXI (1952), p. 226.

they will not include any statements he would treat as nonlaw statements.[1] And normally there will be many other statements he believes to be true which he will deliberately exclude from his presuppositions. The peculiar problem of interpreting ordinary counterfactual statements is that of specifying which, among the statements the asserter believes, he intends to *exclude* from his presuppositions. What statements he will exclude will depend upon what it is he is concerned to call attention to, emphasize, or convey.

Let us suppose a man accepts the following statements, taking the universal statements to be law statements : (1) All gold is malleable ; (2) No cast-iron is malleable ; (3) Nothing is both gold and cast-iron ; (4) Nothing is both malleable and not malleable ; (5) That is cast-iron ; (6) That is not gold ; and (7) That is not malleable. We may contrast three different situations in which he asserts three different counterfactuals having the same antecedents.

First, he asserts, pointing to an object his hearers don't know to be gold and don't know not to be gold, " If that *were* gold, it would be malleable ". In this case, he is supposing the denial of (6) ; he is excluding from his presuppositions (5), (6), and (7) ; and he is concerned to emphasize (1).

Secondly, he asserts, pointing to an object he and his hearers agree to be cast-iron, " If *that* were gold, then some gold things would not be malleable ". He is again supposing the denial of (6) ; he is excluding (1) and (6), but he is no longer excluding (5) or (7) ; and he is concerned to emphasize either (5) or (2).

Thirdly, he asserts, " If that were gold, then some things would be both malleable and not malleable ". He is again supposing the denial of (6) ; he is now excluding (3) and no longer excluding (1), (5), (6), or (7) ; and he is now concerned to emphasize (1), (2), or (5).

Still other possibilities readily suggest themselves.

If, then, we were to ask " What if that were gold? " our question would have a number of possible answers—e.g., the subjunctive forms of the denial of (7), the denial of (1), and the denial of (4). Any one of these three answers might be appropriate, but they would not *all* be appropriate in conjunction. Which answer is the appropriate one will depend upon what we wish to know. If, in asking " What if that were gold ?", we wish to know of some law statement describing gold, the denial

---

[1] Instead of saying his presuppositions include no statement he treats as a law statement, it might be more accurate to say this : if his presuppositions include any statement N he would interpret as a nonlaw statement, then N and the man's supposition cannot be so formulated that the supposition constitutes a substitution-instance of N's antecedent.

of (7) is appropriate ; if we wish to know what are the properties of the thing in question, the denial of (1) is appropriate ; and if we wish to know whether the thing has properties such that a statement saying nothing gold has those properties is a law statement, the denial of (4) is appropriate. The counterfactual question, " What if that were gold? ", is, therefore, clearly ambiguous. But in each case, the question could be formulated clearly and unambiguously.

Counterfactuals are similar to *probability* statements in that each type of statement is, in a certain sense, elliptical. If we ask, " What is the probability that this man will survive ? ", our question is incompletely formulated ; a more explicit formulation would be, " With respect to such-and-such evidence, what is the probability that this man will survive? " Similarly, if we ask, "What would American policy in Asia be if Stevenson were President ", our question is incompletely formulated ; a more explicit formulation would be, " Supposing that Stevenson were President, and presupposing so-and-so, but not so-and-so, what would be the consequences with respect to American policy in Asia? " But there is an important respect in which counterfactual statements *differ* from such probability statements. If a man wishes to know what is the probability of a certain statement, i.e., if he wishes to know the truth of a categorical probability statement, then, we may say, he should take into consideration *all* the relevant evidence available to him ; the premises of his probability inference should omit no relevant statement which he is justified in believing.[1] But this " require-ment of total evidence " cannot be assumed to hold in the case of counterfactual inference. If a man asks, " What would American policy in Asia be, if Stevenson were President? ", and if his question may be interpreted in the way in which it ordinarily would be interpreted, then there are many facts included in his store of knowledge which we would expect him to *overlook*, or *ignore*, in answering his question ; i.e., there are many facts which we would expect him deliberately to *exclude* from his presuppositions. Normally we would expect him to exclude the fact that Eisenhower's program is the one which has been followed since 1953 ; another is the fact that Mr. Dulles is Secretary of State. But there are other facts, which may also be included in the man's store of knowledge, whose status is more questionable. Does he intend to exclude the fact that Congress was Republican; does he intend to exclude those Asiatic events which have occurred as a result of Eisenhower's policies ; does

[1] Compare Rudolf Carnap, *Logical Foundations of Probability*, p. 211 ff.

he intend to exclude the fact that Stevenson went to Asia in 1953? There is no point in insisting either that he consider or that he exclude these facts. But, if he wishes to be understood, he should tell us which are the facts that he is considering, or presupposing, and which are the ones he is excluding.

Bradley suggested the ambiguity of some counterfactual statements may be attributed to the fact that " the supposition is not made evident " (*op. cit.* p. 89). In our terminology, it would be more accurate to say that the *presupposition* is not made evident ; for the supposition is usually formulated explicitly in the antecedent of the counterfactual statement. (But when a man says, " If that thing, which is not S, were S . . . . . , " the subordinate indicative clause expresses neither a supposition nor a presupposition.) Ideally it might be desirable to formulate our counterfactuals in somewhat the following way : " Supposing that that is S, and presupposing so-and-so, then it follows that that is P." In practice, however, it is often easy to tell, from the context in which a counterfactual is asserted, just what it is that is being presupposed and what it is that is being excluded[1]

Although I have been using the terms " counterfactual " and " contrary-to-fact " throughout this discussion, it is important to note that, when a man arrives at a conditional statement in the manner we have been discussing, his supposition—and thus also the antecedent of his conditional—need *not* be anything he believes to be false. For example, a man in deliberating will consider the consequences of a supposition, taken along with certain presuppositions, and he will also consider the consequences of its denial, taken along with the same presuppositions. It is misleading to say, therefore, that the conditionals he may then affirm are " counterfactual ", or " contrary-to-fact ", for he may have no beliefs about the truth or falsity of the respective antecedents and one of these antecedents will in fact be true.[2] A better term might be " suppositional conditional " or, indeed, " hypothetical statement ".

Brown University, U.S.A.

[1] " The Contrary-to-fact Conditional " (pp. 303-304 ; Feigl-Sellars, p. 494) I discuss what I take to be certain conventions of ordinary language pertaining to this point.

[2] Compare Alan Ross Anderson, "A Note on Subjunctive and Counterfactual Conditionals ", ANALYSIS, Vol. 12 (1951), pp. 35-38; Roderick M. Chisholm, review of David Pears' " Hypotheticals ", *Journal of Symbolic Logic*, Vol. 15 (1950), pp. 215-216.

The Problem of Counterfactual Conditionals[1]

1. The Problem in General

THE analysis of counterfactual conditionals is no fussy little grammatical exercise. Indeed, if we lack the means for interpreting counterfactual conditionals, we can hardly claim to have any adequate philosophy of science. A satisfactory definition of scientific law, a satisfactory theory of confirmation or of disposition terms (and this includes not only predicates ending in "ible" and "able" but almost every objective predicate, such as "is red"), would solve a large part of the problem of counterfactuals. Conversely, a solution to the problem of counterfactuals would give us the answer to critical questions about law, confirmation, and the meaning of potentiality.

I am not at all contending that the problem of counterfactuals is logically or psychologically the first of these related problems. It makes little difference where we start if we can go ahead If the study of counterfactuals has up to now failed this pragmatic test, the alternative approaches are little better off.

What, then, is the *problem* about counterfactual conditionals? Let us confine ourselves to those in which antecedent and consequent are inalterably false—as, for example, when I say of a piece of butter that was eaten yesterday, and that had never been heated,

If that piece of butter had been heated to 150° F., it would have melted.

Considered as truth-functional compounds, all counterfactuals are of course true, since their antecedents are false. Hence

If that piece of butter had been heated to 150° F., it would not have melted

would also hold. Obviously something different is intended, and the problem is to define the circumstances under which a given counterfactual holds while the opposing conditional with the contradictory consequent fails to hold. And this criterion of truth must be set up in the face of the fact that a counterfactual by its nature can never be subjected to any direct empirical test by realizing its antecedent.

In one sense the name "problem of counterfactuals" is misleading, because the problem is independent of the form in which a given statement happens to be expressed. The problem of counterfactuals is equally a problem of factual conditionals, for any counterfactual can be transposed into a conditional with a true antecedent and consequent; e.g.,

Since that butter did not melt, it wasn't heated to 150° F.

The possibility of such transformation is of no great importance except to clarify the nature of our problem. That "since" occurs in the contrapositive shows that what is in question is a certain kind of connection between the two component sentences; and the truth of statements of this kind—whether they have the form of counterfactual or factual conditionals or some other form—depends not upon the truth or falsity of the components but upon whether the intended connection obtains. Recognizing the possibility of transformation serves mainly to focus attention on the central problem and to discourage speculation as to the nature of counterfacts. Although I shall begin my study by considering counterfactuals as such, it must be borne in mind that a general solution would explain the kind of connection involved irrespective of any assumption as to the truth or falsity of the components.

The effect of transposition upon conditionals of another kind, which I call "semifactuals", is worth noticing briefly. Should we assert

Even if the match had been scratched, it still would not have lighted,

we would uncompromisingly reject as an equally good expression of our meaning the contrapositive,

Even if the match lighted, it still wasn't scratched.

Our original intention was to affirm not that the non-lighting could be inferred from the scratching, but simply that the lighting could not be inferred from the scratching. Ordinarily a semifactual conditional has the force of denying what is affirmed by the opposite, fully counterfactual conditional. The sentence

Even had that match been scratched, it still wouldn't have lighted

is normally meant as the direct negation of

Had the match been scratched, it would have lighted.

That is to say, in practice full counterfactuals affirm, while semifactuals deny, that a certain connection obtains between antecedent and consequent.[2] Thus it is clear why a semifactual generally has not the same meaning as its contrapositive.

There are various special kinds of counterfactuals that present special problems. An example is the case of 'counteridenticals', illustrated by the statements

If I were Julius Caesar, I wouldn't be alive in the twentieth century,

and

If Julius Caesar were I, he would be alive in the twentieth century.

Here, although the antecedent in the two cases is a statement of the same identity, we attach two different consequents which, on the very assumption of that identity, are incompatible. Another special class of counterfactuals is that of the 'countercomparatives', with antecedents such as

If I had more money, ...

The trouble with these is that when we try to translate the counterfactual into a statement about a relation between two tenseless, non-modal sentences, we get as an antecedent something like

If "I have more money than I have" were true, . . . ,

which wrongly represents the original antecedent as self-contradictory. Again there are the 'counterlegals', conditionals with antecedents that either deny general laws directly, as in

If triangles were squares, . . . ,

or else make a supposition of particular fact that is not merely false but impossible, as in

If this cube of sugar were also spherical, ...

Counterfactuals of all these kinds offer interesting but not insurmountable special difficulties.[3] In order to concentrate upon the major problems concerning counterfactuals in general, I shall usually choose my examples in such a way as to avoid these more special complications.

As I see it, there are two major problems, though they are not independent and may even be regarded as aspects of a single problem. A counterfactual is true if a certain connection obtains between the antecedent and the consequent. But as is obvious from examples already given, the consequent seldom follows from the antecedent by logic alone. (1) In the first place, the assertion that a connection holds is made on the presumption that certain circumstances not stated in the antecedent obtain. When we say

If that match had been scratched, it would have lighted,

we mean that conditions are such—i.e. the match is well made, is dry enough, oxygen enough is present, etc.—that "That match lights" can be inferred from "That match is scratched". Thus the connection we affirm may be regarded as joining the consequent with the conjunction of the antecedent and other statements that truly describe relevant conditions. Notice especially that our assertion of the counterfactual is *not* conditioned upon these circumstances obtaining. We do not assert that the counterfactual is true *if* the circumstances obtain; rather, in asserting the counterfactual we commit ourselves to the actual truth of the statements describing the requisite relevant conditions. The first major problem is to define relevant conditions: to specify what sentences are meant to be taken in conjunction with an antecedent as a basis for inferring the consequent. (2) But even after the particular relevant conditions are specified, the connection obtaining will not ordinarily be a logical one. The principle that permits inference of

That match lights

from

That match is scratched. That match is dry enough. Enough oxygen is present. Etc.

is not a law of logic but what we call a natural or physical or causal law. The second major problem concerns the definition of such laws.

2. *The Problem of Relevant Conditions*

It might seem natural to propose that the consequent follows by law from the antecedent and a description of the actual state-of-affairs of the world, that we need hardly

B

235

define relevant conditions because it will do no harm to include irrelevant ones. But if we say that the consequent follows by law from the antecedent and *all* true statements, we encounter an immediate difficulty:—among true sentences is the negate of the antecedent, so that from the antecedent and all true sentences everything follows. Certainly this gives us no way of distinguishing true from false counterfactuals.

We are plainly no better off if we say that the consequent must follow from *some* set of true statements conjoined with the antecedent; for given any counterfactual antecedent A, there will always be a set S—namely, the set consisting of *not-A*—such that from $A \cdot S$ any consequent follows. (Hereafter I shall regularly use the following symbols: "A" for the antecedent; "C" for the consequent; "S" for the set of statements of the relevant conditions or, indifferently, for the conjunction of these statements.)

Perhaps then we must exclude statements logically incompatible with the antecedent. But this is insufficient; for a parallel difficulty arises with respect to true statements which are not logically but are otherwise incompatible with the antecedent. For example, take

If that radiator had frozen, it would have broken.

Among true sentences may well be (S)

That radiator never reached a temperature below 33° F.

Now we have as true generalizations both

All radiators that freeze but never reach below 33° F. break,

and also

All radiators that freeze but never reach below 33° F. fail to break;

for there are no such radiators. Thus from the antecedent of the counterfactual and the given S, we can infer any consequent.

The natural proposal to remedy this difficulty is to rule that counterfactuals can not depend upon empty laws; that the connection can be established only by a principle of the form "All x's are y's" when there are some x's. But this is ineffectual. For if empty principles are excluded, the following non-empty principles may be used in the case given with the same result:

Everything that is either a radiator that freezes but does not reach below 33° F., or that is a soap bubble, breaks;

Everything that is either a radiator that freezes but does not reach below 33° F., or is powder, does not break.

By these principles we can infer any consequent from the A and S in question.

The only course left open to us seems to be to define relevant conditions as the set of all true statements each of which is both logically and non-logically compatible with A where non-logical incompatibility means violation of a non-logical law.[4] But another difficulty immediately appears. In a counterfactual beginning

If Jones were in Carolina, . . .

the antecedent is entirely compatible with

Jones is not in South Carolina

and with

Jones is not in North Carolina

and with

North Carolina plus South Carolina is identical with Carolina;

but all these taken together with the antecedent make a set that is self-incompatible, so that again any consequent would be forthcoming.

Clearly it will not help to require only that for *some* set S of true sentences, $A \cdot S$ be self-compatible and lead by law to the consequent; for this would make a true counterfactual of

If Jones were in Carolina, he would be in South Carolina,

and also of

If Jones were in Carolina, he would be in North Carolina,

which cannot both be true.

It seems that we must elaborate our criterion still further, to characterize a counterfactual as true if and only if there is some set S of true statements such that $A \cdot S$ is self-compatible and leads by law to the consequent, while there is no such set S' such that $A \cdot S'$ is self-compatible and leads by law to the negate of the consequent.[5] Unfortunately even this is not enough. For among true sentences will be the negate of the consequent: $-C$. Is $-C$ compatible with A or not? If not, then A alone without any additional conditions must lead by law to C. But if $-C$ is compatible with A (as in most cases), then if we take $-C$ as our S, the conjunction $A \cdot S$ will give us $-C$. Thus the criterion we have set up will seldom be satisfied; for since $-C$ will normally be compatible with A, as the need for introducing the relevant conditions testifies, there will normally be an S (namely, $-C$) such that $A \cdot S$ is self-compatible and leads by law to $-C$.

Part of our trouble lies in taking too narrow a view of our problem. We have been trying to lay down conditions under which an A that is known to be false leads to a C that is known to be false; but it is equally important to make sure that our criterion does not establish a similar connection between our A and the (true) negate of C. Because our S together with A was to be so chosen as to give us C, it seemed gratuitous to specify that S must be compatible with C; and because $-C$ is true by supposition, S would necessarily be compatible with it. But we are testing whether our criterion not only admits the true counterfactual we are concerned with but also excludes the opposing conditional. Accordingly, our criterion must be modified by specifying that S be compatible with both C and $-C$.[6] In other words, S by itself must not decide between C and $-C$, but S together with A must lead to C but not to $-C$. We need not know whether C is true or false.

Our rule thus reads that a counterfactual is true if and only if there is some set S of true sentences such that S is compatible with C and with $-C$, and such that $A \cdot S$ is self-compatible and leads by law to C; while there is no set S' compatible with C and with $-C$, and such that $A \cdot S'$ is self-compatible and leads by law to $-C$. As thus stated, the rule involves a certain redundancy; but simplification is not in point here, for the criterion is still inadequate.

The requirement that $A \cdot S$ be self-compatible is not strong enough; for S might comprise true sentences that although *compatible with A*, were such that *they would not be true if A were true*. For this reason, many statements that we would regard as definitely false would be true according to the stated criterion. As an example, consider the familiar case where for a given match M, we would affirm

(i) If match M had been scratched, it would have lighted,

but deny

(ii) If match M had been scratched, it would not have been dry.[7]

According to our tentative criterion, statement (ii) would be quite as true as statement (i). For in the case of (ii), we may take as an element in our S the true sentence

Match M did not light,

which is presumably compatible with A (otherwise nothing would be required along with A to reach the opposite as the consequent of the true counterfactual statement (i)). As our total $A \cdot S$ we may have

Match M is scratched. It does not light. It is well made. Oxygen enough is present . . . etc.;

and from this, by means of a legitimate general law, we can infer

It was not dry.

And there would seem to be no suitable set of sentences S' such that $A \cdot S'$ leads by law to the negate of this consequent. Hence the unwanted counterfactual is established in accord with our rule. The trouble is caused by including in our S a true statement which though compatible with A would not be true if A were. Accordingly we must exclude such statements from the set of relevant conditions; S, in addition to satisfying the other requirements already laid down, must be not merely compatible with A but 'jointly tenable' or *cotenable* with A. A is cotenable with S, and the conjunction $A \cdot S$ self-cotenable, if it is not the case that S would not be true if A were.[8]

Parenthetically it may be noted that the relative fixity of conditions is often unclear, so that the speaker or writer has to make explicit additional provisos or give subtle verbal clues as to his meaning. For example, each of the following two counterfactuals would normally be accepted:

If New York City were in Georgia, then New York City would be in the South.

If Georgia included New York City, then Georgia would not be entirely in the South.

Yet the antecedents are logically indistinguishable. What happens is that the direction of expression becomes important, because in the former case the meaning is

If New York City were in Georgia, and the boundaries of Georgia remained unchanged, then . . . ,

while in the latter case the meaning is

If Georgia included New York City, and the boundaries of New York City remained unchanged, then . . . .

Without some such cue to the meaning as is covertly given

by the word-order, we should be quite uncertain which of the two consequents in question could be truly attached. The same kind of explanation accounts for the paradoxical pairs of counteridenticals mentioned earlier.

Returning now to the proposed rule, I shall neither offer further corrections of detail nor discuss whether the requirement that S be cotenable with A makes superfluous some other provisions of the criterion; for such matters become rather unimportant beside the really serious difficulty that now confronts us. In order to determine the truth of a given counterfactual it seems that we have to determine, among other things, whether there is a suitable S that is cotenable with A and meets certain further requirements. But in order to determine whether or not a given S is cotenable with A, we have to determine whether or not the counterfactual "If A were true, then S would not be true" is itself true. But this means determining whether or not there is a suitable S_1, cotenable with A, that leads to $-S$ and so on. Thus we find ourselves involved in an infinite regressus or a circle; for cotenability is defined in terms of counterfactuals, yet the meaning of counterfactuals is defined in terms of cotenability. In other words to establish any counterfactual, it seems that we first have to determine the truth of another. If so, we can never explain a counterfactual except in terms of others, so that the problem of counterfactuals must remain unsolved.

Though unwilling to accept this conclusion, I do not at present see any way of meeting the difficulty. One naturally thinks of revising the whole treatment of counterfactuals in such a way as to admit first those that depend on no conditions other than the antecedent, and then use these counterfactuals as the criteria for the cotenability of relevant conditions with antecedents of other counterfactuals, and so on. But this idea seems initially rather unpromising in view of the formidable difficulties of accounting by such a step-by-step method for even so simple a counterfactual as

If the match had been scratched, it would have lighted.

3. The Problem of Law

Even more serious is the second of the problems mentioned earlier: the nature of the general statements that enable us to infer the consequent upon the basis of the antecedent and the statement of relevant conditions. The distinction between these connecting principles and relevant conditions is imprecise and arbitrary; the 'connecting principles' might be conjoined to the condition-statements, and the relation of the antecedent-conjunction $(A \cdot S)$ to the consequent thus made a matter of logic. But the same problems would arise as to the kind of principle that is capable of sustaining a counterfactual; and it is convenient to consider the connecting principles separately.

In order to infer the consequent of a counterfactual from the antecedent A and a suitable statement of relevant conditions S, we make use of a general statement; namely, the generalization[9] of the conditional having $A \cdot S$ for antecedent and C for consequent. For example, in the case of

If the match had been scratched, it would have lighted

the connecting principle is

Every match that is scratched, well made, dry enough, in enough oxygen, etc., lights.

But notice that *not* every counterfactual is actually sustained by the principle thus arrived at, *even* if that principle is *true*. Suppose, for example, that all I had in my right pocket on VE day was a group of silver coins. Now we would not under normal circumstances affirm of a given penny P

If P had been in my pocket on VE day, P would have been silver,[10]

242

even though from

P was in my pocket on VE day

we can infer the consequent by means of the general statement

Everything in my pocket on VE day was silver.

On the contrary, we would assert that if P had been in my pocket, then this general statement would not be true. The general statement will *not* permit us to infer the given consequent from the counterfactual assumption that P was in my pocket, because the general statement will not itself withstand that counterfactual assumption. Though the supposed connecting principle is indeed general, true, and perhaps even fully confirmed by observation of all cases, it is incapable of sustaining a counterfactual because it remains a description of accidental fact, not a law. The truth of a counterfactual conditional thus seems to depend on whether the general sentence required for the inference is a law or not. If so, our problem is to distinguish accurately between causal laws and casual facts.[11]

The problem illustrated by the example of the coins is closely related to that which led us earlier to require the cotenability of the antecedent and the relevant conditions, in order to avoid resting a counterfactual on any statement that would not be true if the antecedent were true. For decision as to the cotenability of two sentences depends partly upon decisions as to whether certain general statements are laws, and we are now concerned directly with the latter problem. Is there some way of so distinguishing laws from non-laws, among true universal statements of the kind in question, that laws will be the principles that will sustain counterfactual conditionals?

Any attempt to draw the distinction by reference to a notion of causative force can be dismissed at once as unscientific. And it is clear that no purely syntactical criterion can be adequate, for even the most special descriptions of particular facts can be cast in a form having any desired degree of syntactical universality. "Book B is small" becomes "Everything that is Q is small" if "Q" stands for some predicate that applies uniquely to B. What then does distinguish a law like

All butter melts at 150° F.

from a true and general non-law like

All the coins in my pocket are silver ?

Primarily, I would like to suggest, the fact that the first is accepted as true while many cases of it remain to be determined, the further, unexamined cases being predicted to conform with it. The second sentence, on the contrary, is accepted as a description of contingent fact *after* the determination of all cases, no prediction of any of its instances being based upon it. This proposal raises innumerable problems, some of which I shall consider presently; but the idea behind it is just that the principle we use to decide counterfactual cases is a principle we are willing to commit ourselves to in deciding unrealized cases that are still subject to direct observation.

As a first approximation then, we might say that a law is a true sentence used for making predictions. That laws are used predictively is of course a simple truism, and I am not proposing it as a novelty. I want only to emphasize the Humean idea that rather than a sentence being used for prediction because it is a law, it is called a law because it is used for prediction; and that rather than the law being used for prediction because it describes a causal connection, the meaning of the causal connection is to be interpreted in terms of predictively used laws.

By the determination of all instances, I mean simply the examination or testing by other means of all things that satisfy the antecedent, to decide whether all satisfy the consequent also. There are difficult questions about the meaning of "instance", many of which Professor Hempel has investigated. Most of these are avoided in our present study by the fact that we are concerned with a very narrow class of sentences: those arrived at by generalizing conditionals of a certain kind. Remaining problems about the meaning of "instance" I shall have to ignore here. As for "determination", I do not mean final discovery of truth, but only enough examination to reach a decision as to whether a given statement or its negate is to be admitted as evidence for the hypothesis in question.

Our criterion excludes vacuous principles as laws. The generalizations needed for sustaining counterfactual conditionals cannot be vacuous, for they must be supported by evidence.[12] The limited scope of our present problem makes it unimportant that our criterion, if applied generally to all statements, would classify as laws many statements—e.g., true singular predictions—that we would not normally call laws.

For convenience, I shall use the term "lawlike" for sentences that, whether they are true or not, satisfy the other requirements in the definition of law. A law is thus a sentence that is both lawlike and true, but a sentence may be true without being lawlike, as I have illustrated, or lawlike without being true, as we are always learning to our dismay.

Now if we were to leave our definition as it stands, lawlikeness would be a rather accidental and ephemeral property. Only statements that happen actually to have been used for prediction would be lawlike. And a true sentence that had been used predictively would cease to be a law when it became fully tested—i.e., when none of its instances remained undetermined. The definition, then, must be restated in some such way as this: A general statement is lawlike if and only if it is acceptable prior to the determination of all its instances. This is immediately objectionable because "acceptable" itself is plainly a dispositional term; but I propose to use it only tentatively, with the idea of eliminating it eventually by means of a non-dispositional definition. Before trying to accomplish that, however, we must face another difficulty in our tentative criterion of lawlikeness.

Suppose that the appropriate generalization fails to sustain a given counterfactual because that generalization, while true, is unlawlike, as is

Everything in my pocket is silver.

All we would need to do to get a law would be to broaden the antecedent strategically. Consider, for example, the sentence

Everything that is in my pocket or is a dime is silver.

Since we have not examined all dimes, this is a predictive statement and—since presumably true—would be a law. Now if we consider our original counterfactual and choose our S so that $A \cdot S$ is

P is in my pocket. P is in my pocket or is a dime,

then the pseudo-law just constructed can be used to infer from this the sentence "P is silver". Thus the untrue counterfactual is established. If one prefers to avoid an alternation as a condition-statement, the same result can be obtained by using a new predicate such as "dimo" to mean "is in my pocket or is a dime".[13]

The change called for, I think, will make the definition of lawlikeness read as follows: A sentence is lawlike if its acceptance does not depend upon the determination of any given instance.[14] Naturally this does not mean that acceptance is to be independent of all determination of instances, but only that there is no particular instance on the determination of which acceptance depends. This criterion excludes from the class of laws a statement like

That book is black and oranges are spherical

on the ground that acceptance requires knowing whether the book is black; it excludes

Everything that is in my pocket or is a dime is silver

on the ground that acceptance demands examination of all things in my pocket. Moreover, it excludes a statement like

All the marbles in this bag except Number 19 are red, and Number 19 is black

on the ground that acceptance would depend on examination of or knowledge gained otherwise concerning marble Number 19. In fact the principle involved in the proposed criterion is a rather powerful one and seems to exclude most of the troublesome cases.

We must still, however, replace the notion of the acceptability of a sentence, or of its acceptance *depending* or *not depending* on some given knowledge, by a positive definition of such dependence. It is clear that to say that the acceptance of a given statement depends upon a certain kind and amount of evidence is to say that given such evidence, acceptance of the statement is in accord with certain general standards for the acceptance of statements that are not fully tested. So one turns naturally to theories of induction and confirmation to learn the distinguishing factors or circumstances that determine whether or not a sentence is acceptable without complete evidence. But publications on confirmation not only have failed to make clear the distinction between confirmable and non-confirmable statements, but show little recognition that such a problem exists.[15] Yet obviously in the case of some sentences like

Everything in my pocket is silver

or

No twentieth-century president of the United States will be between 6 feet 1 inch and 6 feet 1½ inches tall,

not even the testing with positive results of all but a single instance is likely to lead us to accept the sentence and predict that the one remaining instance will conform to it; while for other sentences such as

All dimes are silver

or

All butter melts at 150° F.

or

All flowers of plants descended from this seed will be yellow

positive determination of even a few instances may lead us to accept the sentence with confidence and make predictions in accordance with it.

There is some hope that cases like these can be dealt with by a sufficiently careful and intricate elaboration of current confirmation theories; but inattention to the problem of distinguishing between confirmable and non-confirmable sentences has left most confirmation theories open to more damaging counterexamples of an elementary kind.

Suppose we designate the 26 marbles in a bag by the letters of the alphabet, using these merely as proper names having no ordinal significance. Suppose further that we are told that all the marbles except d are red, but we are not told what color d is. By the usual kind of confirmation theory this gives strong confirmation for the statement

$Ra. Rb. Rc. Rd. \ldots Rz$

because 25 of the 26 cases are known to be favorable while none is known to be unfavorable. But unfortunately the same argument would show that the very same evidence would equally confirm

Ra. Rb. Rc. Re. . . . Rz. $-$Rd,

for again we have 25 favorable and no unfavorable cases. Thus "Rd" and "$-$Rd" are equally and strongly confirmed by the same evidence. If I am required to use a single predicate instead of both "R" and "$-$R" in the second case, I will use "P" to mean:

is in the bag and either is not d and is red, or is d and is not red.

Then the evidence will be 25 positive cases for

All the marbles are P

from which it follows that d is P, and thus that d is not red. The problem of what statements are confirmable merely becomes the equivalent problem of what predicates are projectible from known to unknown cases.

So far, I have discovered no way of meeting these difficulties. Yet as we have seen, some solution is urgently wanted for our present purpose; for only where willingness to accept a statement involves predictions of instances that may be tested does acceptance endow that statement with the authority to govern counterfactual cases, which cannot be directly tested.

In conclusion, then, some problems about counterfactuals depend upon the definition of cotenability, which in turn seems to depend upon the prior solution of those problems. Other problems require an adequate definition of law. The tentative criterion of law here proposed is reasonably satisfactory in excluding unwanted kinds of statements, and in effect, reduces one aspect of our problem to the question how to define the circumstances under which a statement is acceptable independently of the determination of any given instance. But this question I do not know how to answer.

NOTES: I

[1] (page 13) My indebtedness in several matters to the work of C. I. Lewis has seemed too obvious to call for detailed mention.

[2] (page 15) The practical import of a semifactual is thus different from its literal import. Literally a semifactual and the corresponding counterfactual are not contradictories but contraries, and both may be false (cf. Note I.8 below). The presence of the auxiliary terms "even" and "still", or either of them, is perhaps the idiomatic indication that a not quite literal meaning is intended.

[3] (page 16) Of the special kinds of counterfactuals mentioned, I shall have something to say later about counter-identicals and counterlegals. As for countercomparatives, the following procedure is appropriate: — Given "If I had arrived one minute later, I would have missed the train", first expand this to "$(\exists t)$. t is a time. I arrived at t. If I had arrived one minute later than t, I would have missed the train". The counterfactual conditional constituting the final clause of this conjunction can then be treated, within the quantified whole, in the usual way. Translation into "If 'I arrive one minute later than t' were true, then 'I miss the train' would have been true" does not give us a self-contradictory component.

[4] (page 19) This of course raises very serious questions, which I shall come to presently, about the nature of non-logical law.

[5] (page 20) Note that the requirement that $A \cdot S$ be self-compatible can be fulfilled only if the antecedent is self-compatible; hence the conditionals I have called "counterlegals" will all be false. This is convenient for our present purpose of investigating counterfactuals that are not counterlegals. If it later appears desirable to regard all or some counterlegals as true, special provisions may be introduced.

⁶ (page 21) It is natural to inquire whether for similar reasons we should stipulate that S must be compatible with both A and $-A$, but this is unnecessary. For if S is incompatible with $-A$, then A follows from S; therefore if S is compatible with both C and $-C$, then $A \cdot S$ cannot lead by law to one but not the other. Hence no sentence incompatible with $-A$ can satisfy the other requirements for a suitable S.

⁷ (page 21) Of course, some sentences similar to (ii), referring to other matches under special conditions may be true; but the objection to the proposed criterion is that it would commit us to many such statements that are patently false. I am indebted to Morton G. White for a suggestion concerning the exposition of this point.

⁸ (page 22) The double negative cannot be eliminated here; for ". . . if S would be true if A were" actually constitutes a stronger requirement. As we noted earlier (Note I.2), if two conditionals having the same counterfactual antecedent are such that the consequent of one is the negate of the consequent of the other, the conditionals are contraries and both may be false. This will be the case, for example, if every otherwise suitable set of relevant conditions that in conjunction with the antecedent leads by law either to a given consequent or its negate leads also to the other.

⁹ (page 24) The sense of "generalization" intended here is that explained by C. G. Hempel in 'A Purely Syntactical Definition of Confirmation', *Journal of Symbolic Logic*, vol. 8 (1943), pp. 122–43. See also III, Sect. 3, below.

¹⁰ (page 24) The antecedent in this example is intended to mean "If P, while remaining distinct from the things that were in fact in my pocket on VE day, had also been in my pocket then", and *not* the quite different, counteridentical "If P had been identical with one of the things that were in my pocket on VE day". While the antecedents of most counterfactuals (as, again, our familiar one about the match) are — literally speaking — open to both sorts of interpretation, ordinary usage normally calls for some explicit indication when the counteridentical meaning is intended.

C

251

[11] (page 25) The importance of distinguishing laws from non-laws is too often overlooked. If a clear distinction can be defined, it may serve not only the purposes explained in the present paper but also many of those for which the increasingly dubious distinction between analytic and synthetic statements is ordinarily supposed to be needed.

[12] (page 27) Had it been sufficient in the preceding section to require only that $A \cdot S$ be self-*compatible*, this requirement might now be eliminated in favor of the stipulation that the generalization of the conditional having $A \cdot S$ as antecedent and C as consequent should be non-vacuous; but this stipulation would not guarantee the self-*cotenability* of $A \cdot S$.

[13] (page 28) Apart from the special class of connecting principles we are concerned with, note that under the stated criterion of lawlikeness, any statement could be expanded into a lawlike one; for example: given "This book is black" we could use the predictive sentence "This book is black and all oranges are spherical" to argue that the blackness of the book is the consequence of a law.

[14] (page 28) So stated, the definition counts vacuous principles as laws. If we read instead "given class of instances", vacuous principles will be non-laws since their acceptance depends upon examination of the null class of instances. For my present purposes the one formulation is as good as the other.

[15] (page 29) The points discussed in this and the following paragraph have been dealt with a little more fully in my 'A Query on Confirmation', *Journal of Philosophy*, vol. xliii (1946), pp. 383–5.

3.3 The Metalinguistic Theory: Laws of Nature

Metalinguistic theorists commonly give a special place among cotenable factual premises to laws of nature. A law is thought to be cotenable with any antecedent, except an antecedent that is logically inconsistent with that law, or perhaps with some other law. On this view, if the antecedent of a counterfactual, together with some laws, implies the consequent, and if the antecedent is logically consistent with all laws, then the counterfactual is true. (Or: if that is thought to be the case, then the counterfactual is assertable.) On this view also, there can be no true counterfactual saying that if so-and-so particular state of affairs were to hold, then such-and-such law would be violated.

I could, if I wished, incorporate this special status of laws into my theory by imposing the following constraint on systems of spheres: the set of all and only those worlds that do not violate the laws prevailing at a world i is one of the spheres around i. Equivalently, in terms of comparative similarity: whenever the laws prevailing at i are violated at a world k but not at a world j, j is closer than k to i. This would mean that any violation of the laws of i, however slight, would outweigh any amount of difference from i in respect of particular states of affairs.

I have not chosen to impose any such constraint. I doubt that laws of nature have as much of a special status as has been thought. Such special status as they do have, they need not have by fiat. I think I can explain, within the theory already given, why laws tend to be cotenable, unless inconsistent, with counterfactual suppositions.

I adopt as a working hypothesis a theory of lawhood held by F. P. Ramsey in 1928: that laws are 'consequences of those propositions which we should take as axioms if we knew everything and organized it as simply as possible in a deductive system'.* We need not state Ramsey's theory as a counterfactual about omniscience. Whatever we may or may not ever come to know, there exist (as abstract objects) innumerable true deductive systems: deductively closed, axiomatizable sets of true sentences. Of these true deductive systems, some can be axiomatized more *simply* than others. Also, some of them have more *strength*, or *information content*, than others. The virtues of simplicity and strength tend to conflict. Simplicity without strength can be had from pure logic, strength without simplicity from (the deductive closure of) an almanac. Some deductive systems, of course, are neither simple nor strong. What we value in a deductive system is a properly balanced combination of simplicity and strength—as much of both as truth and our way of balancing will permit. We can restate Ramsey's 1928 theory of lawhood as follows: a contingent generalization is a *law of nature* if and only if it appears as a theorem (or axiom) in each of the true deductive systems that achieves a best combination of simplicity and strength.‡ A generalization is a law at a world i, likewise, if and only if it appears as a theorem in each of the best deductive systems true at i.

In science we have standards—vague ones, to be sure—for assessing

---

* R. B. Braithwaite has kindly permitted me to read a short unpublished note, written by Ramsey in March, 1928, stating this theory of lawhood. Ramsey regarded this note as superseded by 'General Propositions and Causality', published in *The Foundations of Mathematics*: 237–255. He there alludes to his theory of 1928 in the words I have quoted (page 242); rejects it on the ground that we never will know everything; and goes on to develop a different theory. See also Braithwaite's mention of the 1928 note in his editorial introduction, *The Foundations of Mathematics*: xiii.

‡ I doubt that our standards of simplicity would permit an infinite ascent of better and better systems; but if they do, we should say that a law must appear as a theorem in all sufficiently good true systems.

the combinations of strength and simplicity offered by deductive systems. We trade off these virtues against each other and against probability of truth on the available evidence. If we knew everything, probability of truth would no longer be a consideration. The false systems would drop out, leaving the true ones to compete in simplicity-cum-strength. (Imagine that God has decided to provide mankind with a *Concise Encyclopedia of Unified Science*, chosen according to His standards of truthfulness and our standards of simplicity and strength.) Our standards of simplicity and strength, and of the proper balance between them, apply—though we who are not omniscient have no occasion so to apply them—to the set of all true deductive systems. Thus it makes sense to speak of the best true systems, and of the theorems common to all the best true systems.

I adopt Ramsey's 1928 theory of lawhood, glossed as above, because of its success in explaining some facts about laws of nature. (1) It explains why lawhood is not just a matter of the generality, syntactically or semantically defined, of a single sentence. It may happen that two true sentences are alike general, but one is a law of nature and the other is not. That can happen because the first does, and the second does not, fit together with other truths to make a best system. (2) It explains why lawhood is a contingent property. A generalization may be true as a law at one world, and true but not as a law at another, because the first world but not the second provides other truths with which it makes a best system. (3) It therefore explains how we can know by exhausting the instances that a generalization—say, Bode's 'Law'—is true, but not yet know if it is a law. (4) It explains why *being* a law is not the same as being regarded as a law—being projected, and so forth—and not the same as being regarded as a law and also being true. It allows there to be laws of which we have no inkling. (5) It explains why we have reason to take the theorems of well-established scientific theories provisionally as laws. Our scientific theorizing is an attempt to approximate, as best we can, the true deductive systems with the best combination of simplicity and strength. (6) It explains why lawhood has seemed a rather vague and difficult concept: our standards of simplicity and strength, and of the proper balance between them, are only roughly fixed. That may or may not matter. We may hope, or take as an item of faith, that our world is one where certain true deductive systems come out as best, and certain generalizations come out as laws, by *any* remotely reasonable standards—but we might be unlucky.

On the working hypothesis that the laws of a world are the generalizations that fit into the best deductive systems true there, we can also say that the laws are generalizations which (given suitable companions) are highly informative about that world in a simple way. Such generalizations are important to us. It makes a big difference to the character of a world which generalizations enjoy the status of lawhood there. Therefore similarity and difference of worlds in respect of their laws is an important respect of similarity and difference, contributing weightily to overall similarity and difference. Since a difference in laws would be a big difference between worlds, we can expect that worlds with the same laws as a world *i* will tend to be closer to *i* than worlds at which the laws of *i* hold only as accidental generalizations, or are violated, or—worse still—are replaced by contrary laws. In other words, the laws of *i* will hold throughout many of the spheres around *i*, and thus will tend to be cotenable with counterfactual suppositions. That is so simply because laws are especially important to us, compared with particular facts or true generalizations that are not laws.

Though similarities or differences in laws have some tendency to outweigh differences or similarities in particular facts, I do not think they invariably do so. Suppose that the laws prevailing at a world *i* are deterministic, as we used to think the laws of our own world were. Suppose a certain roulette wheel in this deterministic world *i* stops on black at a time *t*, and consider the counterfactual antecedent that it stopped on red. What sort of antecedent-worlds are closest to *i*? On the one hand, we have antecedent-worlds where the deterministic laws of *i* hold without exception, but where the wheel is determined to stop on red by particular facts different from those of *i*. Since the laws are deterministic, the particular facts must be different at all times before *t*, no matter how far back. (Nor can we assume that the differences of particular fact diminish as we go back in time. Assume for the sake of argument that *i* and its laws are such that any antecedent-world where the laws hold without exception differs more and more from *i* as we go back.) On the other hand, we have antecedent-worlds that are exactly like *i* until *t* or shortly before; where the laws of *i* hold *almost* without exception; but where a small, localized, inconspicuous miracle at *t* or just before permits the wheel to stop on red in violation of the laws. Laws are very important, but great masses of particular fact count for something too; and a localized violation is not the most serious sort of difference of law. The violated deterministic law has presumably not been replaced by a contrary law. Indeed, a version of the violated law, complicated and weakened by a clause to permit the one exception, may still be simple and strong enough to survive as a law. Therefore some of the antecedent-worlds where the law is violated may be closer to *i* than any of the ones where the particular facts are different at all times before *t*. At least, this seems plausible enough to deter me from decreeing the opposite. I therefore proceed on the assumption that the preeminence of laws of nature among cotenable factual premises is a matter only of degree.

My example of the deterministic roulette wheel raises a problem for me: what about differences of particular fact at times *after* *t*? Among the antecedent worlds I prefer—those where the wheel stops on red by a minor miracle and the particular facts are just as they are at *i* until *t* or shortly before—there are two sorts. There are some where the determin- istic laws of *i* are unviolated after *t* and the particular facts after *t* diverge more and more from those of *i*. (I now assume that the deter- ministic laws are deterministic both forward and backward, so that they do not permit a reconvergence.) There are others where a second minor miracle occurs just after *t*, erasing all traces of the first miracle, so that we have two violations of law instead of one but the particular facts from that time on are just as they are at *i*. If I have decided that a small miracle *before* *t* makes less of a difference from *i* than a big difference of particular fact at all times *before* *t*, then why do I not also think that a small miracle *after* *t* makes less of a difference from *i* than a big difference of particular fact at all times *after* *t*? That is not what I do think: the worlds with no second miracle and divergence must be regarded as closer, since I certainly think it true (at *i*) that if the wheel had stopped on red at *t*, all sorts of particular facts afterward would have been otherwise than they are at *i*. The stopping on red would have plenty of traces and consequences from that time on.

Perhaps it is just brute fact that we put more weight on earlier similarities of particular fact than on later ones. Divergence of par- ticular fact throughout the past might make more of a difference than a small violation of law, but a small violation of law might make more of a difference than divergence of particular fact throughout the future. Then the closest antecedent-worlds to *i* would be those with a miracle and with no difference of particular fact before *t*, but with no miracle and with divergence of particular fact after *t*. Such discrimination be- tween the two directions of time seems anthropocentric; but we are understandably given to just such anthropocentric discrimination, and it would be no surprise if it turns out to infect our standards of com- parative similarity and our truth conditions for counterfactuals.

But perhaps my standards are less discriminatory than they seem. For some reason—something to do with the *de facto* or nomological asymmetries of time that prevail at *i* if *i* is a world something like ours— it seems to take less of a miracle to give us an antecedent-world exactly like *i* in the past than it does to give us one exactly like *i* in the future. For the first, all we need is one little miraculous shove, applied to the wheel at the right moment. For the second, we need much more. All kinds of traces of the wheel's having stopped on red must be falsified. The rest position of the wheel; the distribution of light, heat, and sound in the vicinity; the memories of the spectators—all must be changed to bring about a reconvergence of particular fact between the antecedent-world and *i*. One shove will not do it; many of the laws of *i* must be violated in many ways at many places. Small wonder if the closest antecedent-worlds to *i* are worlds where the particular facts before *t* are preserved at the cost of a small miracle, but the particular facts after *t* are not preserved at the cost of a bigger, more complicated miracle.

CANADIAN JOURNAL OF PHILOSOPHY
Volume VII, Number 4, December 1977

The Nature of Laws

MICHAEL TOOLEY, Australian National University

This paper is concerned with the question of the truth conditions of nomological statements. My fundamental thesis is that it is possible to set out an acceptable, noncircular account of the truth conditions of laws and nomological statements if and only if relations among universals — that is, among properties and relations, construed realistically — are taken as the truth-makers for such statements.

My discussion will be restricted to strictly universal, nonstatistical laws. The reason for this limitation is not that I feel there is anything dubious about the concept of a statistical law, nor that I feel that basic laws cannot be statistical. The reason is methodological. The case of strictly universal, nonstatistical laws would seem to be the simplest case. If the problem of the truth conditions of laws can be solved for this simple subcase, one can then investigate whether the solution can be extended to the more complex cases. I believe that the solution I propose here does have that property, though I shall not pursue that question here.[1]

---

1   I am indebted to a number of people, especially David Armstrong, David Bennett, Mendel Cohen, Michael Dunn, Richard Routley, and the editors of this *Journal*, for helpful comments on earlier versions of this paper.

Michael Tooley

1. Some Unsatisfactory Accounts

The thesis that relations among universals are the truth-makers for laws may strike some philosophers as rather unappealing, for a variety of reasons. Perhaps the two most important are these. First, it entails a strong version of realism with regard to universals. Secondly, traditional semantical accounts of the concept of truth have generally been nominalistic in flavor. Not in the sense that acceptance of them involves commitment to nominalism, but in the sense that they involve no reference to universals. This seems, in part, an historical accident. Semantical accounts of truth in which a concept such as an object's exemplifying a property plays a central role can certainly be set out. For most types of sentences, accounts involving such explicit reference to universals may well introduce additional conceptual apparatus without any gain in philosophical illumination. However I will attempt to show that there is at least one class of statements for which this is not the case, namely, nomological statements.

I shall begin by considering some important alternative accounts of the nature of laws. I think that getting clear about how these accounts are defective will both point to certain conditions that any adequate account must satisfy, and provide strong support for the thesis that the truth-makers for laws must be relations among universals.

Perhaps the most popular account of the nature of laws is that a generalization expresses a law if and only if it is both lawlike and true, where lawlikeness is a property that a statement has, or lacks, simply in virtue of its meaning. Different accounts of lawlikeness have been advanced, but one requirement is invariably taken to be essential: a lawlike statement cannot contain any essential reference to specific individuals. Consider, for example, the generalization: "All the fruit in Smith's garden are apples." Since this statement entails the existence of a particular object — Smith's garden — it lacks the property of lawlikeness. So unless it is entailed by other true statements which are lawlike, it will be at best an accidentally true generalization.

There are at least three serious objections to this approach. First, consider the statement that all the fruit in any garden with property *P* are apples. This generalization is free of all essential reference to specific individuals. Thus, unless it is unsatisfactory in some other way, it is lawlike. But *P* may be quite a complex property, so chosen that there is, as a matter of fact, only one garden possessing that property, namely Smith's. If that were so, one might well want to question whether the generalization that all fruit in any garden with property *P* are apples was a law. It would seem that statements can be both lawlike and true, yet fail to be laws.

A second objection to this approach is that it cannot deal in a satisfactory manner with generalizations that are vacuously true, that is, which lack "positive" confirming instances.[2] Consider the statement: "Whenever two spheres of gold more than eight miles in diameter come into contact, they turn red." The statement is presumably lawlike, and is true under the standard interpretation. Is it then a law? The usual response is that a vacuously true generalization is a law only if it is derivable from generalizations that are not vacuously true. But this seems wrong. Imagine a world containing ten different types of fundamental particles. Suppose further that the behavior of particles in interactions depends upon the types of the interacting particles. Considering only interactions involving two particles, there are 55 possibilities with respect to the types of the two particles. Suppose that 54 of these possible interactions have been carefully studied, with the result that 54 laws have been discovered, one for each case, which are not interrelated in any way. Suppose finally that the world is sufficiently deterministic that, given the way particles of types X and Y are currently distributed, it is impossible for them ever to interact at any time, past, present, or future. In such a situation it would seem very reasonable to believe that there is some *underived* law dealing with the interaction of particles of types X and Y. Yet precisely this view would have to be rejected if one were to accept the claim that a vacuously true generalization can be a law only if derivable from laws that are not vacuously true.

A third objection is this. Assuming that there can be statistical laws, let us suppose that it is a law that the probability that something with property P has property Q is 0.999999999. Suppose further that there are, as a matter of fact, very few things in the world with property P, and, as would then be expected, it happens that all of these things have property Q. Then the statement that everything with property P has property Q would be both lawlike and true, yet it would not be a law.

One might even have excellent grounds for holding that it was not a law. There might be some powerful and very well established theory which entailed that the probability that something with property P would have property Q was not 1.0, but 0.999999999, thus implying that it was not a law that everything with property P would have property Q.

If this argument is correct, it shows something quite important. Namely, that there are statements that would be laws in some worlds,

---

2 A vacuously true generalization is often characterized as a conditional statement whose antecedent is not satisfied by anything. This formulation is not entirely satisfactory, since it follows that there can be two logically equivalent generalizations, only one of which is vacuously true. A sound account would construe being vacuously true as a property of the content of a generalization, rather than as a property of the form of the sentence expressing the generalization.

but only accidentally true generalizations in others. So there cannot be any property of lawlikeness which a statement has simply in virtue of its meaning, and which together with truth is sufficient to make a statement a law.

A second attempt to explain what it is for a statement to express a law appeals to the fact tha laws entail some counterfactuals, and support others, while accidentally true generalizations do neither. If this approach is to provide a noncircular analysis, it must be possible to give a satisfactory account of the truth conditions of subjunctive conditional statements which does not involve the concept of a law. This does not seem possible. The traditional, consequence analysis of subjunctive conditionals explicitly employs the concept of a law. And the principal alternative, according to which truth conditions for subjunctive conditionals are formulated in terms of comparative similarity relations among possible worlds, involves implicit reference to laws, since possession of the same laws is one of the factors that weighs most heavily in judgments concerning the similarity of different possible worlds. The latter theory is also exposed to very serious objections.[3] As a result, it appears unlikely that any noncircular analysis of the concept of a law in terms of subjunctive conditional statements is possible.

A third approach to the problem of analyzing the concept of a law is the view, advanced by Ramsey, that laws are "consequences of those propositions which we should take as axioms if we knew everything and organized it as simply as possible in a deductive system."[4] My earlier example of the universe in which there are ten different types of fundamental particles, two of which never interact, shows that this account does not provide an adequate description of the truth conditions of laws. In the world where particles of types X and Y never meet, there will be many true generalizations about their behavior when they interact. Unfortunately, none of these generalizations will have any positive instances; they will all be only vacuously true. So knowledge of everything that happens in such a universe will not enable one to furmulate a *unique* axiomatic system containing

---

3 See, for example, the incisive discussions by Jonathan Bennett in his article, "Counterfactuals and Possible Worlds," *Canadian Journal of Philosophy*, **4** (1974), pages 381-402, and, more recently, by Frank Jackson in his article, "A Causal Theory of Counterfactuals", *Australasian Journal of Philosophy*, **55** (1977), pages 3-21.

4 F. P. Ramsey, "General Propositions and Causality," in *The Foundations of Mathematics*, edited by R. B. Braithwaite, Paterson, New Jersey, 1960, page 242. The view described in the passage is one which Ramsey had previously held, rather than the view he was setting out in the paper itself. For a sympathetic discussion of Ramsey's earlier position, see pages 72-77 of David Lewis's *Counterfactuals*, Cambridge, Massachusetts, 1973.

theorems about the manner of interaction of particles of types X and Y. Adopting Ramsey's approach would force one to say that in such a universe there could not be any law describing how particles of types X and Y would behave if they were to interact. I have argued that this is unacceptabl~

2. Universals as the Truth-Makers for Nomological Statements

What, then, is it that makes a generalization a law? I want to suggest that a fruitful place to begin is with the possibility of underived laws having no positive instances. This possibility brings the question of what makes a generalization a law into very sharp focus, and it shows that an answer that might initially seem somewhat metaphysical is not only plausible, but unavoidable.

Consider, then, the universe containing two types of particles that never meet. What in that world could possibly make true some specific law concerning the interaction of particles of types X and Y? All the events that constitute the universe throughout all time are perfectly compatible with different, and conflicting laws concerning the interaction of these two types of particles. At this point one may begin to feel the pull of the view that laws are not statements, but inference tickets. For in the universe envisaged, there is nothing informative that one would be justified in inferring from the supposition that an X type particle has interacted with a Y type particle. So if laws are inference tickets, there are, in our imaginary universe, no laws dealing with the interaction of particles of types X and Y.

But what if, in the universe envisaged, there could be underived laws dealing with the interaction of particles of types X and Y? Can one draw any conclusions from the assumption that such basic laws without positive instances are possible — specifically, conclusions about the truth-makers for laws? I would suggest that there are two very plausible conclusions. First, *nonnomological* facts about particulars cannot serve as the truth-makers for *all* laws. In the universe in which particles of types X and Y never interact, it might be a law that when they do, an event of type P occurs. But equally, it might be a law that an event of type Q occurs. These two generalizations will not be without instances, but none of them will be of the positive variety. And in the absence of positive instances, there is no basis for holding that one generalization is a law, and the other not. So at least in the case of underived laws without positive instances, nonnomological facts about particulars cannot serve as the truth-makers.

What, then, are the facts about the world that make such statements laws? A possible answer is that the truth-makers are facts about particulars that can only be expressed in *nomological* language. Thus, in the case we are considering, one might try saying that what makes it a law that particles of types *X* and *Y* interact in a specific way are the nomological facts that particles of types *X* and *Y* have certain dispositional properties. But this is not to make any progress. The question of the truth-makers of underived laws without positive instances has merely been replaced by that of the truth-makers of statements attributing unactualized dispositional properties to objects, and if one is willing in the latter case to say that such statements are semantically basic, and that no further account can be given of the fact that an object has a dispositional property, one might equally well say the same thing of laws, that is, that there just are basic facts to the effect that there are specific laws applying to certain types of objects, and no further account of this can be given. In either case one is abandoning the project of providing an account of the truth conditions of nomological statements in nonnomological terms, and thus also the more general program of providing truth conditions for intensional statements in purely extensional terms.

The upshot, then, is that an account of the truth conditions of underived laws without positive instances in terms of nomological facts about particulars is unilluminating, while an account in terms of nonnomological facts about particulars seems impossible. This, then, is the second conclusion: no facts about particulars can provide a satisfactory account of the truth conditions of such laws.

But how then can there be such laws? The only possible answer would seem to be that it must be facts about *universals* that serve as the truth-makers for basic laws without positive instances. But if facts about universals constitute the truth-makers for some laws, why shouldn't they constitute the truth-makers for all laws? This would provide a uniform account of the truth conditions of laws, and one, moreover, that explains in a straightforward fashion the difference between laws and accidentally true generalizations.

Let us now consider how this idea that facts about universals can be the truth-makers for laws is to be developed. Facts about universals will consist of universals' having properties and standing in relations to other universals. How can such facts serve as truth-makers for laws? My basic suggestion here is that the fact that universals stand in certain relationships may *logically necessitate* some corresponding generalization about particulars, and that when this is the case, the generalization in question expresses a law.

This idea of a statement about particulars being entailed by a statement about a relation among universals is familiar enough in another context, since some philosophers have maintained that analytical statements are true in virtue of relations among universals. In this latter case, the relations must be necessary ones, in order for the statement about particulars which is entailed to be itself logically necessary. Nomological statements, on the other hand, are not logically necessary, and because of this the relations among universals involved here must be *contingent* ones.

The idea of contingent relations among universals logically necessitating corresponding statements about particulars is admittedly less familiar. But why should it be more problematic? Given the relationship that exists between universals and particulars exemplifying them, any property of a universal, or relation among universals, regardless of whether it is necessary or merely contingent, will be reflected in corresponding facts involving the particulars exemplifying the universal.

It might be suggested, though, that what is problematic is rather the idea of a contingent relation among universals. Perhaps this idea is, like the notion of a necessarily existent being, ultimately incoherent? This possibility certainly deserves to be examined. Ideally, one would like to be able to prove that the concept of contingent relations among universals is coherent. Nevertheless, one generally assumes that a concept is coherent unless there are definite grounds for thinking otherwise. So unless some reason can be offered for supposing that the concept of contingent relations among universals is incoherent, one would seem to be justified in assuming that this is not the case.

Let us refer to properties of universals, and relations among universals, as *nomological* if they are contingent properties or relations which logically necessitate corresponding facts about particulars. How can one *specify* such nomological properties and relations? If the properties or relations were observable ones, there would be no problem. But in our world, at least, the facts about universals which are the truth-makers for laws appear to be unobservable. One is dealing, then, with theoretical relations among universals, and the problem of specifying nomological relations and properties is just a special case of the problem of specifying the meaning of statements involving theoretical terms.

Theoretical statements cannot be analyzed in purely observational terms. From this, many have concluded that theoretical statements cannot, in the strict sense, be analyzed at all in terms of statements free of theoretical vocabulary. But it is clear that this does not follow, since the class of statements that are free of theoretical vocabulary does not coincide with the class of observation statements. Thus the statement, "This table has parts too small to be observed," although it contains no theoretical vocabulary, is not a pure observation statement, since it refers to something beyond what is observable. This situation can arise because, in addition to observational vocabulary and theoretical vocabulary, one also has logical and quasi-logical vocabulary — including expressions such as "part", "property", "event", "state", "particular", and so on — and statements containing such vocabulary together with observational vocabulary can refer to unobservable states of affairs.

This suggests the possibility, which I believe to be correct, that theoretical statements, though not analyzable in terms of observation statements, are analyzable in terms of statements that contain nothing beyond observational, logical, and quasi-logical vocabulary. The natural and straightforward way of doing this is suggested by the method of Ramsey sentences, and has been carefully worked out and defended by David Lewis in his article, "How to Define Theoretical Terms."[5] Essentially, the idea is this. Let T be any theory. If T contains any singular theoretical terms, eliminate them by paraphrase. Then replace all theoretical predicates and functors by names of corresponding entities, so that, for example, an expression such as "...is a neutrino" is replaced by an expression such as "...has the property of neutrino-hood." The result can be represented by $T(P_1, P_2, ..., P_n)$, where each P_i is the theoretical name of some property or relation. All such theoretical names are then replaced by distinct variables, and the corresponding existential quantifiers prefixed to the formula. The resulting sentence — $\exists x_1 \exists x_2 ... \exists x_n \ T(x_1, x_2, ..., x_n)$ — is a Ramsey sentence for the theory T. Suppose now that there is only one ordered n-tuple that satisfies $T(x_1, x_2, ..., x_n)$. It will then be possible to define the theoretical name P_i by identifying the property or relation in question with the i_{th} member of the unique n-tuple which satisfies $T(x_1, x_2, ..., x_n)$.

Expressed in intuitive terms, the underlying idea is this. The meaning of theoretical terms is to be analyzed by viewing them as referring to properties (or relations) by characterizing them as properties (or relations) that stand in certain logical or quasi-logical relations to other properties and relations, both theoretical and observable, the logical and quasi-logical relations being specified by the relevant theory. One might compare here the way in which the mind is characterized in central state materialism: the mind is that entity, or collection of states and processes, that stands in certain specified relations to behavior. The above approach to the meaning of theoretical statements involves, in effect, a similar relational analysis of theoretical terms.

Some possible objections to this approach deserve to be at least briefly noted. One is that the procedure presupposes that the theoretical names will not name anything unless there is a unique n-tuple that satisfies the appropriate formula. I think that Lewis makes

---

5 *Journal of Philosophy*, **67** (1970), pages 427-446.

out a plausible case for this view. However the requirement can be weakened slightly. One might, for example, take the view that P_i is definable even where there is more than one n-tuple that satisfies $T(x_1, x_2, ..., x_n)$, provided that every n-tuple has the same ith element.

A second objection is that replacing predicates and functors by names is not automatic, as Lewis supposes. For unless there are disjunctive properties, negative properties, etc., there is no reason for thinking that there will be a one-to-one correspondence between predicates on the one hand, and properties and relations on the other. This point is surely correct. However it shows only that the initial paraphrase has to be carried out in a metaphysically more sophisticated way.

A slightly more serious difficulty becomes apparent if one considers some very attenuated theories. Suppose, for example, that T consists of a single statement $(x)(Mx \supset Px)$, where M is theoretical and P is observational. This theory will have the peculiarity that the corresponding Ramsey sentence is logically true,[6] and given Lewis's approach, this means that the property M-hood exists only if it is identical with P-hood.

One response is to refuse to count a set of sentences as a theory if, like T here, it has no observational consequences. However this requirement seems overly stringent. Even if a theory does not entail any observational statements, it may have probabilistic implications: the likelihood of R given Q together with T may differ from the likelihood of R given Q alone, where Q and R are observational statements, even though there are no observational statements entailed by T.

An alternative response is to adopt the view that the Ramsey sentence for a theory should be replaced by a slightly different sentence which asserts not merely that there is some ordered n-tuple $(x_1, x_2, ..., x_n)$ that satisfies the formula $T(x_1, x_2, ..., x_n)$, but that there is some n-tuple that satisfies the formula *and* whose existence is not entailed simply by the existence of the observable properties involved in the theory. Then, if there is a unique ordered n-tuple with those two properties, one can define the theoretical terms $P_1, P_2, ..., P_n$.

In any case, I believe that one is justified in thinking that difficulties such as the above can be dealt with, and I shall, in my attempt to state truth conditions for laws, assume that theoretical statements can be adequately analyzed along the Lewis-type lines outlined above.

There are three further ideas that are needed for my account of the truth conditions of nomological statements. But the account will be

---

6   This problem was pointed out by Israel Scheffler in his book, *The Anatomy of Inquiry*, New York, 1963. See section 21 of part II, pages 218ff.

more perspicuous if the motivation underlying the introduction of these ideas is first made clear. This can perhaps best be done by outlining a simpler version of the basic account, and then considering some possible problems that it encounters.[7]

The simpler version can be seen as attempting to specify explicitly a small number of relations among universals that will serve as the truth-makers for all possible laws. One relation which seems clearly essential is that of *nomic necessitation*. This relation can be characterized — though *not defined* — as that relation which holds between two properties P and Q if and only if it is a law that for all x, if x has property P then x has property Q. Is this one relation of nomic necessitation sufficient to handle all possible laws? The answer to this depends in part upon certain metaphysical matters. Consider a law expressed by a statement of the form $(x)(Px \supset -Qx)$. If this type of law is to be handled via the relation of nomic necessitation, one has to say that the property P stands in the relation of nomic necessitation to the property of not having property Q, and this commits one to the existence of negative properties. Since negative properties are widely thought suspect, and with good reason, another relation has to be introduced to handle laws of this form: the relation of *nomic exclusion*.

Are these two relations jointly sufficient? Consider some problematic cases. First, laws of the form $(x)Mx$. Is it possible to state truth conditions for such laws in terms of the relations of nomic necessitation and nomic exclusion? Perhaps. A first try would be to treat its being a law that everything has property M as equivalent to its being true, of every property P, that it is a law that anything that has property P also has property M. But whether this will do depends upon certain issues about the existence of properties. If different properties would have existed if the world of particulars had been different in certain ways, the suggested analysis will not be adequate. One will have to say instead that its being a law that everything has property M is equivalent to its being a law that, for every property P, anything with property P has property M — in order to exclude the possibility of there being some property Q, not possessed by any object in the world as it actually is, but which is such that if an object had property Q, it would lack property M. This revision, since it involves the occurrence of a universal quantifier ranging over properties within the scope of a nomological operator, means that laws apparently about particulars are being analyzed in terms of laws about universals. This, however,

---

7   This version of the general theory is essentially that set out by David Armstrong in his forthcoming book, *Universals and Scientific Realism*. In revising the present paper I have profited from discussions with Armstrong about the general theory, and the merits of our competing versions of it.

would not seem to be a decisive objection to this way of viewing laws expressed by statements of the form $(x)Mx$.

A much more serious objection concerns laws expressed by statements of the form $(x)[Px \supset (Qx \lor Rx)]$. If the world were partially indeterministic, there might well be laws that stated, for example, that if an object has property P, then it has either property Q or property R, and yet no laws that specified *which* of those properties an object will have on any given occasion. Can the truth conditions for laws of this form be expressed in terms of the relations of nomic necessitation and nomic exclusion? The answer depends on whether there are disjunctive properties, that is, on whether, if Q and R are properties, there is a property, Q or R, which is possessed by objects that have property Q and by objects that have property R. If, as many philosophers have maintained, there are no disjunctive properties, then the relations of nomic necessitation and nomic exclusion will not suffice to provide truth conditions for laws of the form $(x)[Px \supset (Qx \lor Rx)]$.

A third case that poses difficulties concerns laws expressed by statements of the form $(x)(-Px \supset Qx)$. If negative properties are rejected, such laws cannot be handled in any immediate fashion by the relations of nomic necessitation and nomic exclusion. Nevertheless, this third case does not appear to raise any new issues. For if one can handle laws expressed by sentences of the form $(x)Mx$ in the way suggested above, one can rewrite laws of the form $(x)(-Px \supset Qx)$ in the form $(x)(Px \lor Qx)$, and then apply the method of analysis suggested for laws of the form $(x)Mx$. The result will be a law that is conditional in form, with a positive antecedent and a disjunctive consequent, which is the case just considered.

The conclusion seems to be this. The relation of nomic necessitation by itself does not provide a satisfactory account unless there are both negative and disjunctive properties. Supplementing it with the relation of nomic exclusion may allow one to dispense with negative properties, but not with disjunctive ones. It would seem best to try to set out a more general account that will allow one to avoid all dubious metaphysical assumptions. Let us now turn to such an account.

The first concept required is that of the *universals involved in a proposition*. This notion is a very intuitive one, though how best to explicate it is far from clear. One approach would be to attempt to show that propositions can be identified with set theoretical constructs out of universals. This treatment of propositions is not without its difficulties, but it is a reasonably natural one if propositions are viewed as nonlinguistic entities.

The second, and related idea, is that of the logical form or structure of a proposition. One can view this form as specified by a *construction function* which maps ordered n-tuples of universals into propositions.

Michael Tooley

Thus one could have, for example, a construction function K such that K(redness, roundness) = the proposition that all red things are round. Conceived in the most general way, some construction functions will map ordered n-tuples of universals into propositions that involve as constituents universals not contained in the original n-tuple. Thus G could be a function so defined that G(property P) = the proposition that everything with property P is green. Other construction functions will map ordered n-tuples of universals into propositions which do not contain, as constituents, all the universals belonging to the n-tuple. H could be a function so defined that H(property P, property Q) = the proposition that everything has property Q. In order to capture the notion of logical form, one needs a narrower notion of construction function, namely, one in which something is a construction function if and only if it is a mapping from ordered n-tuples of universals into propositions that contain, as constituents, all and only those universals belonging to the ordered n-tuple. In this narrower sense, K is a construction function, but G and H are not.

The final idea required is that of a *universal being irreducibly of order k*. Properties of, and relations among particulars, are universals of order one. If nominalism is false, they are irreducibly so. A universal is of order two if it is a property of universals of order one, or a relation among things, some of which are universals of order one, and all of which are either universals of order one or particulars. It is irreducibly so if it cannot be analyzed in terms of universals of order one. And in general, a universal is of order $(k + 1)$ if it is a property of universals of order k, or a relation among things, some of which are universals of order k, and all of which are either particulars or universals of order k or less. It is irreducibly of order $(k + 1)$ if it cannot be analyzed in terms of particulars and universals of order k or less.

Given these notions, it is possible to explain the general concept of a *nomological relation* — which will include, but not be restricted to, the relations of nomic necessitation and nomic exclusion. Thus, as a first approximation:

R is a *nomological relation* if and only if

(1) R is an n-ary relation among universals;

(2) R is irreducibly of order $(k + 1)$, where k is the order of the highest order type of element that can enter into relation R;

(3) R is a contingent relation among universals, in the sense that there are universals U_1, U_2, ..., U_n such that it is neither necessary that $R(U_1, U_2, ..., U_n)$ nor necessary that not $R(U_1, U_2, ..., U_n)$;

272

(4) there is a construction function K such that
   (i) if $P_1, P_2, ..., P_n$ are either properties or relations, and of the appropriate types, then $K(P_1, P_2, ..., P_n)$ is a proposition about particulars, and
   (ii) the proposition that $R(P_1, P_2, ..., P_n)$ logically entails the proposition which is the value of $K(P_1, P_2, ..., P_n)$.

This characterization of the theoretical concept of a nomological relation in logical and quasi-logical terms can in turn be used to state truth conditions for nomological statements:

S is a true nomological statement if and only if there exists a proposition p which is expressed by S, and there exists a nomological relation R and an associated construction function K, and universals $P_1, P_2, ..., P_n$ such that

(1) it is not logically necessary that p;

(2) the proposition that p is identical with the value of $K(P_1, P_2, ..., P_n)$;

(3) it is true that $R(P_1, P_2, ..., P_n)$;

(4) it is not logically necessary that $R(P_1, P_2, ..., P_n)$;

(5) the proposition that $R(P_1, P_2, ..., P_n)$ logically entails the proposition that p.

The basic idea, then, is that a statement expresses a nomological state of affairs if is true in virtue of a contingent, nomological relation holding among universals. Different types of nomological relations are specified by different construction functions. A relation is a relation of nomic necessitation if it is of the type specified by the construction function which maps ordered couples (P, Q) of universals into propositions of the form $(x)(Px \supset Qx)$. It is a relation of nomic exclusion if it is of the type determined by the construction function mapping ordered couples (P, Q) of universals into propositions of the form $(x)(Px \supset -Qx)$.

It is critical to this account that a nomological relation be genuinely a relation among universals and nothing else, as contrasted, for example, with a relation that is apparently among universals, but which can be analyzed in terms of properties of, and relations among, particulars. Hence the requirement that a relation, to be nomological, always be *irreducibly* of an order greater than the order of the universals that enter into it. If this requirement were not imposed, every true generalization would get classified as nomological. For suppose that everything with property *P* just happens to have property *Q*, and consider the relation *R* which holds between properties *A* and *B* if and only if everything with property *A* has property *B*. Properties *P* and *Q* stand in relation *R*. The relation is a contingent one. Its holding between properties *P* and *Q* entails the proposition that everything with property *P* has property *Q*. So if condition (2) were dropped from the definition of the concept of a nomological relation, *R* would qualify as a nomological relation, and it would be a nomological truth that everything with property *P* has property *Q*.

But while condition (2) is essential, it is not quite adequate. For suppose that it is a law that everything with property *S* has property *T*, and that the truth-maker for this law is the fact that *S* and *T* stand in a certain relation *W*, where *W* is irreducibly of order two. Then one can define a relation *R* as follows: Properties *P* and *Q* stand in relation *R* if and only if everything with property *P* has property *Q*, and properties *S* and *T* stand in relation *W*. So defined, relation *R* will not be analyzable in terms of universals of order one, so condition (2) will not be violated. But if relations such as *R* were admitted as nomological, then, provided that there was at least one true nomological statement, all generalizations about particulars would get classified as nomological statements.

There are alternative ways of coping with this difficulty. One is to replace condition (2) by:

(2*) If $R(U_1, U_2, ..., U_n)$ is analytically equivalent to $C_1 \wedge C_2 \wedge ... \wedge C_m$, then every nonredundant C_i — that is, every C_i not entailed by the remainder of the conjunctive formula — is irreducibly of order $(k + 1)$.

This condition blocks the above counterexample. But given the somewhat ad hoc way in which it does so, one might wonder whether there may not be related counterexamples which it fails to exclude. What is one to say, for example, about a relation *R* defined as follows: Properties *P* and *Q* stand in relation *R* if and only if either everything with property *P* has property *Q* or properties *P* and *Q* stand in relation *W*? I would hold that this is not a counterexample, on the ground that there cannot be disjunctive relations. But this is to appeal to a view that some philosophers would reject.

A second, and more radical approach, involves replacing condition (2) by:

(2**) Relation *R* is not analyzable in terms of other universals of *any* order.

This more radical approach, which I believe is preferable, does necessitate another change in the account. For suppose that there is a nomological relation R_1 holding between properties P and Q, in virtue of which it is a law that everything with property P has property Q, and a nomological relation R_2 holding between properties Q and S, in virtue of which it is a law that everything with property Q has property S. It will then be a law that everything with property P has property S, and this may be so simply in virtue of the relations R_1 and R_2 which hold between properties P and Q, and Q and S, respectively, and not because of any additional relation holding between properties P and S. Given the revised account of nomological relations, and hence of nomological statements, the generalization that everything with property P has property S could not be classified as nomological. But this consequence can be avoided by viewing the revised account as concerned only with *basic* or *underived* nomological statements, and then defining nomological statements as those entailed by the class of basic nomological statements.

3. Laws and Nomological Statements

The class of nomological statements characterized in the preceding section does not seem to coincide with the class of laws. Suppose it is a nomological truth that $(x)(Px \supset Qx)$. Any statement entailed by this must also be nomological, so it will be a nomological truth that $(x)[(Px \wedge Rx) \supset Qx]$, regardless of what property R is. Now it is certainly true that the latter statement will be nomologically necessary, and thus, in a broad sense of "law", it will express a law. Nevertheless it is important for some purposes — such as the analysis of causal statements and subjunctive conditionals — to define a subclass of nomological statements to which such statements will not belong. Consider, for example, the nomological statement that all salt, when in water, dissolves. If this is true, it will also be a nomological truth that all salt, when both in water and in the vicinity of a piece of gold, dissolves. But one does not want to say that the cause of a piece of salt's dissolving was that it was in water and in the vicinity of some gold. In the description of causes one wants to exclude irrelevant facts. Or consider the counterfactual: "If this piece of salt were in water and were not dissolving, it would not be in the vicinity of a piece of gold." If one says that all nomological statements support counterfactuals, and that it is a nomological truth that all salt when both in water and near gold dissolves, one will be forced to accept the preceding counterfactual, whereas it is clear that there is good reason not to accept it.

Intuitively, what one wants to do is to define a subclass of nomological statements in the broad sense, containing only those involving no irrelevant conditions. Nomological statements belonging to this subclass will be laws.[8] But how is this class to be defined? It cannot be identified with the class of underived nomological statements, since it is certainly possible for laws in the strict sense to be entailed by other, more comprehensive laws. The most plausible answer, I think, emerges if one rewrites nomological statements in full disjunctive normal form.[9] Thus, if this is done for the statement $(x)[(Px \land Rx) \supset Qx]$, one has:

$$(x) \, [\, (Px \land Rx \land Qx) \lor (Px \land -Rx \land Qx) \lor (-Px \land Rx \land Qx) \lor$$
$$(-Px \land -Rx \land Qx) \lor (Px \land -Rx \land -Qx)$$
$$\lor (-Px \land Rx \land -Qx) \lor (-Px \land -Rx \land -Qx)].$$

Of the seven disjuncts that compose the matrix of the statement rewritten in this way, one is of special interest: $Px \land -Rx \land -Qx$. For if it is a nomological truth that $(x)(Px \supset Qx)$, it is nomologically impossible for anything to satisfy that disjunct. It is this feature, I suggest, that distinguishes between nomological statements in general and laws in the narrow sense. If so, the following is a natural analysis of the concept of a law:

S expresses a law if and only if

(1)   S is a nomological statement, and

(2)   there is no nomological statement T such that, when S is rewritten in full disjunctive normal form, there is a disjunct D in the matrix such that T entails that there will be nothing that satisfies D.

---

8   This interpretation of the expressions "nomological statement" and "law" follows that of Hans Reichenbach in his book, *Nomological Statements and Admissible Operations*, Amsterdam, 1954. It may be that the term "law" is ordinarily used in a less restricted sense. However there is an important distinction to be drawn here, and it seems natural to use the term "law" in perhaps a slightly narrower sense in order to have convenient labels for these two classes of statements.

9   This method of handling the problem was employed by Hans Reichenbach in *Nomological Statements and Admissible Operations*.

This account is not, however, entirely satisfactory. Suppose that the statements $(x)(Px \supset Qx)$ and $(x)(Rx \supset Qx)$ both express laws. Since the statement $(x)[(Px \supset Qx) \land (Rx \supset Rx)]$ is logically equivalent to the former statment, it presumably expresses the same law. But when the last statement is written in full disjunctive normal form, it contains the disjunct $-Px \land -Qx \land Rx$, which cannot be satisfied by anything if it is a law that $(x)(Rx \supset Qx)$.

So some revision is necessary. The natural move is to distinguish between essential and inessential occurrences of terms in a statement: a term occurs essentially in a statement S if and only if there is no logically equivalent statement S^* which does not contain an occurrence of the same term. The account can thus be revised to read:

S expresses a law if and only if

(1) S is a nomological statement, and

(2)   there are no nomological statements S^* and T such that S^* is logically equivalent to S, all constant terms in S^* occur essentially, and when S^* is rewritten in full disjunctive normal form, there is a disjunct D in the matrix such that T entails that nothing will satisfy D.

4. Objections

In this section I shall consider three objections to the approach advocated here. The first is that the account offered of the truth conditions of nomological statements is in some sense ad hoc and unilluminating. The second objection is that the analysis commits one to a very strong version of realism with respect to universals. The third is that the account offered places an unjustifiable restriction upon the class of nomological statements.

The basic thrust of the first objection is this. There is a serious problem about the truth conditions of laws. The solution offered here is that there are relations — referred to as nomological — which hold among universals, and which function as truth-makers for laws. How does this solution differ from simply saying that there are special facts — call them nomological — which are the facts which make laws true? How is the one approach any more illuminating than the other?

The answer is two-fold. First, to speak simply of nomological facts does nothing to *locate* those facts, that is, to specify the individuals that are the constituents of the facts in question. In contrast, the view advanced here does locate the relevant facts: they are facts about

277

universals, rather than facts about particulars. And support was offered for this contention, viz., that otherwise no satisfactory analysis of the truth conditions of basic laws without positive instances is forthcoming. Secondly, the relevant facts were not merely located, but *specified*, since not only the individuals involved, but their relevant attributes, were described. It is true that the attributes had to be specified theoretically, and hence in a sense indirectly, but this is also the case when one is dealing with theoretical terms attributing properties to particulars. The Lewis-type account of the meaning of theoretical terms that was appealed to is just as applicable to terms that refer to relations among universals — including nomological relations — as it is to terms that refer to properties of, and relations among, particulars.

Still, the feeling that there is something unilluminating about the account may persist. How does one *determine*, after all, that there is, in any given case, a nomological relation holding among universals? In what sense have truth conditions for nomological statements really been supplied if it remains a mystery how one answers the epistemological question?

I think this is a legitimate issue, even though I do not accept the verificationist claim that a statement has factual meaning only if it is in principle verifiable. In the next section I will attempt to show that, given my account of the truth conditions of nomological statements, it is possible to have evidence that makes it reasonable to accept generalizations as nomological.

The second objection is that the analysis offered involves a very strong metaphysical commitment. It is not enough to reject nominalism. For one must, in the first place, also hold that there are higher order universals which are not reducible to properties of, and relations among, particulars. And secondly, although the account may not entail that Platonic realism — construed minimally as the claim that there are some uninstantiated universals — is true, it does entail that if the world had been slightly different, it *would* have been true.

The first part of this objection does not seem to have much force as an objection. For what reasons are there for holding that there are no irreducible, higher order relations? On the other hand, it does point to a source of uneasiness which many are likely to feel. Semantics is usually done in a way that is compatible with nominalism. Truth conditions of sentences are formulated in terms of particulars, sets of particulars, sets of sets of particulars, and so on. The choice, as a metalanguage for semantics, of a language containing terms whose referents are either universals, or else intensional entities, such as concepts, though rather favored by the later Carnap[10] and others, has not been generally accepted. Once irreducible higher order relations enter into the account, a shift to a metalanguage containing terms referring to entities other than particulars appears unavoidable, and I suspect that this shift in metalanguages may make the solution proposed here difficult for many to accept.

The second part of this objection raises a deeper, and more serious point. Does it follow from the analysis offered that, if the world had been somewhat different, Platonic realism would have been true? The relevant argument is this. Suppose that materialism is false, and that there is, for example, a nonphysical property of being an experience of the red variety. Then consider what our world would have been like if the earth had been slightly closer to the sun, and if conditions in other parts of the universe had been such that life evolved nowhere else. The universe would have contained no sentient organism, and hence no experiences of the red variety. But wouldn't it have been true in *that* world that if the earth had been a bit farther from the sun, life would have evolved, and there would have been experiences of the red variety? If so, in virtue of what would this subjunctive conditional have been true? Surely an essential part of what would have made it true is the existence of a certain psychophysical law linking complex physical states to experiences of the red variety. But if the truth-makers for laws are relations among universals, it could not be a law in that world that whenever a complex physical system is in a certain state, there is an experience of the red variety, unless the property of being an experience of the red variety exists in that world. Thus, if the account of laws offered above is correct, one can describe a slightly altered version of our world in which there would be uninstantiated, and hence transcendent, universals.

The argument, as stated, is hardly conclusive. It does depend, for example, on the assumption that materialism is false. However this assumption is not really necessary. All that is required is the assumption that there are emergent properties. It makes no difference to the argument whether such emergent properties are physical or nonphysical.

If our world does not contain any emergent properties, it will not be possible to argue that if our world had been slightly different, there would have been uninstantiated universals. However one can argue for a different conclusion that may also seem disturbing. For if it is granted, not that there are emergent properties, but only that the

---

10 Compare Carnap's discussion in the section entitled "Language, Modal Logic, and Semantics," especially pages 889-905, in *The Philosophy of Rudolf Carnap*, edited by Paul A. Schilpp, La Salle, Illinois, 1963.

concept of an emergent property is coherent, then it can be argued that the existence of uninstantiated, and hence transcendent universals, is logically possible. A conclusion that will commend itself to few philosophers who reject Platonic realism, since the arguments usually directed against Platonic realism, if sound, show that it is *necessarily* false.

Another way of attempting to avoid the conclusion is by holding that if the world had been different in the way indicated, there would have been no psychophysical laws. This view may be tenable, although it strikes me as no more plausible than the stronger contention that there cannot be basic laws that lack positive instances. As a result, I am inclined to accept the contention that if the account of laws set out above is correct, there is reason to believe that Platonic realism, construed only as the doctrine that there are uninstantiated universals, is not incoherent.

The final objection to be considered is that there are statements that it would be natural, in some possible worlds, to view as nomological, but which would not be so classified on the account given here. Suppose, for example, the world were as follows. All the fruit in Smith's garden at any time are apples. When one attempts to take an orange into the garden, it turns into an elephant. Bananas so treated become apples as they cross the boundary, while pears are resisted by a force that cannot be overcome. Cherry trees planted in the garden bear apples, or they bear nothing at all. If all these things were true, there would be a very strong case for its being a law that all the fruit in Smith's garden are apples. And this case would be in no way undermined if it were found that no other gardens, however similar to Smith's in all other respects, exhibited behavior of the sort just described.

Given the account of laws and nomological statements set out above, it cannot, in the world described, be a law, or even a nomological statement, that all the fruit in Smith's garden are apples. If relations among universals are the truth-makers for nomological statements, a statement that contains essential reference to a specific particular can be nomological only if entailed by a corresponding, universally quantified statement free of such reference. And since, by hypothesis, other gardens do not behave as Smith's does, such an entailment does not exist in the case in question.

What view, then, is one to take of the generalization about the fruit in Smith's garden, in the world envisaged? One approach is to say that although it cannot be a law, in that world, that all the fruit in Smith's garden are apples, it can be the case that there is some property P such that Smith's garden has property P, and it is a law that all the fruit in any garden with property P are apples. So that even though it is not a nomological truth that all the fruit in Smith's garden are apples, one can, in a loose sense, speak of it as "derived" from a nomological statement.

This would certainly seem the most natural way of regarding the generalization about Smith's garden. The critical question, though, is whether it would be reasonable to maintain this view in the face of any conceivable evidence. Suppose that careful investigation, over thousands of years, has not uncovered any difference in intrinsic properties between Smith's garden and other gardens, and that no experimental attempt to produce a garden that will behave as Smith's does has been successful. Would it still be reasonable to postulate a theoretical property P such that it is a law that all the fruit in gardens with property P are apples? This issue strikes me as far from clear, but I incline slightly to a negative answer. For it would seem that, given repeated failures to produce gardens that behave as Smith's garden does, one might well be justified in concluding that if there is such a property P, it is one whose exemplification outside of Smith's garden is nomologically impossible. And this seems like a strange sort of property to be postulating.

I am inclined to think, then, that it is logically possible for there to be laws and nomological statements, in the strict sense, that involve ineliminable reference to specific individuals. But it does not matter, with regard to the general view of laws advanced here, whether that is so. If the notion of nomological statements involving ineliminable reference to specific individuals turns out to be conceptually incoherent, the present objection will be mistaken. Whereas if there can be such nomological statements, my account requires only minor revision to accommodate them. The definition of a construction function will have to be changed so that instances of universals, and not merely universals, can be elements of the ordered n-tuples that it takes as arguments, and the definition of a nomological relation will have to be similarly altered, so that it can be a relation among both universals and instances of universals. These alterations will result in an analysis of the truth conditions of nomological statements that allows for the possibility of ineliminable reference to specific particulars. And they will do so, moreover, without opening the door to accidentally true generalizations. Both condition (2*) and condition (2**), as set out in section 2 above, appear sufficiently strong to block such counterexamples.

5. The Epistemological Question

In the previous section I mentioned, but did not discuss, the contention that the analysis of nomological statements advanced here is unilluminating because it does not provide any account of how one determines whether a given nomological relation holds among specific universals. I shall argue that this objection is mistaken, and that in fact one of the merits of the present view is that it does make possible an answer to the epistemological question.

Suppose, then, that a statement is nomological if there is an irreducible theoretical relation holding among certain universals which necessitates the statement's being true. If this is right, there should not be any special epistemological problem about the grounds for accepting a given statement as nomological. To assert that a statement is nomological will be to advance a theory claiming that certain universals stand in some nomological relation. So one would think that whatever account is to be offered of the grounds for accepting theories as true should also be applicable to the case of nomological statements.

I should like, however, to make plausible, in a more direct fashion, this claim that the sorts of considerations that guide our choices among theories also provide adequate grounds for preferring some hypotheses about nomological statements to others. Consider, then, a familiar sort of example. John buys a pair of pants, and is careful to put only silver coins in the right hand pocket. This he does for several years, and then destroys the pants. What is the status of the generalization: "Every coin in the right hand pocket of John's pants is silver"? Even on the limited evidence described, one would not be justified in accepting it as nomological. One of the central types of considerations in support of a theory is that it in some sense provides the best explanation of certain observed states. And while the hypothesis that there is a theoretical relation *R* which holds among the universals involved in the proposition that every coin in the right hand pocket of John's pants is silver, and which necessitates that proposition's being true, does explain why there were only silver coins in the pocket, this explanation is unnecessary. Another explanation is available, namely, that John wanted to put only silver coins in that pocket, and carefully inspected every coin to ensure that it was silver before putting it in.

If the evidence is expanded in certain ways, the grounds for rejecting the hypothesis that the statement is nomological become even stronger. Suppose that one has made a number of tests on other pants, ostensibly similar to John's, and found that the right hand pockets accept copper coins as readily as silver ones. In the light of this additional evidence, there are two main hypotheses to be considered:

H_1: It is nomologically possible for the right hand pocket of any pair of pants of type *T* to contain a nonsilver coin;

H$_2$:   There is a pair of pants of type *T*, namely John's, such that it is nomologically impossible for the right hand pocket to contain a nonsilver coin; however all other pairs of pants of type *T* are such that it is nomologically possible for the right hand pocket to contain a nonsilver coin.

H$_1$ and H$_2$ are conflicting hypotheses, each compatible with all the evidence. But it is clear that H$_1$ is to be preferred to H$_2$. First, because H$_1$ is simpler than H$_2$, and secondly, because the generalization explained by H$_2$ is one that we already have an explanation for.

Let us now try to get clearer, however, about the sort of evidence that provides the strongest support for the hypothesis that a given generalization is nomological. I think the best way of doing this is to consider a single generalization, and to ask, first, what a world would be like in which one would feel that the generalization was merely accidentally true, and then, what changes in the world might tempt one to say that the generalization was not accidental, but nomological. I have already sketched a case of this sort. In our world, if all the fruit in Smith's garden are apples, it is only an accidentally true generalization. But if the world were different in certain ways, one might classify the generalization as nomological. If one never succeeded in getting pears into Smith's garden, if bananas changed into apples, and oranges into elephants, as they crossed the boundary, etc., one might well be tempted to view the generalization as a law. What we now need to do is to characterize the evidence that seems to make a critical difference. What is it, about the sort of events described, that makes them significant? The answer, I suggest, is that they are events which determine which of "conflicting" generalizations are true. Imagine that one has just encountered Smith's garden. There are many generalizations that one accepts — generalizations that are supported by many positive instances, and for which no counterexamples are known, such as "Pears thrown with sufficient force towards nearby gardens wind up inside them," "Bananas never disappear, nor change into other things such as apples," "Cherry trees bear only cherries as fruit." One notices that there are many apples in the garden, and no other fruit, so the generalization that all the fruit in Smith's garden are apples is also supported by positive instances, and is without counterexamples. Suppose now that a banana is moving in the direction of Smith's garden. A partial conflict situation exists, in that there are some events which will, if they occur, falsify the generalization that all the fruit in Smith's garden are apples, and other events which will falsify the generalization that bananas never change into other objects. There are, of course, other possible events which will falsify neither generalization: the banana may simply stop moving as it reaches the boundary. However there may well be other gneralizations that one accepts which will make the situation one of inevitable conflict, so that whatever the outcome, at least one generalization will be falsified. Situations of inevitable conflict can arise even for two generalizations, if related in the proper way. Thus, given the generalizations that $(x)(Px \supset Rx)$ and that $(x)(Qx \supset -Rx)$, discovery of an object b such that both Pb and Qb would be a situation of inevitable conflict.

Many philosophers have felt that a generalization's surviving such situations of conflict, or potential falsification, provides strong evidence in support of the hypothesis that the generalization is nomological. The problem, however, is to *justify* this view. One of the merits of the account of the nature of laws offered here is that it provides such a justification.

The justification runs as follows. Suppose that one's total evidence contains a number of supporting instances of the generalization that $(x)(Px \supset Rx)$, and of the generalization that $(x)(Qx \supset -Rx)$, and no evidence against either. Even such meager evidence may provide some support for the hypothesis that these generalizations are nomological, since the situation may be such that the only available explanation for the absence of counterexamples to the generalizations is that there are theoretical relations holding among universals which necessitate those generalizations. Suppose now that a conflict situation arises: an object b is discovered which is both P and Q. This new piece of evidence will reduce somewhat the likelihood of both hypotheses, since it shows that at least one of them must be false. Still, the total evidence now available surely lends some support to both hypotheses. Let us assume that it is possible to make at least a rough estimate of that support. Let m be the probability, given the available evidence, that the generalization $(x)(Px \supset Rx)$ is nomological, and n the probability that the generalization $(x)(Qx \supset -Rx)$ is nomological — where $(m + n)$ must be less than or equal to one. Suppose finally that b, which has property P and property Q, turns out to have property R, thus falsifying the second generalization. What we are now interested in is the effect this has upon the probability that the first generalization is nomological. This can be calculated by means of Bayes' Theorem:

> Let S be: It is a nomological truth that $(x)(Px \supset Rx)$.
> Let T be: It is a nomological truth that $(x)(Qx \supset -Rx)$.
> Let H_1 be: S and not-T.
> Let H_2 be: Not-S and T.
> Let H_3 be: S and T.
> Let H_4 be: Not-S and not-T.
> Let E describe the total antecedent evidence, including the fact that Pb and Qb
> Let E^* be: E and Rb.

Then Bayes' Theorem states:

Probability(H_1, given that E^* and E) =

$$\frac{\text{Probability}(E^*, \text{ given that } H_1 \text{ and } E) \times \text{Probability } (H_1 \text{ and } E)}{\sum_{i=1}^{n}[\text{Probability}(E^*, \text{ given that } H_i \text{ and } E) \times \text{Probability } (H_1 \text{ and } E)]}$$

Taking the antecedent evidence as given, we can set the probability of E equal to one. This implies that the probability of H_i and E will be equal to the probability of H_i given that E.

Probability(H_3, given that E)
$\qquad$ = Probability (S and T, given that E)
$\qquad$ = O, since E entails that either not-S or not-T.

Probability(H_1, given that E)
$\qquad$ = Probability (S and not-T, given that E)
$\qquad$ = Probability (S, given that E) -
$\qquad$ Probability (S and T, given that E)
$\qquad$ = m - O = m.

Similarly, probability (H_2, given that E) = n.

Probability(H_4, given that E)
$\qquad$ = 1 - [Probability (H_1, given that E) +
$\qquad$ Probability (H_2, given that E) +
$\qquad$ Probability (H_3, given that E)]
$\qquad$ = 1 - $(m + n)$

Probability(E^*, given that H_i and E)
$\qquad$ = 1, if i = 1 or i = 3
$\qquad$ = 0, if i = 2
$\qquad$ = some value k, if i = 4.

Bayes' Theorem then gives:

Probability(H_1, given that E^* and E)

Michael Tooley

$$= \frac{1 \times m}{(1 \times m) + (0 \times n) + (1 \times 0) + (k \times (1 - (m + n)))}$$

$$= \frac{m}{m + k(1 - m - n)}$$

In view of the fact that, given that E^*, it is not possible that T, this value is also the probability that S, that is, the likelihood that the generalization that $(x)(Px \supset Rx)$ is nomological. Let us consider some of the properties of this result. First, it is easily seen that, provided neither m nor n is equal to zero, the likelihood that the generalization is nomological will be greater after its survival of the conflict situation than it was before. The value of $\frac{m}{m + k(1 - m - n)}$ will be smallest when k is largest. Setting k equal to one gives the value $\frac{m}{m + 1 - m - n}$ i.e., $\frac{m}{1 - n}$, and this is greater than m if neither m nor n is equal to zero.

Secondly, the value of $\frac{m}{m + k(1 - m - n)}$ increases as the value of k decreases, and this too is desirable. If the event that falsified the one generalization were one that would have been very likely if neither generalization had been nomological, one would not expect it to lend as much support to the surviving generalization as it would if it were an antecedently improbable event.

Thirdly, the value of $\frac{m}{m + k(1 - m - n)}$ increases as n increases. This means that survival of a conflict with a well supported generalization results in a greater increase in the likelihood that a generalization is nomological than survival of a conflict with a less well supported generalization. This is also an intuitively desirable result.

Finally, it can be seen that the evidence provided by survival of conflict situations can quickly raise the likelihood that a generalization is nomological to quite high values. Suppose, for example, that $m = n$, and that $k = 0.5$. Then $\frac{m}{m + k(1 - m - n)}$ will be equal to $2m$. This

286

result agrees with the view that laws, rather than being difficult or impossible to confirm, can acquire a high degree of confirmation on the basis of relatively few observations, provided that those observations are of the right sort.

But how is this justification related to the account I have advanced as to the truth conditions of nomological statements? The answer is that there is a crucial assumption that seems reasonable if relations among universals are the truth-makers for laws, but not if facts about particulars are the truth-makers. This is the assumption that m and n are not equal to zero. If one takes the view that it is facts about the particulars falling under a generalization that make it a law, then, if one is dealing with an infinite universe, it is hard to see how one can be justified in assigning any non-zero probability to a generalization, given evidence concerning only a finite number of instances. For surely there is some non-zero probability that any given particular will falsify the generalization, and this entails, given standard assumptions, that as the number of particulars becomes infinite, the probability that the generalization will be true is, in the limit, equal to zero.[11]

In contrast, if relations among universals are the truth-makers for laws, the truth-maker for a given law is, in a sense, an "atomic" fact, and it would seem perfectly justified, given standard principles of confirmation theory, to assign some non-zero probability to this fact's obtaining. So not only is there an answer to the epistemological question; it is one that is only available given the type of account of the truth conditions of laws advocated here.

6. Advantages of this Account of Nomological Statements

Let me conclude by mentioning some attractive features of the general approach set out here. First, it answers the challenge advanced by Chisholm in his article "Law Statements and Counterfactual Inference":[12]

---

11 See, for example, Rudolf Carnap's discussion of the problem of the confirmation of universally quantified statements in section F of the appendix of his book, *The Logical Foundations of Probability*, 2nd edition, Chicago, 1962, pages 570-1.

12 Roderick M. Chisholm, "Law Statements and Counterfactual Inference," *Analysis*, **15** (1955), pages 97-105, reprinted in *Causation and Conditionals*, edited by Ernest Sosa, London, 1975. See page 149.

Can the relevant difference between law and non-law statements be described in familiar terminology without reference to counterfactuals, without use of modal terms such as 'causal necessity', 'necessary condition', 'physical possibility', and the like, and without use of metaphysical terms such as 'real connections between matters of fact'?

The account offered does precisely this. There is no reference to counterfactuals. The notions of logical necessity and logical entailment are used, but no nomological modal terms are employed. Nor are there any metaphysical notions, unless a notion such as a contingent relation among universals is to be counted as metaphysical. The analysis given involves nothing beyond the concepts of logical entailment, irreducible higher order universals, propositions, and functions from ordered sets of universals into propositions.

A second advantage of the account is that it contains no reference to possible worlds. What makes it true that statements are nomological are not facts about dubious entities called possible worlds, but facts about the actual world. True, these facts are facts about universals, not about particulars, and they are theoretical facts, not observable ones. But neither of these things should worry one unless one is either a reductionist, at least with regard to higher order universals, or a rather strict verificationist.

Thirdly, the account provides a clear and straightforward answer to the question of the difference between nomological statements and accidentally true generalizations: a generalization is accidentally true in virtue of facts about particulars; it is a nomological truth in virtue of a relation among universals.

Fourthly, this view of the truth conditions of nomological statements explains the relationships between different types of generalizations and counterfactuals. Suppose it is a law that $(x)(Px \supset Qx)$. This will be so in virtue of a certain irreducible relation between the universals P and Q. If now one asks what would be the case if some object b which at present lacks property P were to have P, the answer will be that this supposition about the particular b does not give one any reason for supposing that the universals P and Q no longer stand in a relation of nomic necessitation, so one can conjoin the supposition that b has property P with the proposition that the nomological relation in question holds between P and Q, from which it will follow that b has property Q. And this is why one is justified in asserting the counterfactual "If b were to have property P, it would also have property Q".

Suppose instead that it is only an accidentally true generalization that $(x)(Px \supset Qx)$. Here it is facts about particulars that make the generalization true. So if one asks what would be the case if some particular *b* which lacks property *P* were to have *P*, the situation is very different. Now one is supposing an alteration in facts that may be relevant to the truth conditions of the generalization. So if object *b* lacks property *Q*, the appropriate conclusion may be that if *b* were to have property *P*, the generalization that $(x)(Px \supset Qx)$ would no longer be true, and thus that one would not be justified in conjoining that generalization with the supposition that *b* has property *P* in order to support the conclusion that *b* would, in those circumstances, have property *Q*. And this is why accidentally true generalizations, unlike laws, do not support the corresponding counterfactuals.

Fifthly, this account of nomological statements allows for the possibility of even basic laws that lack positive instances. And this accords well with our intuitions about what laws there would be in cases such as a slightly altered version of our own world, in which life never evolves, and in that of the universe with the two types of fundamental particles that never meet.

Sixthly, it is a consequence of the account given that if *S* and *T* are logically equivalent sentences, they must express the same law, since there cannot be a nomological relation among universals that would make the one true without making the other true. I believe that this is a desirable consequence. However some philosophers have contended that logically equivalent sentences do not always express the same law. Rescher, for example, in his book *Hypothetical Reasoning*, claims that the statement that it is a law that all *X*'s are *Y*'s makes a different assertion from the statement that it is a law that all non-*Y*'s are non-*X*'s, on the grounds that the former asserts "All *X*'s are *Y*'s *and further* if *z* (which isn't an *X*) were an *X*, then *z* would be a *Y*", while the latter asserts "All non-*Y*'s are non-*X*'s *and further* if *z* (which isn't a non-*Y*) were a non-*Y*, then *z* would be a non-*X*."[13] But it would seem that the answer to this is simply that the statement that it is a law that all *X*'s are *Y*'s *also* entails that if *z* (which isn't a non-*Y*) were a non-*Y*, then *z* would be a non-*X*. So Rescher has not given us any reason for supposing that logically equivalent sentences can express different laws.

The view that sentences which would normally be taken as logically equivalent may, when used to express laws, not be equivalent, has also been advanced by Stalnaker and Thomason. Their argument is this. First, laws can be viewed as generalized subjunctive conditionals. "All *P*'s are *Q*'s", when stating a law, can be analyzed as "For all *x*, if *x* were a *P* then *x* would be a *Q*". Secondly, contraposition does not hold for subjunctive conditionals. It may be true that if *a* were *P* (at time t_1),

---

13 Nicholas Rescher, *Hypothetical Reasoning*, Amsterdam, 1964, page 81.

then a would be Q (at time t_2), yet false that if a were not Q (at time t_2), then a would not have been P (at time t_1). Whence it follows that its being a law that all P's are Q's is not equivalent to its being a law that all non-Q's are non-P's.[14]

The flaw in this argument lies in the assumption that laws can be analyzed as generalized subjunctive conditionals. The untenability of the latter claim can be seen by considering any possible world W which satisfies the following conditions:

(1) The only elementary properties in W are P, Q, F, and G;

(2) There is some time when at least one individual in W has properties F and P, and some time when at least one individual in W has properties F and Q;

(3) It is true, but not a law, that everything has either property P or property Q;

(4) It is a law that for any time t, anything possessing properties F and P at time t will come to have property G at a slightly later time t^*;

(5) It is a law that for any time t, anything possessing properties F and Q at time t will come to have property G at a slightly later time t^*;

(6) No laws are true in W beyond those entailed by the previous two laws.

Consider now the generalized subjunctive conditional, "For all x, and for any time t, if x were to have property F at time t, x would come to have property G at a slightly later time t^*". This is surely true in W, on any plausible account of the truth conditions of subjunctive conditionals. For let x be any individual in W at any time t. If x has P at time t, then in view of the law referred to at (4), it will be true that if x were to have F at t, it would come to have G at t^*. While if x has Q at t, the conditional will be true in virtue of that fact together with the law referred to at (5). But given (3), x will, at time t, have either property P or property Q. So it will be true in W, for any x whatsoever, that if x were to have property F at time t, it would come to have property G at a slightly later time t^*.

If now it were true that laws are equivalent to generalized subjunctive conditionals, it would follow that it is a law in W that for every x, and every time t, if x has F at t, then x will come to have G at t^*.

---

14 Robert C. Stalnaker and Richmond H. Thomason, "A Semantical Analysis of Conditional Logic," *Theoria*, **36** (1970), pages 39-40.

But this law does not follow from the laws referred to at (4) and (5), and hence is excluded by condition (6). The possibility of worlds such as *W* shows that laws are not equivalent to generalized subjunctive conditionals. As a result, the Stalnaker-Thomason argument is unsound, and there is no reason for thinking that logically equivalent statements can express non-equivalent laws.

Seventhly, given the above account of laws and nomological statements, it is easy to show that such statements have the logical properties one would naturally attribute to them. Contraposition holds for laws and nomological statements, in view of the fact that logically equivalent statements express the same law. Transitivity also holds: if it is a law (or nomological statement) that (x) $(Px \supset Qx)$, and that (x) $(Qx \supset Rx)$, then it is also a law (or nomological statement) that (x) $(Px \supset Rx)$. Moreover, if it is a law (or nomological statement) that (x) $(Px \supset Qx)$, and that (x) $(Px \supset Rx)$, then it is a law (or nomological statement) that (x) $[Px \supset (Qx \wedge Rx)]$, and conversely. Also, if it is a law (or nomological statement) that (x) $[(Px \vee Qx) \supset Rx]$, then it is a law (or nomological statement) both that (x) $(Px \supset Rx)$, and that $(x)(Qx \supset Rx)$, and conversely. And in general, I think that laws and nomological statements can be shown, on the basis of the analysis proposed here, to have all the formal properties they are commonly thought to have.

Eighthly, the account offered provides a straightforward explanation of the nonextensionality of nomological contexts. The reason that it can be a law that (x) $(Px \supset Rx)$, and yet not a law that $(x)(Qx \supset Rx)$, even if it is true that (x) $(Px \equiv Qx)$, is that, in view of the fact that the truth-makers for laws are relations among universals, the referent of the predicate "P" in the sentence "It is a law that (x) $(Px \supset Rx)$" is, at least in the simplest case, a universal, rather than the set of particulars falling under the predicate. As a result, interchange of co-extensive predicates in nomological contexts may alter the referent of part of the sentence, and with it, the truth of the whole.

Finally, various epistemological issues can be resolved given this account of the truth conditions of nomological statements. How can one establish that a generalization is a law, rather than merely accidentally true? The general answer is that if laws hold in virtue of theoretical relations among universals, then whatever account is to be given of the grounds for accepting theories as true will also be applicable to laws. The latter will not pose any independent problems. Why is it that the results of a few carefully designed experiments can apparently provide very strong support for a law? The answer is that if the truth-makers for laws are relations among universals, rather than facts about particulars, the assignment of nonzero initial probability to a law ceases to be unreasonable, and one can then employ standard theorems of probability theory, such as Bayes' theorem, to show how a few observations of the right sort will result in a probability assignment that quickly takes on quite high values.

To sum up, the view that the truth-makers for laws are irreducible relations among universals appears to have much to recommend it. For it provides not only a noncircular account of the truth conditions of nomological statements, but an explanation of the formal properties of such statements, and a solution to the epistemological problem for laws.

XII*—FUNDAMENTALISM vs. THE PATCHWORK OF LAWS

by Nancy Cartwright

I

For Realism. A number of years ago I wrote *How The Laws of Physics Lie*. That book was generally perceived to be an attack on realism. Nowadays I think that I was deluded about the enemy: it is not *realism* but *fundamentalism* that we need to combat.

My advocacy of realism—local realism about a variety of different kinds of knowledge in a variety of different domains across a range of highly differentiated situations—is Kantian in structure. Kant frequently used what should be a puzzling argument form to establish quite abstruse philosophical positions (∅): We have X —perceptual knowledge, freedom of the will, whatever. But without ∅ (the transcendental unity of apperception, or the kingdom of ends) X would be impossible, or inconceivable. Hence ∅. The objectivity of local knowledge is my ∅; X is the possibility of planning, prediction, manipulation, control and policy setting. Unless our claims about the expected consequences of our actions are reliable, our plans are for nought. Hence knowledge is possible.

What might be found puzzling about the Kantian argument form are the X's from which it starts. These are generally facts that appear in the clean and orderly world of pure reason as refugees with neither proper papers nor proper introductions, of suspect worth and suspicious origin. The facts that I take to ground objectivity are similarly alien in the clear, well lighted streets of reason, where properties have exact boundaries, rules are unambiguous, and behaviour is precisely ordained. I know that I can get an oak tree from an acorn, but not from a pine cone; that nurturing will make my child more secure; that feeding the hungry and housing

*Meeting of the Aristotelian Society, held in the Senior Common Room, Birkbeck College, London, on Monday, May 9th 1994 at 8.15 p.m.

the homeless will make for less misery; and that giving more smear tests will lessen the incidence of vaginal cancer. Getting closer to physics, which is ultimately our topic here, I also know that I can drop a pound coin from the upstairs window into the hands of my daughter below, but probably not a paper tissue; that I can head north by following my compass needle (so long as I am on foot and not in my car), that...

I know these facts even though they are vague and imprecise, and I have no reason to assume that that can be improved on. Nor, in many cases, am I sure of the strength or frequency of the link between cause and effect, nor of the range of its reliability. And I certainly do not know in any of the cases which plans or policies would constitute an optimal strategy. But I want to insist that these items are items of knowledge. They are, of course, like all genuine items of knowledge (as opposed to fictional items like sense data or the synthetic *a priori*) defeasible and open to revision in the light of further evidence and argument. If I do not know these things, what do I know and how can I come to know anything?

Besides this odd assortment of inexact facts, we also have a great deal of very precise and exact knowledge, chiefly supplied by the natural sciences. I am not thinking here of abstract laws, which as an empiricist I take to be of considerable remove from the world to which they are supposed to apply, but rather of the precise behaviour of specific kinds of concrete systems, knowledge of, say, what happens when neutral K-mesons decay, which allows us to establish c–p violation, or of the behaviour of SQUIDS (super conducting quantum interference devices) in a shielded fluctuating magnetic field, which allows us to detect the victims of strokes. This knowledge is generally regimented within a highly articulated, highly abstract theoretical scheme.

One cannot do positive science without the use of induction, and where those concrete phenomena can be legitimately derived from an abstract scheme, they serve as a kind of inductive base for that scheme. *How The Laws of Physics Lie* challenged the soundness of these derivations and hence of the empirical support for the abstract laws. I still maintain that these derivations are generally shaky, but that is not the point I want to make here. So let us for the sake of argument assume the contrary: the derivations are deductively correct and they use only true premises. Then, granting the validity of the appropriate inductions,[1] we have reason to be realists about the laws in question. But that does not give us reason to be fundamentalists. To grant that a law is true—even a law of 'basic' physics or a law about the so-called 'fundamental particles'—is far from admitting that it is universal, that it holds everywhere and governs in all domains.

II

Against Fundamentalism. Return to my rough division of law-like items of knowledge into two categories: (1) those that are legitimately regimented into theoretical schemes, these generally, though not always, being facts about behaviour in highly structured, manufactured environments like a spark chamber; (2) those that are not. There is a tendency to think that all facts must belong to one grand scheme, and moreover that this is a scheme in which the facts in the first category have a special and privileged status. They are exemplary of the way nature is supposed to work. The others must be made to conform to them. This is the kind of fundamentalist doctrine that I think we must resist. Biologists are clearly already doing so on behalf of their own special items of knowledge. Reductionism has long been out of fashion in biology and now emergentism is again a real possibility. But the long-debated relations between biology and physics are not good paradigms for the kind of anti-fundamentalism I urge. Biologists used to talk about how new laws emerge with the appearance of 'life'; nowadays they talk, not about life, but about levels of complexity and organization. Still in both cases the relation in question is that between larger, richly endowed, complex systems, on the one hand, and fundamental laws of physics on the other: it is the possibility of 'downwards' reduction that is at stake.

I want to go beyond this. Not only do I want to challenge the possibility of downwards reduction but also the possibility of 'cross-wise reduction'. Do the laws of physics that are true of systems (literally true, we may imagine for the sake of argument) in the highly contrived environments of a laboratory or inside the

---

1 These will depend on the circumstance and on our general understanding of the similarities and structures or kinds and essences that obtain in those circumstances.

housing of a modern technological device, do these laws carry across to systems, even systems of very much the same kind, in different and less regulated settings? Can our refugee facts always, with sufficient effort and attention, be remoulded into proper members of the physics community, behaving tidily in accord with the fundamental code. Or must—and should—they be admitted into the body of knowledge on their own merit?

In moving from the physics experiment to the facts of more everyday experience, we are not only changing from controlled to uncontrolled environments, but often from micro to macro as well. In order to keep separate the issues which arise from these two different shifts, I am going to choose for illustration a case from classical mechanics, and will try to keep the scale constant. Classical electricity and magnetism would serve as well. Moreover in order to make my claims as clear as possible, I shall consider the simplest and most well-known example, that of Newton's third law and its application to falling bodies: $F = ma$. Most of us, brought up within the fundamentalist canon, read this with a universal quantifier in front: for any body in any situation, the acceleration it undergoes will be equal to the force exerted on it in that situation divided by its inertial mass. I want instead to read it, as indeed I believe we should read *all* nomologicals, as a *ceteris paribus* law: for any body in any situation, *if nothing interferes*, its acceleration will equal the force exerted on it divided by its mass. But what can interfere with a force in the production of motion other than another force? Surely there is no problem: the acceleration will always be equal to the *total* force divided by the mass. That is just what I want to question.

Think again about how we construct a theoretical treatment of a real situation. Before we can apply the abstract concepts of basic theory—assign a quantum field, a tensor, a Hamiltonian, or in the case of our discussion, write down a force function—we must first produce a model of the situation in terms the theory can handle. From that point the theory itself provides 'language-entry rules' for introducing the terms of its own abstract vocabulary, and thereby for bringing its laws into play. *How The Laws of Physics Lie* illustrated this for the cases of the Hamiltonian—which is roughly the quantum analogue of the classical force function. Part of learning quantum mechanics is learning how to write the Hamiltonian for canonical models, for example, for systems in free motion, for a square well potential, for a linear harmonic oscillator, and so forth. Ronald Giere has made the same point for classical mechanics.[2]

The basic strategy for treating a real situation is to piece together a model from these fixed components; and then to determine the prescribed composite Hamiltonian from the Hamiltonians for the parts. Questions of realism arise when the model is compared with the situation it is supposed to represent. *How the Laws of Physics Lie* argued that even in the best cases, the fit between the two is not very good. I concentrated there on the best cases because I was trying to answer the question 'Do the explanatory successes of modern theories argue for their truth?' Here I want to focus on the multitude of 'bad' cases, where the models, if available at all, provide a very poor image of the situation. These are not cases that disconfirm the theory. You can't show that the predictions of a theory for a given situation are false until you have managed to describe the situation in the language of the theory. When the models are too bad a fit, the theory is not disconfirmed; it is just inapplicable.[3]

Now consider a falling object. Not Galileo's from the leaning tower, nor the pound coin I earlier described dropping from the upstairs window, but rather something more vulnerable to non-gravitational influence. Otto Neurath has a nice example. My doctrine about the case is much like his.

> In some cases a physicist is a worse prophet than a [behaviourist psychologist], as when he is supposed to specify where in St. Stephen's Square a thousand dollar bill swept away by the wind will land, whereas a [behaviourist] can specify the result of a conditioning experiment rather accurately. ('United Science and Psychology', in B. F. McGuiness, ed., *Unified Science*. Dordrecht: Reidel)

Mechanics provides no model for this situation. We have only a partial model, which describes the 1000 dollar bill as an unsupported

---

2 Giere, R.N. 1988. *Explaining Science: A Cognitive Approach*. Chicago: University of Chicago Press.

3 Here I follow Alan Musgrave: 'We do not falsify a theory containing a domain assumption by showing that this assumption is not true of some situations...; we merely show that that assumption is not applicable so that situation in the first place.' ('On Interpreting Friedman,' KYKLOS, 34 (1981). Fasc. 3, 377–387, 381.)

object in the vicinity of the earth, and thereby introduces the force exerted on it due to gravity. Is that the total force? The fundamentalist will say no: there is in principle (in God's completed theory?) a model in mechanics for the action of the wind, albeit probably a very complicated one that we may never succeed in constructing. This belief is essential for the fundamentalist. If there is no model for the 1000 dollar bill in mechanics, then what happens to the note is not determined by its laws. Some falling objects, indeed a very great number, will be outside the domain of mechanics, or only partially affected by it. But what justifies this fundamentalist belief? The successes of mechanics in situations that it can model accurately do not support it, no matter how precise or surprising they are. They show only that the theory is true in its domain, not that its domain is universal. The alternative to fundamentalism that I want to propose supposes just that: mechanics is true, literally true we may grant, for all those motions whose causes can be adequately represented by the familiar models that get assigned force functions in mechanics. For these motions, mechanics is a powerful and precise tool for prediction. But for other motions, it is a tool of limited serviceability.

Let us set our problem of the 1000 dollar bill in St. Stephen's Square to an expert in fluid dynamics. The expert should immediately complain that the problem is ill defined. What exactly is the bill like: is it folded or flat? straight down the middle, or...? is it crisp or crumpled? how long versus wide? and so forth and so forth and so forth. I do not doubt that when answers can be supplied, fluid dynamics can provide a practicable model. But I do doubt that for every real case, or even for the majority, fluid dynamics has enough of the 'right questions' to ask to allow it to model the full set of causes, or even the dominant ones. I am equally sceptical that the models that work will do so by legitimately bringing Newton's laws (or Lagrange's for that matter) into play.[4] How then do airplanes stay afloat? Two observations are important. First, we do not need to maintain that no laws obtain where mechanics runs out. Fluid dynamics may have loose overlaps and intertwinings with

---

4   And the problem is certainly not that a quantum or relativistic or microscopic treatment is needed instead.

mechanics. But it is in no way a subdiscipline of basic physics; it is a discipline on its own. Its laws can direct the 1000 dollar bill as well as can those of Newton or Lagrange. Second, the 1000 dollar bill comes as it comes, and we have to hunt a model for it. Just the reverse is true of the plane. We build it to fit the models we know work. Indeed, that is how we manage to get so much into the domain of the laws we know.

Many will continue to feel that the wind and other exogenous factors must produce a force. The wind after all is composed of millions of little particles which must exert all the usual forces on the bill, both at a distance and via collisions. That view begs the question. When we have a good-fitting molecular model for the wind, and we have in our theory (either by composition from old principles or by the admission of new principles) systematic rules that assign force functions to the models, and the force functions assigned predict exactly the right motions, then we will have good scientific reason to maintain that the wind operates via a force. Otherwise the assumption is another expression of fundamentalist faith.

III

Ceteris Paribus Laws Versus Ascriptions of Natures. If the laws of mechanics are not universal, but nevertheless true, there are at least two options for them. They could be pure *ceteris paribus* laws: laws that hold only in circumscribed conditions or so long as no factors relevant to the effect besides those specified occur. And that's it. Nothing follows about what happens in different settings or in cases where other causes occur. Presumably this option is too weak for our example of Newtonian mechanics. When a force is exerted on an object, the force will be relevant to the motion of the object even if other causes for its motion not renderable as forces are at work as well; and the exact relevance of the force will be given by the formula $F = ma$: the (total) force will contribute a component to the acceleration determined by this formula. For cases like this, the older language of *natures* is appropriate. It is in the nature of a force to produce an acceleration of the requisite size. That means that *ceteris paribus*, it *will* produce that acceleration. But even when other causes are at work, it will 'try' to do so. The idea is familiar

in the case of forces: *trying* to produce an acceleration, F/m, consists
in actually producing F/m as a vector component to the total
acceleration. In general what counts as 'trying' will differ from one
kind of cause to another. To ascribe a behaviour to the nature of a
feature is to claim that that behaviour is exportable beyond the strict
confines of the *ceteris paribus* conditions, although usually only as
a 'tendency' or a 'trying'. The extent and range of the exportability
will vary. Some natures are highly stable; others are very restricted
in their range. The point here is that we must not confuse a
wide-ranging nature with the universal applicability of the related
ceteris paribus law. To admit that forces tend to cause the prescribed
acceleration (and indeed do so in felicitous conditions) is a long
way from admitting that $F=ma$ is universally true.[5] In the next
sections I will describe two different metaphysical pictures in
which fundamentalism about the experimentally derived laws of
basic physics would be a mistake. The first is wholism; the second,
pluralism. It seems to me that wholism is far more likely to give
rise only to *ceteris paribus* laws, whereas natures are more con-
genial to pluralism.

IV

Wholism. We look at little bits of nature, and we look under a very
limited range of circumstances. This is especially true of the exact
sciences. We can get very precise outcomes, but to do so we need
very tight control over our inputs. Most often we do not control
them directly, one by one, but rather we use some general but
effective form of shielding. I know one experiment that aims for
direct control—the Stanford Gravity Probe. Still, in the end, they
will roll the space ship to average out causes they have not been
able to command. Sometimes we take physics outside the laboratory. Then shielding becomes even more important. SQUIDS
(Superconducting quantum interference devices) can make very
fine measurements of magnetic fluctuations, which helps in the

---

5 I have written more about the two levels of generalization, laws and ascriptions of natures,
   in *Natures, Capacities and Their Measurement*, Oxford University Press (1989). See also
   'Aristotelian Natures and the Modern Experimental Method', in *Inference, Explanation
   & Other Philosophical Frustrations.* ed. John Earman, University of California Press
   (1992).

detection of stroke victims. But for administering the tests the hospital must have a Hertz box—a small metal room to block out magnetism from the environment. Or, for a more homely example, we all know that batteries are not likely to work if their protective casing has been pierced.

We tend to think that shielding cannot matter to the laws we use. The same laws apply both inside and outside the shields, the difference is that inside the shield we know how to calculate what the laws will produce, but outside, it is too complicated. Wholists are wary of these claims: if the events we study are locked together and changes depend on the total structure rather than the arrangement of the pieces, we are likely to be very mistaken by looking at small chunks of special cases.

Consider a scientific example, the revolution in communications technology due to fibre optics. Low-loss optical fibres can carry information at rates of many gigabits per second over spans of tens of kilometres. But the development of fibre bundles which lose only a few decibels per kilometre is not all there is to the story. Pulse broadening effects intrinsic to the fibres can be truly devastating. If the pulses broaden as they travel down the fibre, they will eventually smear into each other and destroy the information. That means that the pulses cannot be sent too close together, and the transmission rate may drop to tens or at most hundreds of megabits per second.

We know that is not what happens—the technology has been successful. That's because the right kind of optical fibre in the right circumstance can transmit solitons—solitary waves that keep their shape across vast distances. I'll explain why. The light intensity of the incoming pulse causes a shift in the index of refraction of the optical fibre, producing a slight non-linearity in the index. The non-linearity leads to what is called a 'chirp' in the pulse. Frequencies in the leading half of the pulse are lowered while those in the trailing half are raised. The effects of the chirp combine with those of dispersion to produce the soliton. Stable pulse shapes are not at all a general phenomenon of low loss optical fibres. They are instead a consequence of two different, oppositely directed processes. The pulse widening due to the dispersion is cancelled by the pulse narrowing due to the non-linearity in the index of refraction. We can indeed produce perfectly stable pulses. But to do so we must use fibres of just the right design, and matched precisely with the power and input frequency of the laser that generates the input pulses. By chance that was not hard to do. When the ideas were first tested in 1980 the glass fibres and lasers readily available were easily suited to each other. Given that very special match, fibre optics was off to an impressive start.

Solitons are indeed a stable phenomenon. They are a feature of nature, but of nature under very special circumstance. Clearly it would be a mistake to suppose that they were a general character-istic of low loss optical fibres. The question is, how many of the scientific phenomena we prize are like solitons, local to the environ-ments we encounter, or—more importantly—to the environments we construct? If nature is more wholistic than we are accustomed to think, the fundamentalist's hopes to export the laws of the laboratory to the far reaches of the world will be dashed.

It is clear that I am not very sanguine about the fundamentalist faith. But that is not really out of the kind of wholist intuitions I have been sketching. After all, the story I just told accounts for the powerful successes of the 'false' local theory—the theory that solitons are characteristic of low loss fibres—by embedding it in a far more general theory about the interaction of light and matter. Metaphysically, the fundamentalist is borne out. It may be the case that the successful theories we have are limited in their authority, but their successes are to be explained by reference to a truly universal authority. I do not see why we need to explain their successes. I am prepared to believe in more general theories when we have direct empirical evidence for them. But not merely because they are the 'best explanation' for something which seems to me to need no explanation to begin with. 'The theory is successful in its domain': the need for explanation is the same whether the domain is small, or large, or very small, or very large. Theories are success-ful where they are successful, and that's that. If we insist on turning this into a metaphysical doctrine, I suppose it will look like meta-physical pluralism, to which I now turn.

V

The Patchwork of Laws. Metaphysical nomological pluralism is the doctrine that nature is governed in different domains by different systems of laws not necessarily related to each other in any systematic or uniform way: by a patchwork of laws. Nomological pluralism opposes any kind of fundamentalism. We are here concerned especially with the attempts of physics to gather all phenomena into its own abstract theories. In *How the Laws of Physics Lie* I argued that most situations are brought under a law of physics only by distortion, whereas they can often be described fairly correctly by concepts from more phenomenological laws. The picture suggested was of a lot of different situations in a continuum from ones that fit not perfectly but not badly to those that fit very badly indeed. I did suggest that at one end fundamental physics might run out entirely ('What is... the value of the electric field vector in the region just at the tip of my pencil'), whereas in transistors it works quite well. But that was not the principal focus. Now I want to draw sharp divides: some features of systems typically studied by physics may get into situations where their behaviour is not governed by the laws of physics at all. But that does not mean that they have no guide for their behaviour or only low-level phenomenological laws. They could fall under a quite different organized set of highly abstract principles.

There are two immediate difficulties that metaphysical pluralism encounters. The first is one we create ourselves, by imagining that it must be joined with views that are vestiges of metaphysical monism. The second is, I believe, a genuine problem that nature must solve.

First. We are inclined to ask, How can there be motions not governed by Newton's laws? The answer: there are causes of motion not included in Newton's theory. Many find this impossible because, although they have forsaken reductionism, they cling to a near-cousin: *supervenience*. Suppose we give a complete 'physics' description of the falling object and its surrounds. Mustn't that fix all the other features of the situation? Why? This is certainly not true at the level of discussion at which we stand now: the wind is cold and gusty; the bill is green and white and crumpled. These properties are independent of the mass of the bill, the mass of the earth, the distance between them.

I suppose though I have the supervenience story wrong. It is the microscopic properties of physics that matter; the rest of reality supervenes on them. Why should I believe that? Supervenience is touted as a step forward over reductionism. Crudely, I take it, the advantage is supposed to be that we can substitute a weaker kind of reductionism, 'token–token' reductionism, for the more traditional 'type–type' reductionism which was proving hard to carry out. But the traditional view had arguments in its favour. Science does sketch a variety of fairly systematic connections between micro-structures and macro-properties. Often the sketch is rough, sometimes it is precise, usually its reliability is confined to very special circumstances. Nevertheless there are striking cases. But these cases support type–type reductionism; they are irrelevant for supervenience. Type–type reductionism has well-known problems: the connections we discover often turn out to look more like causal connections than like reductions; they are limited in their domain; they are rough rather than exact; and often we cannot even find good starting proposals where we had hoped to produce nice reductions. These problems suggest modifying the doctrine in a number of specific ways, or perhaps giving it up altogether. But they certainly do not leave us with token–token reductionism as a fallback position. After all, on the story I have just told, it was the appearance of some degree of systematic connection that argued in the first place for the claim that microstructures fixed macro-properties. But it is just this systematicity that is missing in token–token reductionism.

The view that there are macro-properties that do not supervene on micro-features studied by physics is sometimes labelled 'emergentism'. The suggestion is that where there is no supervenience, macro-properties must miraculously come out of nowhere. But why? There is nothing of the newly landed about these properties. They have been here in the world all along, standing beside the properties of physics. Perhaps we are misled by the feeling that the set of properties studied by physics is complete. Indeed, I think that there is a real sense in which this claim is true, but that sense does not support the charge of emergentism. Consider how the domain of properties studied by physics gets set. Here is one caricature: we begin with an interest in motions—deflections, trajectories, orbits. Then we look for the smallest set of properties that is closed (or, closed enough) under prediction. That is, we expand the set until we get all the factors that are causally relevant to our starting factors, and then everything causally relevant to those and so forth. To succeed does not show that we have gotten all the properties there are. This is a fact we need to keep in mind quite independently of the chief claim of this paper, that the predictive closure itself only obtains in highly restricted circumstances. The immediate point is that predictive closure among a set of properties does not imply descriptive completeness.

Second. The second problem that metaphysical pluralism faces is that of consistency. We do not want colour patches to appear in regions from which the laws of physics have carried away all matter and energy. Here are two stories I have told in teaching the mechanical philosophy of the 17th Century. Both are about how to write the Book of Nature to ensure that a consistent universe can be created. In one story God is very interested in physics. He carefully writes out all of the law of physics and lays down the initial distribution of matter and energy in the universe. He then leaves to St. Peter the tedious but intellectually trivial job of calculating all future happenings, including what, if any, macroscopic properties and macroscopic laws will emerge. That is the story of reductionism. Metaphysical pluralism supposes that God is instead very concerned about laws, and so he writes down each and every regularity that his universe will display. In this case St. Peter is left with the gargantuan task of arranging the initial properties in the universe in some way that will allow all God's laws to be true together. The advantage to reductionism is that it makes St. Peter's job easier. God may nevertheless choose to be a metaphysical pluralist.

VI

Conclusion. I have argued that the laws of our contemporary science are, to the extent that they are true at all, at best true *ceteris paribus*. In the nicest cases we may treat them as claims about natures. But we have no grounds in our experience for taking our laws—even our most fundamental laws of physics—as universal. Indeed I should say '*especially* our most fundamental laws of physics', if these are meant to be the laws of fundamental particles. For we have virtually no inductive reason for counting these laws as true of fundamental particles outside the laboratory setting—if they exist there at all. Ian Hacking is famous for his remark, 'If you can spray them, they exist.' I have always agreed with that. But I would now be more cautious: '*When* you can spray them, they exist.'

The claim that theoretical entities are created by the peculiar conditions and conventions of the laboratory is familiar from the social constructionists. The stable low-loss pulses I described earlier provide an example of how that can happen. Here I want to add a caution, not just about the existence of the theoretical entities outside the laboratory, but about their behaviour.

Hacking's point is not only that when we can use theoretical entities in just the way we want to produce precise and subtle effects, they must exist; but also that it must be the case that we understand their behaviour very well if we are able to get them to do what we want. That argues, I believe, for the truth of some very concrete, context-constrained claims, the claims we use to describe their behaviour and control them. But in all these cases of precise control, we build our circumstances to fit our models. I repeat: that does not show that it must be possible to tailor our models to fit every circumstance.

Perhaps we feel that there could be no real difference between the one kind of circumstance and the other, and hence no principled reason for stopping our inductions at the walls of our laboratories. But there is a difference: some circumstances resemble the models we have; others do not. And it is just the point of scientific activity to build models that get in, under the cover of the laws in question, all and only those circumstances that the laws govern.[6] Fundamentalists see matters differently. They want laws; they want true laws; but most of all, they want their favourite laws to be in force everywhere. I urge that we resist fundamentalism. Reality may well be just a patchwork of laws.

Department of Philosophy, Logic and Scientific Method
The London School of Economics and Political Science
Houghton Street
London WC2A 2AE

6 Or, in a more empiricist formulation that I would prefer, 'that the laws accurately describe'.

AMERICAN PHILOSOPHICAL QUARTERLY
Volume 2, Number 4, October 1965

I. CAUSES AND CONDITIONS

J. L. MACKIE

ASKED what a cause is, we may be tempted to say that it is an event which precedes the event of which it is the cause, and is both necessary and sufficient for the latter's occurrence; briefly, that a cause is a necessary and sufficient preceding condition. There are, however, many difficulties in this account. I shall try to show that what we often speak of as a cause is a condition not of this sort, but of a sort related to this. That is to say, this account needs modification, and can be modified, and when it is modified we can explain much more satisfactorily how we can arrive at much of what we ordinarily take to be causal knowledge; the claims implicit within our causal assertions can be related to the forms of the evidence on which we are often relying when we assert a causal connection.

§ 1. SINGULAR CAUSAL STATEMENTS

Suppose that a fire has broken out in a certain house, but has been extinguished before the house has been completely destroyed. Experts investigate the cause of the fire, and they conclude that it was caused by an electrical short-circuit at a certain place. What is the exact force of their statement that this short-circuit caused this fire? Clearly the experts are not saying that the short-circuit was a necessary condition for this house's catching fire at this time; they know perfectly well that a short-circuit somewhere else, or the overturning of a lighted oil stove, or any one of a number of other things might, if it had occurred, have set the house on fire. Equally, they are not saying that the short-circuit was a sufficient condition for this house's catching fire; for if the short-circuit had occurred, but there had been no inflammable material nearby, the fire would not have broken out, and even given both the short-circuit and the inflammable material, the fire would not have occurred if, say, there had been an efficient automatic sprinkler at just the right spot. Far from being a condition both necessary and sufficient for

the fire, the short-circuit was, and is known to the experts to have been, neither necessary nor sufficient for it. In what sense, then, is it said to have caused the fire?

At least part of the answer is that there is a set of conditions (of which some are positive and some are negative), including the presence of inflammable material, the absence of a suitably placed sprinkler, and no doubt quite a number of others, which combined with the short-circuit constituted a complex condition that was sufficient for the house's catching fire—sufficient, but not necessary, for the fire could have started in other ways. Also, of *this* complex condition, the short-circuit was an indispensable part: the other parts of this condition, conjoined with one another in the absence of the short-circuit, would not have produced the fire. The short-circuit which is said to have caused the fire is thus an indispensable part of a complex sufficient (but not necessary) condition of the fire. In this case, then, the so-called cause is, and is known to be, an *insufficient* but *necessary* part of a condition which is itself *unnecessary* but *sufficient* for the result. The experts are saying, in effect, that the short-circuit is a condition of this sort, that it occurred, that the other conditions which conjoined with it form a sufficient condition were also present, and that no other sufficient condition of the house's catching fire was present on this occasion. I suggest that when we speak of the cause of some particular event, it is often a condition of this sort that we have in mind. In view of the importance of conditions of this sort in our knowledge of and talk about causation, it will be convenient to have a short name for them: let us call such a condition (from the initial letters of the words italicized above), an INUS condition.[1]

This account of the force of the experts' statement about the cause of the fire may be confirmed by reflecting on the way in which they will have reached this conclusion, and the way in which anyone who disagreed with it would have to

---

[1] This term was suggested by D. C. Stove who has also given me a great deal of help by criticizing earlier versions of this article.

A

challenge it. An important part of the investigation will have consisted in tracing the actual course of the fire; the experts will have ascertained that no other condition sufficient for a fire's breaking out and taking this course was present, but that the short-circuit did occur and that conditions were present which in conjunction with it were sufficient for the fire's breaking out and taking the course that it did. Provided that there is some necessary and sufficient condition of the fire—and this is an assumption that we commonly make in such contexts—anyone who wanted to deny the experts' conclusion would have to challenge one or another of these points.

We can give a more formal analysis of the statement that something is an INUS condition. Let 'A' stand for the INUS condition—in our example, the occurrence of a short-circuit at that place—and let 'B' and '$\bar{C}$' (that is, 'not-C', or the absence of C) stand for the other conditions, positive and negative, which were needed along with A to form a sufficient condition of the fire—in our example, B might be the presence of inflammable material, $\bar{C}$ the absence of a suitably placed sprinkler. Then the conjunction '$AB\bar{C}$' represents a sufficient condition of the fire, and one that contains no redundant factors; that is, $AB\bar{C}$ is a minimal sufficient condition for the fire.[1] Similarly, let $D\bar{E}F$, $\bar{G}HI$, etc., be all the other minimal sufficient conditions of this result. Now provided that there is some necessary and sufficient condition for this result, the disjunction of all the minimal sufficient conditions for it constitutes a necessary and sufficient condition.[2] That is, the formula "$AB\bar{C}$ or $D\bar{E}F$ or $\bar{G}HI$ or . . ." represents a necessary and sufficient condition for the fire, each of its disjuncts, such as '$AB\bar{C}$', represents a minimal sufficient condition, and each conjunct in each minimal sufficient con-

dition, such as 'A', represents an INUS condition. To simplify and generalize this, we can replace the conjunction of terms conjoined with 'A' (here '$B\bar{C}$') by the single term 'X', and the formula representing the disjunction of all the other minimal sufficient conditions—here "$D\bar{E}F$ or $\bar{G}HI$ or . . ."—by the single term 'Y'. Then an INUS condition is defined as follows:

A is an INUS condition of a result P if and only if, for some X and for some Y, $(AX$ or $Y)$ is a necessary and sufficient condition of P, but A is not a sufficient condition of P and X is not a sufficient condition of P.

We can indicate this type of relation more briefly if we take the provisos for granted and replace the existentially quantified variables 'X' and 'Y' by dots. That is, we can say that A is an INUS condition of P when $(A$ . . . or . . .) is a necessary and sufficient condition of P.

(To forestall possible misunderstandings, I would fill out this definition as follows.[4] First, there could be a set of minimal sufficient conditions of P, but no necessary conditions, not even a complex one; in such a case, A might be what Marc-Wogau calls a moment in a minimal sufficient condition, but I shall not call it an INUS condition. I shall speak of an INUS condition only where the disjunction of all the minimal sufficient conditions is also a necessary condition. Secondly, the definition leaves it open that the INUS condition A might be a conjunct in each of the minimal sufficient conditions. If so, A would be itself a necessary condition of the result. I shall still call A an INUS condition in these circumstances: it is not part of the definition of an INUS condition that it should *not* be necessary, although in the standard cases, such as that

[1] The phrase "minimal sufficient condition" is borrowed from Konrad Marc-Wogau, "On Historical Explanation," *Theoria*, vol. 28 (1962), pp. 213–233. This article gives an analysis of singular causal statements, with special reference to their use by historians, which is substantially equivalent to the account I am suggesting. Many further references are made to this article, especially in n. 9 below.

[2] Cf. n. 8 on p. 227 of Marc-Wogau's article, where it is pointed out that in order to infer that the disjunction of all the minimal sufficient conditions will be a necessary condition, "it is necessary to presuppose that an arbitrary event C, if it occurs, must have sufficient reason to occur." This presupposition is equivalent to the presupposition that there is some (possibly complex) condition that is both necessary and sufficient for C.

It is of some interest that some common turns of speech embody this presupposition. To say "Nothing but X will do," or "Either X or Y will do, but nothing else will," is a natural way of saying that X, or the disjunction (X or Y), is a *necessary* condition for whatever result we have in mind. But taken literally these remarks say only that there is no sufficient condition for this result other than X, or other than (X or Y). That is, we use to mean "a necessary condition" phrases whose literal meanings would be "the only sufficient condition," or "the disjunction of all sufficient conditions." Similarly, to say that Z is "all that's needed" is a natural way of saying that Z is a sufficient condition, but taken literally this remark says that Z is the only necessary condition. But, once again, that the only necessary condition will also be a sufficient one follows only if we presuppose that some condition is both necessary and sufficient.

[4] I am indebted to the referees for the suggestion that these points should be clarified.

sketched above, it is not in fact necessary.[5] Thirdly, the requirement that X by itself should not be sufficient for P insures that A is a nonredundant part of the sufficient condition AX; but there is a sense in which it may not be strictly necessary or indispensable even as a part of *this* condition, for it may be replaceable: for example KX might be another minimal sufficient condition of P.[6] Fourthly, it *is* part of the definition that the minimal sufficient condition, AX, of which A is a nonredundant part, is not also a necessary condition, that there is another sufficient condition Υ (which may itself be a disjunction of sufficient conditions). Fifthly, and similarly, it *is* part of the definition that A is not by itself sufficient for P. The fourth and fifth of these points amount to this: I shall call A an INUS condition only if there are terms which actually occupy the places occupied by 'X' and 'Υ' in the formula for the necessary and sufficient condition. However, there may be cases where there is only one minimal sufficient condition, say AX. Again, there may be cases where A is itself a minimal sufficient condition, the disjunction of all minimal sufficient conditions being $(A$ or $\Upsilon)$; again, there may be cases where A itself is the only minimal sufficient condition, and is itself both necessary and sufficient for P. In any of these cases, as well as in cases where A is an INUS condition, I shall say that A is *at least an* INUS *condition*. As we shall see, we often have evidence which supports the conclusion that something is *at least* an INUS condition; we may or may not have other evidence which shows that it is *no more than* an INUS condition.)

I suggest that a statement which asserts a singular causal sequence, of such a form as "A caused P," often makes, implicitly, the following claims:

(i) A is at least an INUS condition of P—that is,

there is a necessary and sufficient condition of P which has one of these forms: $(AX$ or $\Upsilon)$, $(A$ or $\Upsilon)$, AX, A.

(ii) A was present on the occasion in question.

(iii) The factors represented by the 'X', if any, in the formula for the necessary and sufficient condition were present on the occasion in question.

(iv) Every disjunct in 'Υ' which does not contain 'A' as a conjunct was absent on the occasion in question. (As a rule, this means that whatever 'Υ' represents was absent on this occasion. If 'Υ' represents a single conjunction of factors, then it was absent if at least one of its conjuncts was absent; if it represents a disjunction, then it was absent if each of its disjuncts was absent. But we do not wish to exclude the possibility that 'Υ' should be, or contain as a disjunct, a conjunction one of whose conjuncts is A, or to require that *this* conjunction should have been absent.[7])

I do not suggest that this is the whole of what is meant by "A caused P" on any occasion, or even that it is a part of what is meant on every occasion: some additional and alternative parts of the meaning of such statements are indicated below.[8] But I am suggesting that this is an important part of the concept of causation; the proof of this suggestion would be that in many cases the falsifying of any one of the above-mentioned claims would rebut the assertion that A caused P.

This account is in fairly close agreement, in substance if not in terminology, with at least two accounts recently offered of the cause of a single event.

Konrad Marc-Wogau sums up his account thus:

when historians in singular causal statements speak of a cause or the cause of a certain individual event β, then what they are referring to is another individual event α which is a moment in a minimal sufficient and at the same time necessary condition *post factum* β.[9]

---

[5] Special cases where an INUS condition is also a necessary one are mentioned at the end of § 3.

[6] This point, and the term "nonredundant," are taken from Michael Scriven's review of Nagel's *The Structure of Science*, in *Review of Metaphysics*, 1964. See especially the passage on p. 408 quoted below.

[7] See example of the wicket-keeper discussed below.

[8] See §§ 7, 8.

[9] See pp. 226–227 of the article referred to in n. 2 above. Marc-Wogau's full formulation is as follows: "Let 'msc' stand for minimal sufficient condition and 'nc' for necessary condition. Then suppose we have a class K of individual events $a_1, a_2, \ldots a_n$. (It seems reasonable to assume that K is finite; however even if K were infinite the reasoning below would not be affected.) My analysis of the singular causal statement: α is the cause of β, where α and β stand for individual events, can be summarily expressed in the following statements:

(1) (EK) $(K = \{a_1, a_2, \ldots, a_n\})$; (4) (x) $(\,(x \in K\ x \neq a_1) \supset x$ is not fulfilled when α occurs$)$;

(2) (x) $(x \in K \equiv x$ msc $\beta)$; (5) α is a moment in a_1.

(3) $(a_1 \vee a_2 \vee \ldots a_n)$ nc β;

(3) and (4) say that a_1 is a necessary condition *post factum* for β. If a_1 is a necessary condition *post factum* for β, then every moment in a_1 is a necessary condition *post factum* for β, and therefore also α. As has been mentioned before (note 6) there is assumed to be a temporal sequence between α and β; β is not itself an element in K."

He explained his phrase "necessary condition *post factum*" by saying that he will call an event a_1 a necessary condition *post factum* for x if the disjunction "a_1 or a_2 or a_3 . . . or a_n" represents a necessary condition for x, and of these disjuncts only a_1 was present on the particular occasion when x occurred.

Similarly Michael Scriven has said:

Causes are *not* necessary, even contingently so, they are not sufficient—but they are, to talk that language, *contingently sufficient*. . . . They are part of *a* set of conditions that does guarantee the outcome, and they are non-redundant in that the rest of *this* set (which does not include all the other conditions present) is not alone sufficient for the outcome. It is not even true that they are relatively necessary, i.e., necessary with regard to that set of conditions rather than the total circumstances of their occurrence, for there may be several possible replacements for them which happen not to be present. There remains a ghost of necessity; a cause is a factor from a set of possible factors the presence of one of which (*any* one) is necessary in order that a set of conditions actually present be sufficient for the effect.[10]

There are only slight differences between these two accounts, or between each of them and that offered above. Scriven seems to speak too strongly when he says that causes are not necessary: it is, indeed, not part of the definition of a cause of this sort that it should be necessary, but, as noted above, a cause, or an INUS condition, may be necessary, either because there is only one minimal sufficient condition or because the cause is a moment in each of the minimal sufficient conditions. On the other hand, Marc-Wogau's account of a minimal sufficient condition seems too strong. He says that a minimal sufficient condition contains "only those moments relevant to the effect" and that a moment is relevant to an effect if "it is a necessary condition for β: β would not have occurred if this moment had not been present." This is less accurate than Scriven's statement that the cause only needs to be nonredundant.[11] Also, Marc-Wogau's requirement, in his account of a

necessary condition *post factum*, that only one minimal sufficient condition (the one containing a) should be present on the particular occasion, seems a little too strong. If two or more minimal sufficient conditions (say a_1 and a_2) were present, but a was a moment in each of them, then though neither a_1 nor a_2 was necessary *post factum*, a would be so. I shall use this phrase "necessary *post factum*" to include cases of this sort: that is, a is a necessary condition *post factum* if it is a moment in every minimal sufficient condition that was present. For example, in a cricket team the wicket-keeper is also a good batsman. He is injured during a match, and does not bat in the second innings, and the substitute wicket-keeper drops a vital catch that the original wicket-keeper would have taken. The team loses the match, but it would have won if the wicket-keeper had *both* batted *and* taken that catch. His injury was a moment in two minimal sufficient conditions for the loss of the match; either his not batting, or the catch's not being taken, would on its own have insured the loss of the match. But we can certainly say that his injury caused the loss of the match, and that it was a necessary condition *post factum*.

This account may be summed up, briefly and approximately, by saying that the statement "*A* caused *P*" often claims that *A* was necessary and sufficient for *P* in the circumstances. This description applies in the standard cases, but we have already noted that a cause is nonredundant rather than necessary even in the circumstances, and we shall see that there are special cases in which it may be neither necessary nor nonredundant.

§ 2. DIFFICULTIES AND REFINEMENTS[12]

Both Scriven and Marc-Wogau are concerned not only with this basic account, but with certain difficulties and with the refinements and complications that are needed to overcome them. Before dealing with these I shall introduce, as a refinement of my own account, the notion of a causal field.[13]

[10] *Op. cit.*, p. 408.
[11] However, in n. 7 on pp. 222–233, Marc-Wogau draws attention to the difficulty of giving an accurate definition of "a moment in a sufficient condition." Further complications are involved in the account given in § 5 below of "clusters" of factors and the progressive localization of a cause. A condition which is minimally sufficient in relation to one degree of analysis of factors may not be so in relation to another degree of analysis.
[12] This section is something of an aside: the main argument is resumed in § 3.
[13] This notion of a causal field was introduced by John Anderson. He used it, e.g., in "The Problem of Causality," first published in the *Australasian Journal of Psychology and Philosophy*, vol. 16 (1938), and reprinted in *Studies in Empirical Philosophy* (Sydney, 1962), pp. 126–136, to overcome certain difficulties and paradoxes in Mill's account of causation. I have also used this notion to deal with problems of legal and moral responsibility, in "Responsibility and Language," *Australasian Journal of Philosophy*, vol. 33 (1955), pp. 143–159.

This notion is most easily explained if we leave, for a time, singular causal statements and consider general ones. The question "What causes influenza?" is incomplete and partially indeterminate. It may mean "What causes influenza in human beings in general?" If so, the (full) cause that is being sought is a difference that will mark off cases in which human beings contract influenza from cases in which they do not; the causal field is then the region that is to be thus divided, *human beings in general*. But the question may mean, "Given that influenza viruses are present, what makes some people contract the disease whereas others do not?" Here the causal field is *human beings in conditions where influenza viruses are present*. In all such cases, the cause is required to differentiate, within a wider region in which the effect sometimes occurs and sometimes does not, the sub-region in which it occurs: this wider region is the causal field. This notion can now be applied to singular causal questions and statements. "What caused this man's skin cancer?"[14] may mean "Why did this man develop skin cancer now when he did not develop it before?" Here the causal field is the career of this man: it is within this that we are seeking a difference between the time when skin cancer developed and times when it did not. But the same question may mean "Why did this man develop skin cancer, whereas other men who were also exposed to radiation did not?" Here the causal field is the class of men thus exposed to radiation. And what is the cause in relation to one field may not be the cause in relation to another. Exposure to a certain dose of radiation may be the cause in relation to the former field: it cannot be the cause in relation to the latter field since it is part of the description of that field, and being present throughout that field it cannot differentiate one sub-region of it from another. In relation to the latter field, the cause may be, in Scriven's terms, "Some as-yet-unidentified constitutional factor."

In our first example of the house which caught fire, the history of this house is the field in relation to which the experts were looking for the cause of the fire: their question was "Why did this house catch fire on this occasion, and not on others?" However, there may still be some indeterminacy in this choice of a causal field. Does this house,

considered as the causal field, include all its features, or all its relatively permanent features, or only some of these? If we take all its features, or even all of its relatively permanent ones, as constituting the field, then some of the things that we have treated as conditions—for example the presence of inflammable material near the place where the short-circuit occurred—would have to be regarded as parts of the field, and we could not then take them also as conditions which in relation to this field, as additions to it or intrusions into it, are necessary or sufficient for something else. We must therefore take the house, in so far as it constitutes the causal field, as determined only in a fairly general way, by only some of its relatively permanent features, and we shall then be free to treat its other features as conditions which do not constitute the field, and are not parts of it, but which may occur within it or be added to it. It is in general an arbitrary matter whether a particular feature is regarded as a condition (that is, as a possible causal factor) or as part of the field, but it cannot be treated in both ways at once. If we are to say that something happened to this house because of, or partly because of, a certain feature, we are implying that it would still have been *this* house, the house in relation to which we are seeking the cause of this happening, even if it had not had this particular feature.

I now propose to modify the account given above of the claims often made by singular causal statements. A statement of such a form as "*A* caused *P*" is usually elliptical, and is to be expanded into "*A* caused *P* in relation to the field *F*." And then in place of the claim stated in (i) above, we require this:

(ia) *A* is at least an INUS condition of *P* in the field *F*—that is, there is a condition which, given the presence of whatever features characterize *F* throughout, is necessary and sufficient for *P*, and which is of one of these forms: $(AX$ or $Y)$, $(A$ or $Y)$, AX, A.

In analyzing our ordinary causal statements, we must admit that the field is often taken for granted or only roughly indicated, rather than specified precisely. Nevertheless, the field in relation to which we are looking for a cause of this effect, or saying that such-and-such is a cause, may be definite enough for us to be able to say

---

[14] These examples are borrowed from Scriven, *op. cit.*, pp. 409–410. Scriven discusses them with reference to what he calls a "contrast class," the class of cases where the effect did not occur with which the case where it did occur is being contrasted. What I call the causal field is the logical sum of the case (or cases) in which the effect is being said to be caused with what Scriven calls the contrast class.

that certain facts or possibilities are irrelevant to the particular causal problem under consideration, because they would constitute a shift from the intended field to a different one. Thus if we are looking for the cause, or causes, of influenza, meaning its cause(s) in relation to the field *human beings*, we may dismiss, as not directly relevant, evidence which shows that some proposed cause fails to produce influenza in rats. If we are looking for the cause of the fire in *this house*, we may similarly dismiss as irrelevant the fact that a proposed cause would not have produced a fire if the house had been radically different, or had been set in a radically different environment.

This modification enables us to deal with the well-known difficulty that it is impossible, without including in the cause the whole environment, the whole prior state of the universe (and so excluding any likelihood of repetition), to find a genuinely sufficient condition, one which is "by itself, adequate to secure the effect."[15] It may be hard to find even a complex condition which was absolutely sufficient for this fire because we should have to include, as one of the negative conjuncts, such an item as the earth's not being destroyed by a nuclear explosion just after the occurrence of the suggested INUS condition; but it is easy and reasonable to say simply that such an explosion would, in more senses than one, take us outside the field in which we are considering this effect. That is to say, it may be not so difficult to find a condition which is sufficient in relation to the intended field. No doubt this means that causal statements may be vague, in so far as the specification of the field is vague, but this is not a serious obstacle to establishing or using them, either in science or in everyday contexts.[16]

It is a vital feature of the account I am suggesting that we can say that A caused P, in the sense

described, without being able to specify exactly the terms represented by 'X' and 'Y' in our formula. In saying that A is at least an INUS condition for P in F, one is *not* saying what other factors, along with A, were both present and nonredundant, and one is *not* saying what other minimal sufficient conditions there may be for P in F. One is not even claiming to be able to say what they are. This is in no way a difficulty: it is a readily recognizable fact about our ordinary causal statements, and one which this account explicitly and correctly reflects.[17] It will be shown (in § 5 below) that this elliptical or indeterminate character of our causal statements is closely connected with some of our characteristic ways of discovering and confirming causal relationships: it is precisely for statements that are thus "gappy" or indeterminate that we can obtain fairly direct evidence from quite modest ranges of observation. On this analysis, causal statements implicitly contain existential quantifications; one can assert an existentially quantified statement without asserting any instantiation of it, and one can also have good reason for asserting an existentially quantified statement without having the information needed to support any precise instantiation of it. I can know that there is someone at the door even if the question "Who is he?" would floor me

Marc-Wogau is concerned especially with cases where "there are two events, each of which independently of the other is a sufficient condition for another event." There are, that is to say, two minimal sufficient conditions, both of which actually occurred. For example, lightning strikes a barn in which straw is stored, and a tramp throws a burning cigarette butt into the straw at the same place and at the same time. Likewise for an historical event there may be more than one "cause," and each of them may, on its own, be

---

[15] Cf. Bertrand Russell, "On the Notion of Cause," *Mysticism and Logic* (London, 1917), p. 187. Cf. also Scriven's first difficulty, *op. cit.*, p. 409: "First, there are virtually no known sufficient conditions, literally speaking, since human or accidental interference is almost inexhaustibly possible, and hard to exclude by specific qualification without tautology." The introduction of the causal field also automatically covers Scriven's third difficulty and third refinement, that of the contrast class and the relativity of causal statements to contexts.

[16] J. R. Lucas, "Causation," *Analytical Philosophy*, ed. R. J. Butler (Oxford, 1962), pp. 57–59, resolves this kind of difficulty by an informal appeal to what amounts to this notion of a causal field: ". . . these circumstances [cosmic cataclysms, etc.] . . . destroy the whole causal situation in which we had been looking for Z to appear . . . predictions are not expected to come true when quite unforeseen emergencies arise."

[17] This is related to Scriven's second difficulty, *op. cit.*, p. 409: "there still remains the problem of saying what the other factors are which, with the cause, make up the sufficient condition. If they can be stated, causal explanation is then simply a special case of subsumption under a law. If they cannot, the analysis is surely mythological." Scriven correctly replies that "a combination of the thesis of macro-determinism . . . and observation-plus-theory frequently gives us the very best of reasons for saying that a certain factor combines with an unknown sub-set of the conditions present into a sufficient condition for a particular effect." He gives a statistical example of such evidence, but the whole of my account of typical sorts of evidence for causal relationships in §§ 5 and 7 below is an expanded defence of a reply of this sort.

sufficient.[18] Similarly Scriven considers a case where

> . . . conditions (perhaps unusual excitement plus constitutional inadequacies) [are] present at 4.0 P.M. that guarantee a stroke at 4.55 P.M. and consequent death at 5.0 P.M.; but an entirely unrelated heart attack at 4.50 P.M. is still correctly called the cause of death, which, as it happens, does occur at 5.0. P.M.[19]

Before we try to resolve these difficulties let us consider another of Marc-Wogau's problems: Smith and Jones commit a crime, but if they had not done so the head of the criminal organization would have sent other members to perform it in their stead, and so it would have been committed anyway.[20] Now in this case, if 'A' stands for the actions of Smith and Jones, what we have is that AX is one minimal sufficient condition of the result (the crime), but $\bar{A}Z$ is another, and both X and Z are present. A combines with one set of the standing conditions to produce the result by one route: but the absence of A would have combined with another set of the standing conditions to produce the same result by another route. In this case we *can* say that A was a necessary condition *post factum*. This sample satisfies the requirements of Marc-Wogau's analysis, and of mine, of the statement that A caused this result; and this agrees with what we would ordinarily say in such a case. (We might indeed add that there was *also* a deeper cause—the existence of the criminal organization, perhaps—but this does not matter: our formal analyses do not insure that a particular result will have a unique cause, nor does our ordinary causal talk require this.) It is true that in this case we cannot say what will usually serve as an informal substitute for the formal account, that the cause, here A, was necessary (as well as sufficient) in the circumstances; for $\bar{A}$ would have done just as well. We cannot even say that A was nonredundant. But this shows merely that a formal analysis may be superior to its less formal counterparts.

Now in Scriven's example, we might take it that the heart attack prevented the stroke from occurring. If so, then the heart attack *is* a necessary condition *post factum*: it is a moment in the only minimal sufficient condition that was present in full, for the heart attack itself removed some factor

that was a necessary part of the minimal sufficient condition which has the excitement as one of its moments. This is strictly parallel to the Smith and Jones case. Again it is odd to say that the heart attack was in any way necessary, since the absence of the heart attack would have done just as well: this absence would have been a moment in that other minimal sufficient condition, one of whose other moments was the excitement. Nevertheless, the heart attack was necessary *post factum*, and the excitement was not. Scriven draws the distinction, quite correctly, in terms of continuity and discontinuity of causal chains: "the heart attack was, and the excitement was not the cause of death because the 'causal chain' between the latter and death was interrupted, while the former's 'went to completion'." But it is worth noting that a break in the causal chain corresponds to a failure to satisfy the logical requirements of a moment in a minimal sufficient condition that is also necessary *post factum*.

Alternatively, if the heart attack did not prevent the stroke, then we have a case parallel to that of the straw in the barn, or of the man who is shot by a firing squad, and two bullets go through his heart simultaneously. In such cases the requirements of my analysis, or of Marc-Wogau's, or of Scriven's, are not met: each proposed cause *is* redundant and not even necessary *post factum*, though the disjunction of them is necessary *post factum* and nonredundant. But this agrees very well with the fact that we *would* ordinarily hesitate to say, of either bullet, that it caused the man's death, or of either the lightning or the cigarette butt that it caused the fire, or of either the excitement or the heart attack that it was the cause of death. As Marc-Wogau says, "in such a situation as this we are unsure also how to use the word 'cause'." Our ordinary concept of cause does not deal clearly with cases of this sort, and we are free to decide whether or not to add to our ordinary use, and to the various more or less formal descriptions of it, rules which allow us to say that where more than one at-least-INUS-condition, and its conjunct conditions, are present, each of them caused the result.[21]

The account thus far developed of singular causal statements has been expressed in terms of

[18] *Op. cit.*, pp. 228–233.
[19] *Op. cit.*, pp. 410–411: this is Scriven's fourth difficulty and refinement.
[20] *Op. cit.*, p. 232: the example is taken from P. Gardiner, *The Nature of Historical Explanation* (Oxford, 1952), p. 101.
[21] Scriven's fifth difficulty and refinement are concerned with the direction of causation. This is considered briefly in § 8 below.

statements about necessity and sufficiency; it is therefore incomplete until we have added an account of necessity and sufficiency themselves. This question is considered in § 4 below. But the present account is independent of any particular analysis of necessity and sufficiency. Whatever analysis of these we finally adopt, we shall use it to complete the account of what it is to be an INUS condition, or to be at least an INUS condition. But in whatever way this account is completed, we can retain the general principle that at least part of what is often done by a singular causal statement is to pick out, as the cause, something that is claimed to be at least an INUS condition.

§ 3. GENERAL CAUSAL STATEMENTS

Many general causal statements are to be understood in a corresponding way. Suppose, for example, that an economist says that the restriction of credit causes (or produces) unemployment. Again, he will no doubt be speaking with reference to some causal field; this is now not an individual object, but a class, presumably economies of a certain general kind; perhaps their specification will include the feature that each economy of the kind in question contains a large private enterprise sector with free wage-earning employees. The result, unemployment, is something which sometimes occurs and sometimes does not occur within this field, and the same is true of the alleged cause, the restriction of credit. But the economist is not saying that (even in relation to this field) credit restriction is either necessary or sufficient for unemployment, let alone both necessary and sufficient. There may well be other circumstances which must be present along with credit restriction, in an economy of the kind referred to, if unemployment is to result; these other circumstances will no doubt include various negative ones, the absence of various counteracting causal factors which, if they were present, would prevent this result. Also, the economist will probably be quite prepared to admit that in an economy of this kind unemployment could be brought about by other combinations of circumstances in which the restriction of credit plays no part. So once again the claim that he is making is merely that the restriction of credit is, in economies of this kind, a nonredundant part of one sufficient condition for unemployment: that is, an INUS condition. The economist is probably assuming that there is some condition, no doubt a complex one, which is both necessary and

sufficient for unemployment in this field. This being assumed, what he is asserting is that, for some X and for some Y, $(AX$ or $Y)$ is a necessary and sufficient condition for P in F, but neither A nor X is sufficient on its own, where 'A' stands for the restriction of credit, 'P' for unemployment, and 'F' for the field, economies of such-and-such a sort. In a developed economic theory the field F may be specified quite exactly, and so may the relevant combinations of factors represented here by 'X' and 'Y'. (Indeed, the theory may go beyond statements in terms of necessity and sufficiency to ones of functional dependence, but this is a complication which I am leaving aside for the present.) In a preliminary or popular statement, on the other hand, the combinations of factors may either be only roughly indicated or be left quite undetermined. At one extreme we have the statement that $(AX$ or $Y)$ is a necessary and sufficient condition, where 'X' and 'Y' are given definite meanings; at the other extreme we have the merely existentially quantified statement that this holds for *some* pair X and Y. Our knowledge in such cases ordinarily falls somewhere between these two extremes. We can use the same convention as before, deliberately allowing it to be ambiguous between these different interpretations, and say that in any of these cases, where A is an INUS condition of P in F, $(A \ldots$ or $\ldots)$ is a necessary and sufficient condition of P in F.

A great deal of our ordinary causal knowledge is of this form. We know that the eating of sweets causes dental decay. Here the field is human beings who have some of their own teeth. We do not know, indeed it is not true, that the eating of sweets by any such person is a sufficient condition for dental decay: some people have peculiarly resistant teeth, and there are probably measures which, if taken along with the eating of sweets, would protect the eater's teeth from decay. All we know is that sweet-eating combined with a set of positive and negative factors which we can specify, if at all, only roughly and incompletely, constitutes a minimal sufficient condition for dental decay— but not a necessary one, for there are other combinations of factors, which do not include sweet-eating, which would also make teeth decay, but which we can specify, if at all, only roughly and incompletely. That is, if 'A' now represents sweet-eating, 'P' dental decay, and 'F' the class of human beings with some of their own teeth, we can say that, for some X and Y, $(AX$ or $Y)$ is necessary and sufficient for P in F, and we *may* be able to go beyond this merely existentially quantified statement to at least a partial specification of the X and Y in question. That is, we can say that (A ... or ...) is a necessary and sufficient condition, but that A itself is only an INUS condition. And the same holds for many general causal statements of the form "A causes (or produces) P." It is in this sense that the application of a potential difference to the ends of a copper wire produces an electric current in the wire; that a rise in the temperature of a piece of metal makes it expand; that moisture rusts steel; that exposure to various kinds of radiation causes cancer, and so on.

However, it is true that not all ordinary general causal statements are of this sort. Some of them are implicit statements of functional dependence. Functional dependence is a more complicated relationship of which necessity and sufficiency can be regarded as special cases. (It is briefly discussed in § 7 below.) Here too what we commonly single out as causing some result is only one of a number of factors which jointly affect the result. Again, some causal statements pick out something that is not only an INUS condition, but also a necessary condition. Thus we may say that the yellow fever virus is the cause of yellow fever. (This statement is not, as it might appear to be, tautologous, for the yellow fever virus and the disease itself can be independently specified.) In the field in question—human beings—the injection of this virus is not by itself a sufficient condition for this disease, for persons who have once recovered from yellow fever are thereafter immune to it, and other persons can be immunized against it. The injection of the virus, combined with the absence of immunity (natural or artificial), and perhaps combined with some other factors, constitutes a sufficient condition for the disease. Beside this, the injection of the virus is a necessary condition of the disease. If there is more than one complex sufficient condition for yellow fever, the injection of the virus into the patient's bloodstream (either by a mosquito or in some other way) is a factor included in every such sufficient condition. If 'A' stands for this factor, the necessary and sufficient condition has the form (A ... or A ... etc.), where A occurs in every disjunct. We sometimes note the difference between this and the standard case by using the phrase "the cause." We may say not merely that this virus causes yellow fever, but that it is the cause of yellow fever; but we would say only that sweet-eating causes dental decay, not

that it is the cause of dental decay. But about an individual case we could say that sweet-eating was the cause of the decay of this person's teeth, meaning (as in § 1 above) that the only sufficient condition present here was the one of which sweet-eating is a nonredundant part. Nevertheless, there will not in general be any one item which has a unique claim to be regarded as the cause even of an individual event, and even after the causal field has been determined. Each of the moments in the minimal sufficient condition, or in each minimal sufficient condition, that was present can equally be regarded as the cause. They may be distinguished as predisposing causes, triggering causes, and so on, but it is quite arbitrary to pick out as "main" and "secondary," different moments which are equally nonredundant items in a minimal sufficient condition, or which are moments in two minimal sufficient conditions each of which makes the other redundant.[22]

§ 4. NECESSITY AND SUFFICIENCY

One possible account of general statements of the forms "S is a necessary condition of T" and "S is a sufficient condition of T"—where 'S' and 'T' are general terms—is that they are equivalent to simple universal propositions. That is, the former is equivalent to "All T are S" and the latter to "All S are T." Similarly, "S is necessary for T in the field F" would be equivalent to "All FT are S," and "S is sufficient for T in the field F" to "All FS are T." Whether an account of this sort is adequate is, of course, a matter of dispute; but it is not disputed that these statements about necessary and sufficient conditions at least entail the corresponding universals. I shall work on the assumption that this account is adequate, that general statements of necessity and sufficiency are equivalent to universals: it will be worth while to see how far this account will take us, how far we are able, in terms of it, to understand how we use, support, and criticize these statements of necessity and sufficiency.

A directly analogous account of the corresponding singular statements is not satisfactory. Thus it will not do to say that "A short-circuit here was a necessary condition of a fire in this house" is equivalent to "All cases of this house's catching fire are cases of a short-circuit occurring here," because the latter is automatically true if this house has caught fire only once and a short-circuit has

22 Cf. Marc-Wogau's concluding remarks, op. cit., pp. 232–233.

occurred on that occasion, but this is not enough to establish the statement that the short-circuit was a necessary condition of the fire; and there would be an exactly parallel objection to a similar statement about a sufficient condition.

It is much more plausible to relate singular statements about necessity and sufficiency to certain kinds of non-material conditionals. Thus "A short-circuit here was a necessary condition of a fire in this house" is closely related to the counterfactual conditional "If a short-circuit had not occurred here this house would not have caught fire," and "A short-circuit here was a sufficient condition of a fire in this house" is closely related to what Goodman has called the factual conditional, "Since a short-circuit occurred here, this house caught fire."

However, a further account would still have to be given of these non-material conditionals themselves. I have argued elsewhere[23] that they are best considered as condensed or telescoped *arguments*, but that the statements used as premises in these arguments are no more than simple factual universals. To use the above-quoted counterfactual conditional is, in effect, to run through an incomplete argument: "Suppose that a short-circuit did not occur here, then the house did not catch fire." To use the factual conditional is, in effect, to run through a similar incomplete argument, "A short-circuit occurred here; therefore the house caught fire." In each case the argument might in principle be completed by the insertion of other premises which, together with the stated premise, would entail the stated conclusion. Such additional premises may be said to *sustain* the non-material conditional. It is an important point that someone can use a non-material conditional without completing or being able to complete the argument, without being prepared explicitly to assert premises that would sustain it, and similarly that we can understand such a conditional without knowing exactly how the argument would or could be completed. But to say that a short-circuit here was a necessary condition of a fire in this house is to say that there is some set of true propositions which would sustain the above-stated counterfactual, and to say that it was a sufficient condition is to say

that there is some set of true propositions which would sustain the above-stated factual conditional. If this is conceded, then the relating of singular statements about necessity and sufficiency to non-material conditionals leads back to the view that they refer indirectly to certain simple universal propositions. Thus if we said that a short-circuit here was a necessary condition for a fire in this house, we should be saying that there are true universal propositions from which, together with true statements about the characteristics of this house, and together with the supposition that a short-circuit did not occur here, it would follow that the house did not catch fire. From this we could infer the universal proposition which is the more obvious, but unsatisfactory, candidate for the analysis of this statement of necessity, "All cases of this house's catching fire are cases of a short-circuit occurring here," or, in our symbols, "All FP are A." We can use this to represent approximately the statement of necessity, on the understanding that it is to be a consequence of some set of wider universal propositions, and is not to be automatically true merely because there is only this one case of an FP, of this house's catching fire.[24] A statement that A was a sufficient condition may be similarly represented by "All FA are P." Correspondingly, if all that we want to say is that $(A\dots\text{ or }\dots)$ was necessary and sufficient for P in F, this will be represented approximately by the pair of universals "All FP are $(A\dots\text{ or }\dots)$ and all $F(A\dots\text{ or }\dots)$ are P," and more accurately by the statement that there is some set of wider universal propositions from which, together with true statements about the features of F, this pair of universals follows. This, therefore, is the fuller analysis of the claim that in a particular case A is an INUS condition of P in F, and hence of the singular statement that A caused P. (The statement that A is *at least* an INUS condition includes other alternatives, corresponding to cases where the necessary and sufficient condition is $(A\text{ or }\dots)$, $A\dots$, or A.)

Let us go back now to general statements of necessity and sufficiency and take F as a class, not as an individual. On the view that I am adopting, at least provisionally, the statement that Z is a necessary and sufficient condition for P in F is

[23] "Counterfactuals and Causal Laws," *Analytical Philosophy*, ed. R. J. Butler (Oxford, 1962), pp. 66–80.

[24] This restriction may be compared with one which Nagel imposes on laws of nature: "the vacuous truth of an unrestricted universal is not sufficient for counting it a law; it counts as a law only if there is a set of other assumed laws from which the universal is logically derivable" (*The Structure of Science* [New York, 1961], p. 60). It might have been better if he had added "or if there is some other way in which it is supported (ultimately) by empirical evidence." Cf. my remarks in "Counterfactuals and Causal Laws," *op. cit.*, pp. 72–74, 78–80.

equivalent to "All FP are Z and all $F\bar{Z}$ are $\bar{P}$." Similarly, if we cannot completely specify a necessary and sufficient condition for P in F, but can only say that the formula "$(A...$ or $...)$" represents such a condition, this is equivalent to the pair of incomplete universals, "All FP are $(A...$ or $...)$ and all F $(A...$ or $...)$ are P." In saying that our general causal statements often do no more than specify an INUS condition, I am therefore saying that much of our ordinary causal knowledge is knowledge of such pairs of incomplete universals, of what we may call elliptical or *gappy* causal laws.

§ 5. EVIDENCE FOR CAUSAL CONNECTIONS

If we assume that the general causal statement that A causes P, or the singular causal statement that A caused P, often makes the claims set out in §§ 1, 2, 3, and 4, including the claim that A is at least an INUS condition of P, then we can give an account of a combination of reasoning and observation which constitutes evidence for these causal statements.

This account is based on what von Wright calls a complex case[25] of the Method of Difference. Like any other method of eliminative induction, this can be formulated in terms of an assumption, an observation, and a conclusion which follows by a deductively valid argument from the assumption and the observation together. To get any positive conclusion by a process of elimination, we must assume that the result (the phenomenon a cause of which we are going to discover) has *some* cause in the sense that there is some condition the occurrence of which is both necessary and sufficient for the occurrence (as a rule, shortly afterwards) of the result. Also, if we are to get anywhere by elimination, we must assume that the range of possibly relevant causal factors, the items that might in some way constitute this necessary and sufficient condition, is restricted in some way. On the other hand, even if we had specified some such set of possibly relevant factors, it would in most cases be quite implausible to assume that the supposed necessary and sufficient condition is identical with just one of these factors on its own, and fortunately we have no need to do so. If we represent each possibly relevant factor as a single term, the natural assumption to make is merely

that the supposed necessary and sufficient condition will be represented by a formula which is constructed in some way out of some selection of these single terms, by means of negation, conjunction, and disjunction. However, any formula so constructed is equivalent to some formula in disjunctive normal form—that is, one in which negation, if it occurs, is applied only to single terms, and conjunction, if it occurs, only to single terms and/or negations of single terms. So we can assume without loss of generality that the formula of the supposed necessary and sufficient condition is in disjunctive normal form, that it is at most a disjunction of conjunctions in which each conjunct is a single term or the negation of one, that is, a formula such as "$(ABC$ or GH or $J)$." Summing this up, the assumption that we require will have this form:

For some Z, Z is a necessary and sufficient condition for the phenomenon P in the field F, that is, all FP are Z and all $F\bar{Z}$ are $\bar{P}$, and Z is a condition represented by some formula in disjunctive normal form all of whose constituents are taken from the range of possibly relevant factors A, B, C, D, E, etc.

Along with this assumption, we need an observation which has the form of the classical difference observation described by Mill. This we can formulate as follows:

There is an instance I_1, in which P occurs, and there is a negative case N_1, in which P does not occur, such that one of the possibly relevant factors (or the negation of one), say A, is present in I_1 and absent from N_1, but each of the other possibly relevant factors is either present in both I_1 and N_1 or absent both from I_1 and from N_1.

We can set out an example of such an observation as follows, using 'a' and 'p' to stand for "absent" and "present."

	P	A	B	C	D	E	
I_1	p	p	p	a	a	p	etc.
N_1	a	a	p	a	a	p	

Given the above-stated assumption, we can reason in the following way about any such observation:

[25] *A Treatise on Induction and Probability* (New York, 1951), pp. 90 ff. The account that I am here giving of the Method of Difference, and that I would give of the eliminative methods of induction in general, differs, however, in several respects from that of von Wright. An article on "Eliminative Methods of Induction," which sets out my account, is to appear in the *Encyclopedia of Philosophy*, edited by Paul Edwards, to be published by the Free Press of Glencoe, Collier-Macmillan.

Since P is absent from N_1, every sufficient condition for P is absent from N_1, and therefore every disjunct in Z is absent from N_1. Every disjunct in Z which does not contain A is therefore also absent from I_1. But since P is present in I_1, and Z is a necessary condition for P, Z is present in I_1. Therefore at least one disjunct in Z is present in I_1. Therefore at least one disjunct in Z contains A.

What this shows is that Z, the supposed necessary and sufficient condition for P in F, is either A itself, or a conjunction containing A, or a disjunction containing as a disjunct either A itself or a conjunction containing A. That is, Z has one of these four forms: A; $A...$; (A or ...); ($A...$ or ...). We can sum these up by saying that Z has the form $(A$- - - or - - -$)$, where the dashes indicate that these parts of the formula may or may not be filled in. This represents briefly the statement that A is at least an INUS condition. It follows also that if there are in the (unknown) formula which represents the complete necessary and sufficient condition any disjuncts not containing A, none of them was present as a whole in N_1 (but of course some of their component terms may have been present there), and also that in at least one of the disjuncts that contains A, the terms, if any, conjoined with A stand for factors (or negations of factors) that were present in I_1. This is all that follows from this single observation. But in general other observations will show that the dotted spaces do need to be filled in, and that A alone is neither sufficient nor necessary for P in F. We can then infer that the necessary and sufficient condition actually has the form ($A...$ or ...), and that A itself is only an INUS condition.

This analysis is so far merely formal, and we have still to consider whether such a method can be, or is, actually used, whether an assumption of the sort required can be justified and whether an observation of the sort required can ever be made. Even at this stage, however, it is worth noting that the Method of Difference does not require the utterly unrealistic sort of assumption used in what von Wright calls the simple case—namely, that the supposed necessary and sufficient condition is some single factor on its own—but that the much less restrictive assumption used here will still yield information when it is combined with nothing more than the classical difference observation. It is worth

noting also that the information thus obtained, though it falls far short of what von Wright calls absolutely perfect analogy, that is, of a full specification of a necessary and sufficient condition, is information of exactly the form that is implicit in our ordinary causal assertions, both singular and general.[26]

But can observations of the kind required be made? A preliminary answer is that the typical controlled experiment is an attempt to approximate to an observation of this sort. The experimental case corresponds to our I_1, the control case to our N_1, and the experimenter tries to insure that there will be no possibly relevant difference between these two except the one whose effect he is trying to determine, our A. Any differential outcome, present in the experimental case but not in the control case, is what he takes to be this effect, corresponding to our P.

The before-and-after observation is a particularly important variety of this kind. Suppose, for example, that we take a piece of blue litmus paper and dip it in a certain liquid, and it turns red. The situation before it is dipped provides the negative case N_1; the situation after it is dipped provides the instance I_1. As far as we can see, no other possibly relevant feature of the situation has changed, so that I_1 and N_1 are alike with regard to all possibly relevant factors except A, the paper's being dipped in a liquid of this sort, but the result P, the paper's turning red, is present in I_1 but not in N_1. We can take this in either of two ways. First, we may take the field F to be pieces of blue litmus paper, and if we assume that in this field there is some necessary and sufficient condition for P, made up in some way from some selection from the factors we are considering as possibly relevant, we can conclude that (A- - - or - - -) is necessary and sufficient for P in F. Other observations may show that A alone is neither necessary nor sufficient, and hence that the necessary and sufficient condition is ($A...$ or ...). Thus we can establish the gappy causal law, "All FP are ($A...$ or ...) and all $F(A...$ or ...) are P." This amounts to the assertion that in some circumstances being dipped in a liquid of this sort turns blue litmus paper red. Secondly, we can take the field (which we shall here call F_1) to be this particular piece of paper, and what the experiment then establishes

[26] What is established by the present method may be compared with the four claims listed in § 1 above, that A is at least an INUS condition, that A was present on the occasion in question, that the factors represented by 'X'—that is, the other moments in at least one minimal sufficient condition in which A is a moment—were present, and that every disjunct in Y which does not contain A—that is, every minimal sufficient condition which does not contain A—was absent.

is the singular causal statement that on this particular occasion the dipping in this liquid turned this piece of paper red. This is established in accordance with the analysis of singular causal statements completed in § 4. For the experiment, together with the assumption, has established the wider universals indicated by the above-stated gappy causal law. It has shown that for some X and Y all FP are $(AX$ or $Y)$ and all $F(AX$ or $Y)$ are P, and from these, since F_1 is an F (that is, this piece of paper is a piece of blue litmus paper), it follows that for some X and Y all F_1P are $(AX$ or $Y)$ and all $F_1(AX$ or $Y)$ are P. Also, 'X' represents circumstances which were present on this occasion, and 'Y' circumstances which were not present in N_1, the "before" situation. That is to say, the observation, together with the appropriate assumption, entails that there are true propositions which sustain the counterfactual and factual conditionals, "If, in the circumstances, this paper had not been dipped in this liquid it would not have turned red, but since it was dipped it did turn red"; but it does not fully determine what these propositions are, it does not fill in the gaps in the causal laws which sustain these conditionals. The importance of this is that it shows how an observation can reveal not merely a sequence but a causal sequence: what we discover is not merely that the litmus paper was dipped *and then* turned red, but that the dipping *made* it turn red.

It is worth noting that despite the stress traditionally laid, in accounts of the Method of Difference, on the requirement that there should be only *one* point of difference between I_1 and N_1, very little really turns upon this. For suppose that two of our possibly relevant factors, say A and B, were both present in I_1 and both absent from N_1, but that each of the other possibly relevant factors was either present in both or absent from both. Then reasoning parallel to that given above will show that at least one of the disjuncts in Z either contains A or contains B (and may contain both). That is, this observation still serves to show that the cluster of factors (A, B) *contains* something that is at least an INUS condition of P in F, whether this condition turns out in the end to be A alone, or B alone, or the conjunction AB, or the disjunction $(A$ or $B)$. And similar considerations apply if there are more than two points of difference between I_1 and N_1. However many there are, an observation of this form, coupled with our assumption, shows that a cause in our sense (in general an INUS condition) lies somewhere within the cluster of terms,

positive or negative, in respect of which I_1 differs from N_1. (Note that it does *not* show that the other terms, those common to I_1 and N_1, are causally irrelevant; our reasoning does not exclude factors as irrelevant, but positively locates some of the relevant factors within the differentiating cluster.)

This fact rebuts the criticism sometimes leveled against the eliminative methods that they presuppose and require a finally satisfactory analysis of causal factors into their simple components, which we never actually achieve. On the contrary, any distinction of factors, however rough, enables us to start using such a method. We can proceed, and there is no doubt that discovery has often proceeded, by what we may call *the progressive localization of a cause*. Using the Method of Difference in a very rough way, we can discover first, say, that the drinking of wine causes intoxication. That is, the cluster of factors which is crudely summed up in the single term "the drinking of wine" contains somewhere within it an INUS condition of intoxication; and we can subsequently go on to distinguish various possibly relevant factors within this cluster, and by further observations of the same sort locate a cause of intoxication more precisely. In a context in which this cluster is either introduced or excluded as a whole, it is correct to say that the introduction of this cluster was non-redundant or necessary *post factum*, and experiments can establish this, even if, in a different context, in which distinct items in the cluster are introduced or excluded separately, it would be correct to say that only one item, the alcohol, was nonredundant or necessary *post factum*, and this could be established by more exact experimentation.

One merit of this formal analysis is that it shows in what sense a method of eliminative induction, such as the Method of Difference, rests upon a deterministic principle or presupposes the uniformity of nature. In fact, each application of this method requires an assumption which in one respect says much less than this, in another a little more. No sweeping general assumption is needed: we need not assume that every event has a cause, but merely that for events of the kind in question, P, in the field in question, F, there is some necessary and sufficient condition. But—and this is where we need something more than determinism or uniformity in general—we must also assume that this condition is constituted in some way by some selection from a restricted range of possibly relevant factors.

It is this further assumption that raises a doubt

about the use of this method to make causal discoveries. As for the mere deterministic assumption that the phenomenon in question has some necessary and sufficient condition, we may be content to say that this is one which we simply do make in all inquiries of this kind, and leave its justification to be provided by whatever solution we can eventually find for the general problem of induction. But the choice of a range of possibly relevant factors cannot be brushed aside so easily. Also, the wider a range of possibly relevant factors we admit, the harder it will be to defend the claim that I_1 and N_1 are observed to be alike with respect to all the possibly relevant factors except the one, or the indicated cluster of factors, in which they are observed to differ. Alternatively, the more narrowly the range of possibly relevant factors is restricted, the easier it will be to defend the claim that we have made an observation of the required form, but at the same time the less plausible will be our assumption be.

However, this difficulty becomes less formidable if we consider the assumption and the observation together. We want to be able to say that there is no possibly relevant difference, other than the one (or ones) 'noted, between I_1 and N_1. We need not draw up a complete list of possibly relevant factors before we make the observation. In practice we usually assume that a causally relevant factor will be in the spatial neighborhood of the instance of the field in or to which the effect occurs in I_1, or fails to occur in N_1, and it will either occur shortly before or persist throughout the time at which the effect occurs in I_1, or might have occurred, but did not, in N_1. No doubt in a more advanced application of the Method of Difference within an already-developed body of causal knowledge we

can restrict the range of possibly relevant factors much more narrowly and can take deliberate steps to exclude interferences from our experiments; but I am suggesting that even our most elementary and primitive causal knowledge rests upon implicit applications of this method, and the spatio-temporal method of restricting possibly relevant factors is the only one initially available. And perhaps it is all we need. Certainly in terms of it the observer could say, about the litmus paper, for example, "I cannot see any difference, other than the dipping into this liquid, between the situation in which the paper turned red and that in which it did not, that might be relevant to this change."

It may be instructive to compare the Method of Difference as a logical ideal with any actual application of it. If the assumption and the observation were known to be true, then the causal conclusion would be established. Consequently, anything that tells in favor of both the assumption and the observation tells equally in favor of the causal conclusion. No doubt we are never in a position to say that they are known to be true, and therefore that the conclusion is established; but we are often in a position to say that, given the deterministic part of the assumption, we cannot see any respect in which they are not true (since we cannot see any difference that might be relevant between I_1 and N_1), and consequently that we cannot see any escape from the causal conclusion. In this sense at least we can say that an application of this method confirms a causal conclusion: the observer has looked for but failed to find an escape from this conclusion.[27]

In practice we do not rely as much on single observations as this account might suggest. We assure ourselves that it was the dipping in this

[27] An account of how eliminative inductive reasoning supports causal conclusions is given by J. R. Lucas in the article cited in n. 16 above. His account differs from mine in many details, but agrees with it in general outline. Contrast with this the remarks of von Wright, *op. cit.*, p. 135: ". . . in normal scientific practice we have to reckon with plurality rather than singularity, and with complexity rather than simplicity of conditions. This means that the weaker form of the Deterministic Postulate, or the form which may be viewed as a reasonable approximation to what is commonly known as the Law of Universal Causation, is practically useless as a supplementary premiss or 'presupposition' of induction." I hope I have shown that this last remark is misleading.

It has been argued by A. Michotte (*La perception de la causalité* [Louvain, 1946], translated by T. R. and E. Miles as *The Perception of Causality* [London, 1963]) that we have in certain cases an immediate perception or impression of causation. His two basic experimental cases are these. In one, an object A approaches another object B; on reaching B, A stops and B begins to move off in the same direction; here the observer gets the impression that A has "launched" B, has set B in motion. In the other case, A continues to move on reaching B, and B moves at the same speed and in the same direction; here the observer gets the impression that A is carrying B with it. In both cases observers typically report that A has caused the movement of B. Michotte argues that it is an essential feature of observations that give rise to this causal impression that there should be two distinguishable movements, that of the "agent" A and that of the "patient" B, but also that it is essential that the movement of the patient should in some degree copy or duplicate that of the agent.

This would appear to be a radically different account of the way in which we can detect causation by observing a single sequence, for on Michotte's view our awareness of causation can be direct, perceptual, and non-inferential. It must be conceded that not only spatio-temporal continuity, but also qualitative continuity between cause and effect (*l'ampliation du* liquid that turned the litmus paper red by dipping other pieces of litmus paper and seeing them, too, turn red just after they are dipped. This repetition is effective because it serves as a check on the possibility that some other relevant change might have occurred, unnoticed, just at the moment when the first piece of litmus paper was dipped in the liquid. After a few trials it will be most likely that any other relevant change has kept on occurring just as each piece was dipped (or even that there has been a succession of different relevant changes at the right times). Of course, it may be that there is some other relevant change (or set of relevant changes) which keeps on occurring just as each paper is dipped because it is linked with the dipping by what Mill calls "some fact of causation."[28] If so, then this other relevant change may be regarded as part of a cluster of factors which can be grouped together under the title "the dipping of the paper in this liquid," taking this in a broad sense, as possibly including items other than the actual entry of the paper into the liquid. But if this is not so, then it would be a sheer coincidence if this other relevant change kept on occurring just as each piece of paper was dipped, or if there was a succession of relevant changes at the right times. The hypothesis that such co-incidences have continued will soon become implausible, even if it cannot be conclusively falsified.[29] It is an important point that it is not the repetition as such that supports the conclusion that the dipping causes the turning red, but the repetition of a sequence which, on each single occasion, is already *prima facie* a causal one. The repetition tends to disconfirm the set of hypotheses each of which explains a single sequence of a dipping followed by a turning red as a mere

coincidence, and by contrast it confirms the hypothesis that in each such single sequence the dipping is causally connected with the change of color.

The analysis offered here of the Method of Difference has this curious consequence: in employing this method we are liable to use the word "cause" in different senses at different stages. In the assumption, it is said that the phenomenon P has some "cause," meaning some necessary and sufficient condition; but the "cause" actually found—A in our formal example—may be only an inus condition. But we do need to assume that *something* is both necessary and sufficient for P in F to be able to conclude that A is at least an inus condition, that it is a moment in a minimal sufficient condition that was present, and that it was necessary *post factum*.

§ 6. FALSIFICATION OF INCOMPLETE STATEMENTS

A possible objection to this account is that the gappy laws and singular statements used here are so incomplete that they are internally guaranteed against falsification and are therefore not genuine scientific statements at all. However, it is not a satisfactory criterion of a scientific statement that it should be exposed to conclusive falsification: what is important is that to treat a statement as a scientific hypothesis involves handling it in such a way that evidence would be allowed to tell against it. And there are ways in which evidence can be, and is, allowed to tell against a statement which asserts that something is an inus condition.

Suppose, for example, that by using the I_1 and N_1 set out in § 5 above we have concluded that A is at least an inus condition of P—taking this both as a singular causal statement about an individual

mouvement), are important ingredients in the primitive concept of causation; they may contribute to the notion of causal "necessity"; and both these continuities can sometimes be directly perceived. But it is equally clear that these continuities are not in general required either as observed or as postulated features of a causal sequence, and that a sequence which has these continuities may fail to be causal. What is perceived in Michotte's examples is neither necessary nor sufficient for causal relationship as we now understand it, though it may have played an important part in the genesis of the causal concept. It is worth noting that these examples also exhibit the features stressed in my account. They present the observer with an apparently simple and isolated causal field, within which there occurs a marked change, B's beginning to move. The approach of A is the only observed possibly relevant difference between the times when B is stationary and when B begins to move. If B's beginning to move has a cause, then A's approach is a suitable candidate, and nothing else that the observer is allowed to see or encouraged to suspect is so. Thus these examples could *also* give rise to an inferential awareness of causation, though it is true that other examples which would do this equally well would fail, and in Michotte's experiments do fail, to produce a direct impression of causation.

[28] E.g., in the Fifth Canon, *A System of Logic*, Book III, Chapter VIII, § 6.

[29] Cf. J. R. Lucas, *op. cit.*, p. 53: "It might be that two quite independent processes were going on, and we were getting constant concomitance for no reason except the chance fact that the two processes happened to keep in step. If this be so, an arbitrary disturbance in one will reveal the independence of the other. If an arbitrary disturbance in the one is followed by a corresponding alteration in the other, it always could be that it was a genuine coincidence. . . . But to argue this persistently is to make the same illicit extension of 'coincidence' as some phenomenalists do of 'illusion'. . . . It is no longer a practical possibility that we are eliminating but a Cartesian doubt."

field F_1 and as an incomplete law about the general field F. Now suppose that closer examination shows that some other factor, previously unnoticed, say K, was present in I_1 and absent from N_1, and that we also discover (or construct experimentally) further cases I_2 and N_2, such that the observational evidence is now of this form:

	P	A	B	C	D	E	...K	...
I_1	p	p	p	a	a	p	...p	...
N_1	a	a	p	a	a	p	...a	...
I_2	p	a	p	a	a	p	...p	...
N_2	a	p	p	a	a	p	...a	...

Here N_2 shows that for any X which does not contain K, AX is not sufficient: so X must contain K. But any X that contains K is present in I_2, and *may* therefore be sufficient for P on its own, without A. This evidence does not conclusively falsify the hypothesis that A is an INUS condition as stated above, but it takes away all the reason that the previous evidence gave us for this conclusion. Observations of this pattern would tell against this conclusion, and would lead us to replace the view that A causes P, and caused P in I_1, with the view that K causes P, and caused P both in I_2 and in I_1, with A not even forming an indispensable part of the sufficient condition which was present in I_1. (A fuller treatment of this kind of additional evidence would require accounts of the Method of Agreement and of the Joint Method, parallel to that of the Method of Difference given in § 5.)

It remains true that some of the claims made by singular causal statements and by causal laws as here analyzed—that is, claims that some factor is at least an INUS condition of the effect—are not conclusively falsifiable. But ordinary causal laws and singular causal statements are not conclusively falsifiable, as direct consideration will show. It is a merit of the account offered here, not a difficulty for it, that it reproduces this feature of ordinary causal knowledge.[30]

§ 7. FUNCTIONAL DEPENDENCE AND CONCOMITANT VARIATION

As I mentioned in § 3, causal statements sometimes refer not to relations of necessity and sufficiency, nor to any more complex relations based on these, like that of being an INUS condition, but to relations of functional dependence. That is, the effect and the possible causal factors are things which can vary in magnitude, and the cause of

[30] This was pointed out by D. C. Stove.

some effect P is that on whose magnitude the magnitude of P functionally depends. But causal statements of this sort can be expanded and analyzed in an account parallel to that which we have given of causal statements of the previous kinds. Again we speak of a field, individual or general, in relation to which a certain functional dependence holds. Also, we can speak of the *total cause*, the complete set of factors on whose magnitude the magnitude of P, given the field F, wholly depends: that is, variations of P in F are completely covered by a formula which is a function of the magnitudes of all of the factors in this "complete set," and of these alone. This total cause is analogous to a necessary and sufficient condition. It can be distinguished from each of the factors that compose it, each of which is causally relevant to the effect, but it is not the whole cause of its variations: each of these *partial causes* is analogous to an INUS condition.

The problem of finding a cause in this new sense would require, for its full solution, the completion of two tasks. We should have both to identify all the factors in this total cause, and also to discover in what way the effect depends upon them—that is, to discover the law of functional dependence of the effect on the total cause, or the partial differential equations relating it to each of the partial causes. The first—but only the first—of these two tasks can be performed by what is really the Method of Concomitant Variation, developed in a style analogous to that in which the Method of Difference was developed in § 5. That is, we assume that there is something on which the magnitude of P in F functionally depends, and that there is a restricted set of possibly relevant factors; then if while all other possibly relevant factors are held constant one factor, say A, varies and P also varies, it follows that A is at least a partial cause, that it is one of the actually relevant factors. It is this relationship that is commonly asserted by statements of such forms as "A affects P" and "On this occasion A affected P." Some of our causal statements, singular or general, have just this force, and all that I am trying to show here is that these statements can be supported by reasoning along the lines of the Method of Concomitant Variation, developed analogously with the development in § 5 of the Method of Difference. Just as we there assumed that there was some necessary and sufficient condition, and by combining this assumption with our observations discovered something which is at least an INUS condition, so we here assume that there is some total cause and so discover something which is at least a partial cause. However, a complete account of the Method of Concomitant Variation would involve the examination of several other cases besides the one sketched here.[31] For our present purpose, we need note only that there is this functional dependence part of the concept of causation as well as the presence-or-absence part, indeed that the latter can be considered as a special limiting case of the former,[32] but that the two parts are systematically analogous to one another, and that our knowledge of both singular and general causal relationships of these two kinds can be accounted for on corresponding principles.

§ 8. THE DIRECTION OF CAUSATION

This account of causation is still incomplete, in that nothing has yet been said about the direction of causation, about what distinguishes A causing P from P causing A. This is a difficult question, and it is linked with the equally difficult question of the direction of time. I cannot hope to resolve it completely here, but I shall state some of the relevant considerations.[33]

First, it seems that there is a relation which may be called *causal priority*, and that part of what is meant by "A caused P" is that this relation holds in one direction between A and P, not the other. Secondly, this relation is not identical with temporal priority; it is conceivable that there should be evidence for a case of backward causation, for A being causally prior to P whereas P was temporally prior to A. Most of us believe, and I think with good reason, that backward causation does not occur, so that we can and do normally use temporal order to limit the possibilities about causal order; but the connection between the two is synthetic. Thirdly, it could be objected to the analysis of "necessary" and "sufficient" offered in § 4 above that it omits any reference to causal order, whereas our most common use of "necessary" and "sufficient" in causal contexts includes such a reference. Thus "A is (causally) sufficient for B" says "If A, then B, and A is causally prior to B," but "B is (causally) necessary for A" is not equivalent to

this: it says "If A, then B, and B is causally prior to A." However, it is simpler to use "necessary" and "sufficient" in senses which exclude this causal priority, and to introduce the assertion of priority separately into our accounts of "A caused P" and "A causes P." Fourthly, although "A is (at least) an INUS condition of P" is not synonymous with "P is (at least) an INUS condition of A," this difference of meaning cannot exhaust the relation of causal priority. If it did exhaust it, the direction of causation would be a trivial matter, for, given that there is some necessary and sufficient condition of A in the field, it can be proved that if A is (at least) an INUS condition of P, then P is also (at least) an INUS condition of A: we can construct a minimal sufficient condition of A in which P is a moment.[34]

Fifthly, it is often suggested that the direction of causation is linked with controllability. If there is a causal relation between A and B, and we can control A without making use of B to do so, and the relation between A and B still holds, then we decide that B is not causally prior to A and, in general, that A is causally prior to B. But this means only that if one case of causal priority is known, we can use it to determine others: our rejection of the possibility that B is causally prior to A rests on our knowledge that our action is causally prior to A, and the question how we know the latter, and even the question of what causal priority is, have still to be answered. Similarly, if one of the causally related kinds of event, say A, can be randomized, so that occurrences of A are either not caused at all, or are caused by something which enters this causal field *only* in this way, by causing A, we can reject both the possibility that B is causally prior to A and the possibility that some common cause is prior both to A and separately to B, and we can again conclude that A is causally prior to B. But this still means only that we can infer causal priority in one place if we first know that it is absent from another place. It is true that our knowledge of the direction of causation in ordinary cases is thus based on what we find to be controllable, and on what we either find to be random or find that we can randomize; but this cannot without circularity be taken as providing a full account either of what we mean

---

[31] I have given a fuller account of this method in the article cited in n. 25.

[32] Cf. J. R. Lucas, *op. cit.*, p. 65.

[33] As was mentioned in n. 21, Scriven's fifth difficulty and refinement are concerned with this point (*op. cit.*, pp. 411–412), but his answer seems to me inadequate. Lucas touches on it (*op. cit.*, pp. 51–53). The problem of temporal asymmetry is discussed, e.g., by J. J. C. Smart, *Philosophy and Scientific Realism* (London, 1963), pp. 142–148, and by A. Grünbaum in the article cited in n. 36 below.

[34] I am indebted to one of the referees for correcting an inaccurate statement on this point in an earlier version.

B

by causal priority or of how we know about it.

A suggestion put forward by Popper about the direction of time seems to be relevant here.[35] If a stone is dropped into a pool, the entry of the stone will explain the expanding circular waves. But the reverse process, with contracting circular waves, "would demand a vast number of distant coherent generators of waves the coherence of which, to be explicable, would have to be shown . . . as originating from one centre." That is, if B is an occurrence which involves a certain sort of "coherence" between a large number of separated items, whereas A is a single event, and A and B are causally connected, A will explain B in a way in which B will not explain A unless some other single event, say C, first explains the coherence in B. Such examples give us a *direction of explanation*, and it may be that this is the basis, or part of the basis, of the relation I have called causal priority.

§ 9. CONCLUSIONS

Even if Mill was wrong in thinking that science consists mainly of causal knowledge, it can hardly be denied that such knowledge is an indispensable element in science, and that it is worth while to investigate the meaning of causal statements and the ways in which we can arrive at causal knowledge. General causal relationships are among the items which a more advanced kind of scientific theory explains, and is confirmed by its success in explaining. Singular causal assertions are involved in almost every report of an experiment: doing such and such *produced* such and such an effect. Materials are commonly identified by their causal properties: to recognize something as a piece of a certain material, therefore, we must establish singular causal assertions about it, that this object affected that other one, or was affected by it, in such and such a way. Causal assertions are embedded in both the results and the procedures of scientific investigation.

The account that I have offered of the force of various kinds of causal statements agrees both with our informal understanding of them and with

accounts put forward by other writers: at the same time it is formal enough to show how such statements can be supported by observations and experiments, and thus to throw a new light on philosophical questions about the nature of causation and causal explanation and the status of causal knowledge.

One important point is that, leaving aside the question of the direction of causation, the analysis has been given entirely within the limits of what can still be called a regularity theory of causation, in that the causal laws involved in it are no more than straightforward universal propositions, although their terms may be complex and perhaps incompletely specified. Despite this limitation, I have been able to give an account of the meaning of statements about singular causal sequences, regardless of whether such a sequence is or is not of a kind that frequently recurs: repetition is not essential for causal relation, and regularity does not here disappear into the mere fact that this single sequence has occurred. It has, indeed, often been recognized that the regularity theory could cope with single sequences if, say, a unique sequence could be explained as the resultant of a number of laws each of which was exemplified in many other sequences; but my account shows how a singular causal statement can be interpreted, and how the corresponding sequence can be shown to be causal, even if the corresponding complete laws are not known. It shows how even a unique sequence can be directly recognized as causal.

One consequence of this is that it now becomes possible to reconcile what have appeared to be conflicting views about the nature of historical explanation. We are accustomed to contrast the "covering-law" theory adopted by Hempel, Popper, and others with the views of such critics as Dray and Scriven who have argued that explanations and causal statements in history cannot be thus assimilated to the patterns accepted in the physical sciences.[36] But while my basic analysis of singular causal statements in §§ 1 and 2 agrees closely with Scriven's, I have argued in § 4 that this analysis can be developed in terms of complex

[35] "The Arrow of Time," *Nature*, vol. 177 (1956), p. 538; also vol. 178, p. 382 and vol. 179, p. 1297.
[36] See, for example, C. G. Hempel, "The Function of General Laws in History," *Journal of Philosophy*, vol. 39 (1942), reprinted in *Readings in Philosophical Analysis*, ed. by H. Feigl and W. Sellars (New York, 1949), pp. 459–471; C. G. Hempel and P. Oppenheim, "Studies in the Logic of Explanation," *Philosophy of Science*, vol. 15 (1948), reprinted in *Readings in the Philosophy of Science*, ed. by H. Feigl and M. Brodbeck (New York, 1953), pp. 319–352; K. R. Popper, *Logik der Forschung* (Vienna, 1934), translation *The Logic of Scientific Discovery* (London, 1959), pp. 59–60, also *The Open Society* (London, 1952), vol. II, p. 262; W. Dray, *Laws and Explanation in History* (Oxford, 1957); N. Rescher, "On Prediction and Explanation," *British Journal for the Philosophy of Science*, vol. 9 (1958), pp. 281–290; various papers in *Minnesota Studies in the Philosophy of Science*, vol. III, ed. by H. Feigl and G. Maxwell (Minneapolis, 1962); A. Grünbaum, "Temporally-asymmetric Principles,

and elliptical universal propositions, and this means that wherever we have a singular causal statement we shall still have a covering law, albeit a complex and perhaps elliptical one. Also, I have shown in § 5, and indicated briefly, for the functional dependence variants, in § 7, that the evidence which supports singular causal statements also supports general causal statements or covering laws, though again only complex and elliptical ones. Hempel recognized long ago that historical accounts can be interpreted as giving incomplete "explanation sketches," rather than what he would regard as full explanations, which would require fully-stated covering laws, and that such sketches are also common outside history. But in these terms what I am saying is that explanation sketches and the related elliptical laws are often all that we can discover, that they play a part in all sciences, that they can be supported and even established without being completed, and do not serve merely as preliminaries to or summaries of complete deductive explanations. If we modify the notion of a covering law to admit laws which not only are complex but also are known only in an elliptical form, the covering-law theory can accommodate many of the points that have been made in criticism of it, while preserving the structural similarity of explanation in history and in the physical sciences. In this controversy, one point

at issue has been the symmetry of explanation and prediction, and my account may help to resolve this dispute. It shows, in agreement with what Scriven has argued, how the actual occurrence of an event in the observed circumstances—the I_1 of my formal account in § 5—may be a vital part of the evidence which supports an explanation of that event, which shows that it was A that caused P on this occasion. A prediction on the other hand cannot rest on observation of the event predicted. Also, the gappy law which is sufficient for an explanation will not suffice for a prediction (or for a retrodiction): a statement of initial conditions together with a gappy law will not entail the assertion that a specific result will occur, though of course such a law may be, and often is, used to make tentative predictions the failure of which will not necessarily tell against the law. But the recognition of these differences between prediction and explanation does not affect the covering-law theory as modified by the recognition of elliptical laws.

Although what I have given is primarily an account of physical causation, it may be indirectly relevant to the understanding of human action and mental causation. It is sometimes suggested that our ability to recognize a single occurrence as an instance of mental causation is a feature which distinguishes mental causation from physical or "Humean" causation.[37] But this suggestion arises

---

Parity between Explanation and Prediction, and Mechanism versus Teleology," *Philosophy of Science*, vol. 29 (1962), pp. 146–170.

Dray's criticisms of the covering-law theory include the following: we cannot state the law used in an historical explanation without making it so vague as to be vacuous (*op. cit.*, especially pp. 24–37) or so complex that it covers only a single case and is trivial on that account (p. 39); the historian does not come to the task of explaining an event with a sufficient stock of laws already formulated and empirically validated (pp. 42–43); historians do not need to replace judgment about particular cases with deduction from empirically validated laws (pp. 51–52). It will be clear that my account resolves each of these difficulties. Grünbaum draws an important distinction between (1) an asymmetry between explanation and prediction with regard to the grounds on which we claim to know that the explanandum is true, and (2) an asymmetry with respect to the logical relation between the explanans and the explanandum; he thinks that only the former sort of asymmetry obtains. I suggest that my account of the use of gappy laws will clarify both the sense in which Grünbaum is right (since an explanation and a tentative prediction can use similarly gappy laws which are similarly related to the known initial conditions and the result) and the sense in which, in such a case, we may contrast an entirely satisfactory explanation with a merely tentative prediction. Scriven (in his most recent statement, the review cited in n. 10 above) says that "we often pin down a factor as a cause by excluding other possible causes. Simple—but disastrous for the covering-law theory of explanation, because we can eliminate causes only for something *we know has occurred*. And if the grounds for our explanation of an event *have* to include knowledge of that event's occurrence, they cannot be used (without circularity) to predict the occurrence of that event" (p. 414). That is, the observation of this event in these circumstances may be a vital part of the evidence that justifies the particular causal explanation that we give of this event: it may itself go a long way toward establishing the elliptical law in relation to which we explain it (as I have shown in § 5), whereas a law used for prediction cannot thus rest on the observation of the event predicted. But as my account also shows, this does not introduce an asymmetry of Grünbaum's second sort, and is therefore not disastrous for the covering-law theory.

[37] See, for example, G. E. M. Anscombe, *Intention* (Oxford, 1957), especially p. 16; J. Teichmann, "Mental Cause and Effect," *Mind*, vol. 70 (1961), pp. 36–52. Teichmann speaks (p. 36) of "the difference between them and ordinary (or 'Humian') sequences of cause and effect" and says (p. 37) "it is sometimes in order for the person who blinks to say absolutely dogmatically that the cause is such-and-such, and to say this independently of his knowledge of any previously established correlations," and again "if the noise is a cause it seems to be one which is known to be such in a special way. It seems that while it is necessary for an observer to have knowledge of a previously established correlation between noises and Smith's jumpings, before he can assert that one causes the other, it is not necessary for Smith himself to have such knowledge."

from the use of too simple a regularity account of physical causation. If we first see clearly what we mean by singular causal statements in general, and how we can support such a statement by observation of the single sequence itself, even in a physical case, we shall be better able to contrast with this our awareness of mental causes, and to see whether the latter has any really distinctive features.

This account also throws light on both the form and the status of the "causal principle," the deterministic assumption which is used in any application of the methods of eliminative induction. These methods need not presuppose determinism in general, but only that each specific phenomenon investigated by such a method is deterministic. Moreover, they require not only that the phenomenon should have some cause, but that there should be some restriction of the range of possibly relevant factors (at least to spatio-temporally neighboring ones, as explained in § 5). Now the general causal principle, that every event has some

cause, is so general that it is peculiarly difficult either to confirm or to disconfirm, and we might be tempted either to claim for it some *a priori* status, to turn it into a metaphysical absolute presupposition, or to dismiss it as vacuous. But the specific assumption that this phenomenon has some cause based somehow on factors drawn from this range, or even that this phenomenon has some neighboring cause, is much more open to empirical confirmation and disconfirmation: indeed the former can be conclusively falsified by the observation of a positive instance I_1 of P, and a negative case N_1 in which P does not occur, but where each of the factors in the given range is either present in both I_1 and N_1 or absent from both. This account, then, encourages us to regard the assumption as something to be empirically confirmed or disconfirmed. At the same time it shows that there must be some principle of the confirmation of hypotheses other than the eliminative methods themselves, since each such method rests on an empirical assumption.

University of York

THE JOURNAL OF PHILOSOPHY

VOLUME LXX, NO. 8, APRIL 26, 1973

CAUSATION, NOMIC SUBSUMPTION, AND THE
CONCEPT OF EVENT *

IN his celebrated discussion of causation Hume identified four
prima facie constituents in the relation of causation. As everyone knows, they are constant conjunction, contiguity in space
and time, temporal priority, and necessary connection. As ordinarily
understood, the causal relation is a binary relation relating causes
to their effects, and so presumably are the four relations Hume dis-
cerns in it. But what do these four relations tell us about the nature
of the entities they relate?

Constant conjunction is a relation between generic events, that is,
kinds or types of events; constant conjunction makes no clear or
nontrivial sense when directly applied to spatiotemporally bounded
individual events.[1] On the other hand, it is clear that the relation
of temporal priority calls for individual, rather than generic, events
as its relata; there appears to be no useful way of construing 'earlier
than' as a relation between kinds or classes of events in the causal
context.

What of the condition of contiguity? This condition has two parts,
temporal and spatial. Temporal contiguity makes sense when applied to events; two events are contiguous in time if they temporally
overlap. But spatial contiguity makes best sense when applied not
to events but to objects, especially material bodies; intuitively at
least, we surely understand what it is for two bodies to be in contact
or to overlap. For events, however, the very notion of spatial location often becomes fuzzy and indeterminate. When Socrates expired
in the prison, Xantippe became a widow and their three sons became
fatherless. Exactly *where* did these latter events take place? When

---

* I am indebted to Richard Brandt, Alvin and Holly Goldman, and Ernest Sosa
for helpful suggestions.
[1] By 'event' simpliciter I always mean individual events; when I mean generic
events I shall say so.

Hume's two billiard balls collide, what obviously are in spatial contact are the two balls. Are the motions of the balls also in spatial contact? Reflections on these and other cases suggest that the locations of events, and hence their spatial contiguity relations, are parasitic in some intricate ways on the locations of objects.[2] As for the controversial idea of necessary connection, we are clearly more at home with this notion taken in the de dicto sense as applying to sentences, propositions, and the like, than when it is taken in the de re sense as applying directly to objects and events in the world.

Hume's four conditions, therefore, seem at first blush to call for apparently different categories of entities as relata of causal relations. We might say that the four conditions are jointly incongruous ontologically, thereby rendering the causal relation ontologically incoherent. I do not intend these remarks as criticisms of the historical Hume; I am merely pointing up the need for a greater sensitivity to ontological issues in the analysis of causation.

In this paper I want to examine some logical and ontological problems that arise when we try to give a precise characterization of Humean causation.[3] (I call "Humean" any concept of causation that includes the idea that causal relations between individual events somehow involve general regularities.) In fact, my chief concern will be focused not on the full-fledged concept of causation but rather on the concept of nomic subsumption, the idea of bringing individual events under a law, which is at the core of the Humean approach to causation. I begin with an examination of one popular modern formulation of Humean causation, "the nomic-implicational model."

I. "SUBSUMPTION UNDER A LAW"

When we try to explain the notion of subsuming events under a law, a notion of central importance to Humean causation, we immediately face a problem which turns out to be more intractable than it might at first appear: laws are sentences (or statements, propositions, etc.), but events are not. Exactly in what relation must a pair of events stand to a law if the law is to "subsume" the events? Given the categorial difference between laws and events, it would be quite senseless to say that one of the events must be "logically implied" by the other event taken together with the law. However,

---

[2] Zeno Vendler makes the claim that events are primarily temporal entities, whereas objects are primarily spatial, and that the attributions of temporal properties and relations to objects and of spatial properties and relations to events are derivative. See his Linguistics in Philosophy (Ithaca, N.Y.: Cornell, 1967), pp. 143–144.

[3] For a general discussion of Humean causation see Bernard Berofsky, Determinism (Princeton, N.J.: University Press, 1971), esp. chs. IV, VI, and VII.

the temptation to use logico-linguistic constructions is great, and one tries to bring events within the purview of logic by talking about their descriptions.

(1) Law L subsumes events e and e' (in that order) provided there are descriptions D and D' of e and e' respectively such that L and D jointly imply D' (without D alone implying it).[4]

Thus, according to this formulation, the law 'All copper expands upon heating' subsumes the events described by 'This piece of copper was heated at t' and 'This piece of copper expanded at t'. The basic idea is that nomic subsumption is nomic implication between appropriate event descriptions.

Here 'describe' is the key word. The crucial assumption of the nomic-implicational model as embodied in (1) is that *certain sentences describe events*. But how do we explain this notion? There are three important related problems here: (i) What types of sentences describe events? (ii) Given an event-describing sentence, what particular event does it describe? (iii) Under what conditions do two such sentences describe the same event?

Recent investigations [5] have shown that there are no simple answers to these questions and that the intuitive ideas we have about them are full of pitfalls, if not outright contradictions. Let us briefly see how a seemingly natural and promising line of approach runs quickly into a dead end.

Consider a sentence like 'This piece of copper was heated at t', which we would take as a typical event-describing sentence. We may think of the whole sentence as describing the event of this piece of copper being heated at t. An event-describing sentence in this sense has the form 'Object x has property P at time t' and affirms of a concrete object that it has a certain empirical property at a time (let us not worry about polyadic cases). Such a sentence, if true, is thought to describe the event of x's having P at t. Now, once this approach is adopted, the following development is both natural and inescapable: if object a is the very same object as object b, then the event of a's having P at t is the same event as

---

[4] Compare Arthur Pap: "In the scientific sense of 'cause', an event A causes an event B in the sense that there is a law, L, such that from the conjunction of L and a description of A the occurrence of B is logically deducible." *An Introduction to the Philosophy of Science* (New York: Free Press, 1962), p. 271. We shall not consider here the difficulty that, according to (1), undescribed events are not subsumable under any law and as a result cannot enter into causal relations.

[5] See, e.g., Donald Davidson, "The Individuation of Events" in Nicholas Rescher *et al.*, eds., *Essays in Honor of Carl G. Hempel* (Dordrecht: Reidel, 1969); and my "Events and Their Descriptions: Some Considerations," *ibid.*

the event of b's having P at t. Thus, if 'a' and 'b' are coreferential, the sentences 'a has P at t' and 'b has P at t' describe the same event.[6] But now see what happens to the nomic-implicational model (1).

Let the law '$(x)(Fx \rightarrow Gx)$' subsume the two events described by 'c has F' and 'c has G' (we drop 't' for simplicity). Then, if 'b has H' is any true event-describing sentence, the law subsumes the event described by 'b has H' and the event 'c has G'; for the former event is also described by '$(Ix)(x = b \ \& \ c \text{ has } F) \text{ has } H$',[7] which, together with the law '$(x)(Fx \rightarrow Gx)$', but not by itself, implies 'c has G'. In fact, it can be shown that any law that subsumes, in the sense of (1), at least one pair of events subsumes every pair.

The moral of these difficulties for the nomic-implicational model is this: once the description operator 'I' is available, we can pack as much "content" as we like into any singular sentence, and this can likely be done without changing the identity of the event described. Obviously, this is bound to cause trouble for any account of causation or nomic subsumption based on the relation of logical implication, since logical implication essentially depends on the content of sentences.[8]

So far we have examined the difficulties for (1) that arise from the notion of a sentential description of an event. Let us now go on to difficulties of another type arising from the other central idea of (1): that nomic subsumption of events can be linguistically mirrored by nomic implication between their descriptions.

The obvious similarity between the so-called "covering-law model" of explanation and what we have called "the nomic-implicational model" of causation will not have escaped notice. It should then come as no surprise that difficulties for one have counterparts in the difficulties for the other; however, this fact seems not to have been fully appreciated.

A valid argument having the following properties will be called a 'D-N argument' ('D-N' for 'deductive-nomological'): (i) its premises include both laws and singular sentences and its conclusion is singular, and (ii) the argument becomes invalid upon the deletion of the laws from the premises. The covering-law model of explanation, as a first approximation, can be formulated thus: an event

---

[6] For more details see my "Events and Their Descriptions: Some Considerations," *ibid.*

[7] We follow Dana Scott in the use of 'I' as definite description operator. See Scott, "Existence and Description in Formal Logic," in Ralph Schoenman, ed., *Bertrand Russell: Philosopher of the Century* (Boston: Little Brown, 1967).

[8] Thus, the method favored by Davidson for handling event-describing sentences runs afoul of the same difficulties in connection with (1). See his "Causal Relations," this JOURNAL, LXIV, 21 (Nov. 9, 1967): 691–703, esp. p. 699.

is explained when a D-N argument is constructed whose conclusion describes that event. In terms of 'D-N argument', the nomic-implicational model of subsumption under a law comes to this: two events are subsumed under a law just in case there is a D-N argument whose premises are the law and a description of one of the events and whose conclusion is a description of the other event.

It is trivial to show that the notion of D-N argument as characterized cannot coincide with explanation, for the following is easily shown: for any law L and a true event-description D', there is a true singular sentence D such that 'L, D, therefore D'' is a D-N argument.[9] Thus, one law would suffice to explain any event you please. As an example: you want to explain why an object b has property F, for any b and F you choose. So you construct the following D-N argument: 'Copper is an electric conductor, b is F or b is nonconducting copper, therefore b is F'.

With regard to this and similar cases, the proponent of the nomic-implicational model might plead that the singular premise in such an argument (e.g., 'b is F, or b is nonconducting copper'), being a compound sentence of a rather artificial sort, cannot be thought of as an event-description.[10] Apart from the fact that this reply presupposes a satisfactory solution to the problem raised earlier of characterizing 'event-describing sentence', it seems to have a good deal less force against a pseudo-D-N argument like this: 'All crows are black, b is a crow, and c has the color of b. Therefore c is black'.

There is as yet no adequate formulation of the notion of 'D-N argument' that can successfully cope with these and other simple anomalous arguments; and it is unclear how examples of the second sort just described can be handled within the existing scheme of the theory of explanation. In any case, the unsettled state of the formal theory of deductive explanation implies a similar unsettled state for the nomic-implicational approach to Humean causation.

Enough has been said, I think, to justify at least a temporary shift of strategy away from the logico-descriptive approach underlying the nomic-implicational model. In the two sections to follow, we shall explore a direct "ontological approach" which dispenses with talk of descriptions and implications.

[9] For further details see Carl G. Hempel and Paul Oppenheim, "Studies in the Logic of Explanation," reprinted in Hempel, *Aspects of Scientific Explanation* (New York: Free Press, 1965) and the references given in Hempel's "Postscript" to this article.

[10] In fact, a clearer understanding of event-describing sentences is likely to help us with the problem of characterizing the structure of deductive explanation, since many counterexamples to the standard account contain singular premises which are intuitively not event-describing.

II. THE STRUCTURE OF EVENTS

Once we abandon the logico-descriptive approach, we must begin taking events seriously, since the only clear alternative to it is to define the causal relation directly for events without reliance on linguistic intermediaries. But what is an event? What sort of structures do we need as relata of causal relations? In this section I sketch an analysis of events[11] on the basis of which I shall formulate three versions of Humean causation in the next section.

We think of an event as a concrete object (or n-tuple of objects) exemplifying a property (or n-adic relation) at a time. In this sense of 'event', events include states, conditions, and the like, and not only events narrowly conceived as involving changes. Events, therefore, turn out to be complexes of objects and properties, and also time points and segments, and they have something like a propositional structure; the event that consists in the exemplification of property P by an object x at time t bears a structural similarity to the sentence 'x has P at t'. This structural isomorphism is related to the fact that we often take singular sentences of the form 'x has P at t' as referring to, describing, representing, or specifying an event; also we commonly and standardly use gerundial nominals of sentences to refer to events as in 'the sinking of the Titanic', 'this match's being struck', 'this match's lighting', and so forth.

We represent events by expressions of the form

$$'[(x_1, \ldots, x_n, t), P^n]'$$

An expression of this form refers to the event that consists in the ordered n-tuple of concrete objects $(x_1, \ldots, x_n)$ exemplifying the n-adic empirical attribute P^n at time t. Strictly speaking, P^n is $(n + 1)$-adic since we count 't' as an argument place; but we follow the usual procedure of reckoning, for example, redness as a property rather than a relation even though objects are red, or not red, *at a time*. (In fact, there is no reason why time should be limited to a single argument place in an attribute, but let us minimize complexities not directly relevant to our central concerns.) We shall abbreviate '$(x_1, \ldots, x_n)$' as '$(\mathbf{x}_n)$' and '$(x_1, \ldots, x_n, t)$' as '$(\mathbf{x}_n, t)$' re-

[11] This account was adumbrated in my "On the Psycho-Physical Identity Theory," *American Philosophical Quarterly*, III, 3 (July 1966): 231–232. It bears a resemblance to R. M. Martin's analysis in "Events and Descriptions of Events," in J. Margolis, ed., *Fact and Existence* (Oxford: Blackwell, 1969) and also to Alvin I. Goldman's account of action in *A Theory of Human Action* (Englewood Cliffs, N.J.: Prentice-Hall, 1970), ch. 1. Nancy Holmstrom develops a similar notion of event in her doctoral dissertation, *Identities, States, and the Mind-Body Problem*, The University of Michigan, 1970.

spectively, and drop the superscript from 'P^n'. The variable 't' ranges over time instants and intervals; when 't' denotes an interval, 'at t' is to be understood in the sense of 'throughout t'. We call P, $(\mathbf{x}_n)$, and t, respectively, the "constitutive attribute", "the constitutive objects", and "the constitutive time" of the event $[(\mathbf{x}_n, t), P]$.

We adopt the following as the condition of event existence:

Existence condition: $[(\mathbf{x}_n, t), P]$ exists if and only if the n-tuple of concrete objects $(\mathbf{x}_n)$ exemplifies the n-adic empirical attribute P at time t.

Linguistically, we can think of '$[(\mathbf{x}_n, t), P]$' as the gerundive nominalization of the sentence '$(\mathbf{x}_n)$ has P at t'. Thus, '$[(\text{Socrates}, t)$, drinks hemlock$]$' can be read "Socrates' drinking hemlock at t." Notice that $[(x, t), P]$ is not the ordered triple consisting of x, t, and P; the triple exists if x, t, and P exist; the event $[(x, t), P]$ exists only if x has P at t. As property designators we may use ordinary (untensed) predicative expressions; when the order of argument places has to be made explicit we use circled numerals;[12] e.g.,

$$[(a, b, c, t), ② \text{ stands between } ① \text{ and } ③]$$

corresponds, by the existence condition, to the sentence 'b stands between a and c at t'. The proviso that the constitutive attribute of an event be "empirical" is intended to exclude, if one so wishes, tautological, evaluative, and perhaps other kinds of properties; but we must in this paper largely leave open the question of exactly what sorts of attributes are admissible as constitutive attributes of events.

When P is a monadic attribute, that is, when only "monadic events" are considered, the following identity condition is immediate:

Identity condition I_1: $[(x, t), P] = [(y, t'), Q]$ if and only if $x = y$, $t = t'$, and $P = Q$.

Thus, Socrates' drinking hemlock at t is the same event as Xantippe's husband's drinking hemlock at t, and this liquid's turning blue at t is the same event as its turning the color of the sky at t.

Two objections might be voiced at this point. First, it might be contended that the event $[(\text{Brutus}, t), \text{stabs Caesar}]$ is the very same event as $[(\text{Caesar}, t), \text{is stabbed by Brutus}]$, although our identity condition pronounces them to be distinct. Our reply here is

[12] Following W. V. Quine, *Methods of Logic* (New York: Holt, Rinehart & Winston, 1950), pp. 130ff. For formal development property abstracts could be used; see, e.g., Richard Montague, "On the Nature of Certain Philosophical Entities," *Monist*, LIII, 2 (April 1969): 159–194.

that what the critic might have in mind are the dyadic events [(Brutus, Caesar, t), stabs] and [(Caesar, Brutus, t), is stabbed by], and that, according to the identity condition for dyadic events below, these events are indeed one and the same. Generally, we do not allow "mixed universals" [13] such as stabbing Caesar as constitutive attributes of events; only "pure universals" [13] are allowed as such.

Second, it might be objected that the event [(Xantippe's husband, t), dies] is identical with the event [(Xantippe, t), becomes a widow], viz., Xantippe's husband dying at t is the same event as Xantippe's becoming a widow at t, although again I_1 is not satisfied. We answer that these are indeed different events. Consider, for example, their locations: the first obviously took place in the prison in which Socrates took the poison, but it is not clear exactly where the second event occurred. We might want to locate it where Xantippe was at the moment of Socrates' death (and this is the procedure we shall adopt), but clearly not in the prison. To be sure, the two events are connected; in fact, the biconditional '[(Xantippe's husband, t), dies] exists if and only if [(Xantippe, t), becomes a widow] exists' is demonstrable from the existence condition; one might wish to say that necessarily one exists if and only if the other does. But this has no tendency to show that we have one event here and not two. One could just as well argue that since 'The husband of Socrates' wife exists if and only if Socrates' wife exists' is necessarily true, the husband of Socrates' wife is the same as Socrates' wife.

Now for dyadic events: if we want the identity '[(Brutus, Caesar, t), stabs] = [(Caesar, Brutus, t), is stabbed by]', we obviously cannot simply repeat I_1 for dyadic events. But what we should say is equally obvious. For any dyadic relation R, let R^* be its converse. We then have:

Identity condition I_2: $[(x, y, t), R] = [(u, v, t'), Q]$ if and only if either (i) $(x, y) = (u, v)$, $t = t'$, and $R = Q$, or (ii) $(x, y) = (v, u)$, $t = t'$, and $R = Q^*$.

For the general case of n-adic events, we need to generalize the concept of converse to n-adic relations. Any n-termed sequence can be permuted in $n!$ different ways (including the identity permutation). If k is a permutation on n-termed sequences (note that k is a *scheme* of permutation, not a particular permuted sequence), then by '$k(\mathbf{x}_n)$' we denote the sequence resulting from permuting the

[13] For a possible explanation of these terms, see Arthur W. Burks, "Ontological Categories and Language," *Visva-Bharati Journal of Philosophy*, III (1967): 25–46, esp. pp. 28–29.

sequence $(\mathbf{x}_n)$ by k. The $n!$ permutations on n-termed sequences form a group, and for each permutation k there exists an inverse k^{-1} such that $k^{-1}(k(\mathbf{x}_n)) = (\mathbf{x}_n)$. If k is a permutation on n-termed sequences and R is an n-adic relation, $k(R)$ is to be the n-adic relation such that, for every $(\mathbf{x}_n)$, $(\mathbf{x}_n)$ has $k(R)$ if and only if $k^{-1}(\mathbf{x}_n)$ has R.[14] It follows that, for each k, $k(\mathbf{x}_n)$ has $k(R)$ if and only if $(\mathbf{x}_n)$ has R. The $n!$ permutations of an n-adic relation R can be thought of as the converses of R. Just as the converse of a dyadic relation may be identical with the relation itself (that is, the relation is symmetric), some of the converses of an n-adic relation may in fact be identical.

We now state the identity condition for the general case:

Identity condition I_n: $[(\mathbf{x}_n, t), P] = [(\mathbf{y}_m, t'), Q]$ if and only if there exists a permutation k on m-termed sequences such that $(\mathbf{x}_n) = k(\mathbf{y}_m)$, $t = t'$, and $P = k(Q)$.

Obviously, I_n entails I_1 and I_2 for $n = 1, 2$. We can say, for example, that $[(a, b, c, t), ① \text{ gives } ② \text{ to } ③] = [(c, b, a, t), ① \text{ receives } ② \text{ from } ③]$. The permutation involved here is (13) (2), i.e., the permutation whereby the first element is replaced by the third, the second by itself, and the third by the first.

This completes the presentation of what is admittedly a sketchy account of events. And it is only a beginning; many interesting problems remain. First of all, there is the problem of characterizing more precisely the syntactical and semantical properties of the operator '[ ]'. According to our identity condition, Socrates' dying is a different event from Xantippe's becoming a widow. What then is the relationship between the two? What is the relationship between my firing the gun and my killing Jones?[15] How are such notions as "complex events," "compound events," "part-whole" (for events), etc. to be explained? And above all, there is the problem of how the notion of "property" (generally, that of "attribute") is best construed for the purposes of an event theory of this kind,

[14] This is not intended as a definition, but only an informal explanation, of '$k(R)$'. As a definition it would likely be construed as presupposing an extensional interpretation of attributes (whether in the possible-world semantics or in some other scheme), whereas I prefer to be silent on this issue here. It may be useful, however, to point out that we are as much entitled to this informal explanation of '$k(R)$' as we are to the usual informal explanation of the notion of 'converse' of a binary relation.

[15] This problem is extensively discussed in Goldman, *A Theory of Human Action*. See also the APA Symposium on "The Individuation of Action" by Goldman, Judith Jarvis Thomson, and Irving Thalberg, this JOURNAL, LXVIII, 21 (Nov. 4, 1971): 761–787.

and in particular how those properties which can be constitutive properties of events (these properties can be called "generic events") should be characterized. It seems to me that the resolution of these problems about events depends on a satisfactory general account of properties; in fact, many interesting problems about events are likely to remain unresolved until such an account is at hand. In any case, we shall be alluding below to some of these further problems.

III. CAUSATION REVISITED

There appears to be a general agreement that the requirement of constant conjunction for causal relations for individual events is best explained in terms of lawlike correlations between generic events. Constant conjunction obviously makes better sense for repeatedly instantiable universals than for spatiotemporally bounded particulars. But, given a particular causal relation between two individual events, precisely which generic events must be lawfully correlated in order to sustain it?

Our account of events gives a quick answer. Every event has a unique constitutive property (generally, attribute), namely the property an exemplification of which by an object at a time is that event. And, for us, these constitutive properties of events are generic events. It follows that each event falls under exactly one generic event, and that once a particular cause-effect pair is fixed, the generic event that must satisfy the constant conjunction requirement is uniquely fixed. It is important to notice the distinction drawn by our analysis between properties *constitutive of* events and properties *exemplified* by them. An example should make this clear: the property of dying is a constitutive property of the event [(Socrates, t), dying], i.e., Socrates' dying at t, but not a property exemplified by it; the property of occurring in a prison is a property this event exemplifies, but is not constitutive of it. Under our account, then, if Socrates' drinking hemlock (at t) was the cause of his dying (at t'), the two generic events, drinking hemlock and dying, must fulfill the requirement of lawlike constant conjunction.

This procedure, therefore, is in sharp contrast with the procedure in which the inner structure of events is not analyzed and which, as a result, does not associate with each event a unique constitutive property. On that approach no distinction is made between properties constitutive of events and properties exemplified by them; and an individual event is usually thought to fall under many, in fact an indefinite number of, generic events; for example, one and the same event can be the moving of a finger, the pressing of the trigger of a gun, a shooting, and a mercy killing.[16] How, on that view, might one answer the question raised at the outset of this section? Evidently, it would be too strong to require that every generic event under which the cause event falls be lawfully related to every generic event under which the effect event falls. A more reasonable proposal, which seems to be what many have in mind, would be to say that two causally related events are such that there are at least two lawfully correlated generic events under which they respectively fall. Thus, two events, e and e', satisfy the constant-conjunction requirement just in case there are generic events F and G such that e is an F-event, e' is a G-event, and F-events are constantly conjoined with G-events.

Given the considerable freedom permitted by this formula in the choice of the generic events to which the two events belong, the requirement of constant conjunction as stated turns out to be too easy to satisfy. If any grouping of events is allowed as a generic event—or if any property exemplifiable by events is taken as one— then the requirement thus interpreted becomes quite useless; it can be shown that every event satisfies this requirement with respect to any event that satisfies it with respect to at least one event. For let e_1 and e_2 satisfy the requirement in virtue of the constant conjunction between F-events and G-events; that is, e_1 is of kind F, e_2 is of kind G, and whenever an F-event occurs there occurs a corresponding G-event. Let e_3 be any arbitrary event and let R be any relation such that $R(e_3,e_1)$. We explain 'H' to be true of any event e just in case $(\exists f)(R(e,f) \& F(f))$. Then clearly e_3 belongs to the generic event H, and H-events are constantly conjoined with G-events, from which it follows that e_3 and e_2 satisfy the requirement of constant conjunction. This plainly is a result we want to avoid.[17]

In comparison, our procedure will make it a good deal more difficult—too difficult, some will say—to satisfy the constant-conjunction requirement because, as we noted, once cause and effect are fixed, the generic events that must lawfully correlate are also fixed. There may be a way of framing a reasonable condition of constant conjunction without associating a unique generic event with each event, but it is hard to see what it could be. In any case I

[16] Compare Donald Davidson: "I flip the switch, turn on the light, and illuminate the room. Unbeknownst to me I also alert a prowler to the fact that I am home. Here I do not do four things, but only one, of which four descriptions have been given." "Actions, Reasons, and Causes," this JOURNAL, LX, 23 (Nov. 7, 1963): 685–700, p. 686.

[17] This has been adapted from an argument given by J. A. Foster in "Psychophysical Causal Relations," *American Philosophical Quarterly*, V, 1 (January 1968): 65–66.

do not wish to suggest that the foregoing considerations tilt the
the balance decisively in favor of our procedure; as we shall shortly
see, there is a difficulty of a somewhat similar nature for our procedure as well.

What does it mean to say that two generic events are constantly
conjoined or lawfully correlated? It clearly is not enough to repeat
the usual formula that the occurrence of an event of one kind is
always followed by the occurrence of an event of the other kind. We
need to make more specific the relation between the given event of
the first kind and *the* event of the second kind that is to be associated
with it. As an example, the heating of a metallic object and the expansion of a metallic object would be constantly conjoined, according to this formula, provided only that whenever a metallic
object is heated, *some* metallic object *somewhere* expands. In this
particular case, what we have in mind is that whenever a metallic
object is heated *it* expands. But this cannot be made into a general
requirement, since we must allow causal relations between events
whose constitutive objects are different. A similar sort of indeterminacy besets the expression 'whenever' in the above formula; we do
not want to say that a given event of kind F and the particular event
of kind G that follows it must be simultaneous; but to leave this indefinite ("each F-event is followed by a G-event at some time or
other") is to render the requirement vacuous.

What seems needed, then, is a way of relating a particular F-event
to that particular G-event with which it is associated by the constant conjunction of F-events with G-events. Such a relation would
also be useful for correctly pairing a cause with *its* effect and an
effect with *its* cause. If two rifles are fired simultaneously, resulting
in two simultaneous deaths, we need a relation of that kind to pair
each rifle shot with the death it causes and not with the other.[18]
Notice, by the way, that those who would allow for each event a
multiplicity of generic events are faced with the same pairing
problem.

If x's being F at t is causally related to y's being G at t', this must
be so in virtue of some relation R holding for x, t, y, and t'. How else
could the following two facts be explained? First, given that x is F
at t, there are objects other than y that are not G at t'; and there
are times other than t' at which the object y is not G. Second, again
given that x is F at t and this event causes y's being G at t', there
can be (and usually would be) other individual events of kind G

---

[18] Haskell Fain raises a similar problem in "Some Problems of Causal Explanation," *Mind*, LXXII, 288 (October 1963): pp. 519–532.

occurring at t' that are causally unrelated to x's being F at t. Now it seems that there are three different ways in which such a relation could be worked into an analysis of Humean causation: (A) we look for a single "pairing relation" for all cases of constant conjunction (or Humean causal relations); (B) we let the choice of a suitable pairing relation depend on the specific generic events F and G to be correlated (and perhaps the choice may also depend on the specific individual events to be causally related); (C) we build such a pairing relation into the cause event so that the cause is not the event of x's being F at t, but rather the "complex event" of x's being F and also being in relation R to y at t.

In what follows we explore these three possibilities. In addition to their individual strengths and shortcomings, all three will be seen to be subject to one important difficulty. But a close examination and discussion of the comparative merits and faults of these three approaches cannot be attempted here, although of course I shall be making remarks relevant to a comparative evaluation of them. The order in which the three approaches will be considered is this: first (B), then (A), and finally (C).

An analysis of the causal relation that falls under (B) is the following definition of 'causal sufficiency' offered by J. A. Foster (*op. cit.*, p. 67):

a's being F is causally sufficient for b's being G if and only if there exists a relation R such that

(i) $F(a)$, $G(b)$, and $R(a,b)$
(ii) $(x)(F(x) \rightarrow (\exists y)(G(y) \ \& \ R(x,y)))$ [19]
(iii) $(x)(F(x) \ \& \ R(x,b) \supset x = a) \ \& \ (x)(G(x) \ \& \ R(a,x) \supset x = b)$

The condition (ii) of course is the constant-conjunction requirement; and the condition (iii) states that the pairing relation R must be such that at most one thing that is F, namely a, bears R to b and that a bears R to at most one thing that is G, namely b. The choice of R depends not only on F and G but also on a and b.

It seems to me that Foster's (ii) is not the most useful way of stating the lawlike correlation of F and G; there appears to be no simple way of accommodating such mundane examples of causal relations as a's firing a rifle and b's dying, a's having such-and-such mass and b's accelerating with such-and-such rate of acceleration

---

[19] We use the arrow '$\rightarrow$' to denote whatever type of implication the reader deems appropriate for stating laws in something like this form (this in effect is also Foster's practice). We do not consider here the question of precisely what sort of "nomic force" if any, should be carried by a statement of a constant conjunction. For various possible interpretations of causal or nomological implication, see Arthur W. Burks, *Cause, Chance, and Reason* (forthcoming).

(toward a by gravitational attraction), and so on. The problem is simply that the laws in question do not entail a statement of the form (ii) to the effect that if any object has property F *there exists at least one object y* fulfilling the consequent of (ii). (Foster restricts his definition so that a, b, and objects in the range of 'x', 'y', ..., are "momentary particulars" without temporal duration, but this doesn't affect the problem.) It would seem that (ii) is more usefully stated thus: $(x)(y)(F(x) \mathrel{\&} R(x,y) \rightarrow G(y))$.

In any case, let us turn to another problem. Let us assume, as Foster does, that, for any spatiotemporal objects a and b, their exact spatiotemporal relation R satisfies the condition (iii), regardless of what F and G may be; this assumption holds if the identity of spatiotemporal objects is determined completely by their spatiotemporal location. With this assumption at hand we can show the following: If a's being F is causally sufficient for b's being G, then for any object c there exists a property H such that c's being H is causally sufficient for b's being G. For let R_1 be the spatiotemporal relation between c and a, and let R_2 be the spatiotemporal relation between c and b. And we set H to be the property denoted by the expression '$(\exists y)(F(y) \mathrel{\&} R_1(x,y))$'. Then, the law '$(x)(H(x) \rightarrow (\exists y)(G(y) \mathrel{\&} R_2(x, y)))$' holds; and the other conditions are obviously satisfied. To make this more concrete, consider this case: the object b's being heated is causally sufficient for its expanding (here $a = b$ and the relation R can be taken as identity). Let c be an object exactly 50 miles due north of the object that is being heated. The property H in this case is the property an object has in virtue of there being another object 50 miles due south that is being heated. Morevoer, given the law that all objects expand when heated, we have the law that for any object x if x has the property H, then there exists an object 50 miles due south which is expanding. From this it follows that c's having property H is causally sufficient for b's expanding.[17]

Cases like this need not be regarded as necessarily objectionable for Foster's definition, which defines causal sufficiency, not causation. However, they would be clearly objectionable if the relation defined were that of causation. It would be absurd to say that object c's having H caused object a to expand, or that c causally influenced or interacted with a. Notice that Foster's definition can be directly mirrored in our framework of events, since the entities related by his causal sufficiency, a's being F, b's being G, etc., are close analogues of our $[(a, t), F]$, $[(b, t), G]$, etc. The implication of the above example then is that, under a definition of the causal relation similar to Foster's definition of 'causal sufficiency' (notice here that the possible alteration of the condition (ii) does not materially affect the difficulty), if an event is caused by another, then every object is the constituent object in some event which is a cause of the first; that is, there would be no object "causally independent" of that event.

As we shall see, the two remaining ways of handling the pairing problem are open to difficulties of a similar sort. The gist of the difficulties is this: when there is a constant conjunction between F and G, then, for any object you please, we can pick a property H such that the object has H, and H is constantly conjoined with G. Thus, this spurious constant conjunction rides piggyback, so to speak, on the genuine correlation between F and G; we may call this problem "the problem of parasitic constant conjunctions."

We may, I think, question whether the artificially concocted property H can in general be regarded as a constitutive property of an event. A negative answer seems plausible, although a plausible defense of it would be a subtle and difficult matter. We feel that for an object to have this sort of property (recall the special case of H above) is not always for it to undergo, or be disposed to undergo, a "real change"; my being 50 miles east of a burning barn is hardly an event that happens to me.[20] But it would be a mistake to ban all such properties; my being in spatial contact with a burning barn is very much an event that happens to me. Whether a clear distinction between these two kinds of cases can be made that does not beg the question by using causal concepts is an interesting question to which I know of no completely satisfying answer. This is a special case of the more general problem alluded to earlier, namely that of characterizing the properties whose exemplification by an object at a time is an event, i.e., generic events.

We now turn to the approach (A) to the pairing problem. One feature of the event $[(c, t), H]$ which enters into an unwanted causal relation with the event $[(b, t), G]$ is the fact that its constitutive object c, need not be in spatial contact with the constitutive object b, of $[(b, t), G]$. In fact, Hume's condition of spatial contiguity is not mentioned at all in Foster's definition of 'causal sufficiency'. Thus, if we are willing to go along with Hume here, the contiguity relation presents itself as a natural candidate for the pairing relation. This manner of handling the pairing problem differs from the one we have just considered in that there would be a single uniform

---

[20] In this connection see Peter Geach's interesting remarks on "Cambridge changes" in *God and the Soul* (London: Routledge & Kegan Paul, 1969), pp. 71–72.

relation doing the job for all causal relations independent of the particular cause and effect events.

As Hume was aware, however, direct contiguity cannot be generally required for causal relations; following Hume's own suggestion,[21] we shall try first to explain 'direct contiguous causation' and then explain 'causation' as a "chain" of direct contiguous causal relations. Thus, the analysis of causation that follows is not only "Humean"; it is also Hume's.

We first need the contiguity relation for events. It would seem that this relation must be explained in terms of the contiguity relation for objects and times of events (an object is contiguous with another at a time). Thus, if $[(a, T), P]$ is contiguous with $[(b, T'),$ $Q]$, this must be so in virtue of a contiguity relation holding for a, b, T, and T'; and the relevant aspect of the objects a and b is their spatial location at the indicated times. Let '$\text{loc}(x,t)$' denote the spatiotemporal location of x at time t (where x exists at t); where t is an interval, $\text{loc}(x,t)$ will be a spatiotemporal volume. In order not to complicate our problems excessively we consider here only monadic events.

We say that two events $[(a,T), P]$ and $[(b,T'), Q]$ are contiguous just in case $\text{loc}(a,T)$ is contiguous with $\text{loc}(b,T')$—we assume of course that the two events exist. How contiguity for spatiotemporal location is to be explained is a question that depends on the properties of the space-time involved; since nothing in this paper hinges on the exact explanation of this notion, we leave it unanalyzed. We now define 'direct contiguous causation' as follows (we abbreviate 'contiguous with' as "Ct'):

$[(a, T), P]$ is a *direct contiguous cause* of $[(b, T'), Q]$ provided:

(i) $[(a, T), P]$ is contiguous with $[(b, T'), Q]$.

(ii) If $a = b$: $(x)(t)(t')([(x, t), P]$ exists $\&\ \text{Ct}(\text{loc}(x, t), \text{loc}(x, t'))$
$\rightarrow [(x, t'), Q]$ exists$)$.

If $a \neq b$: $(x)(y)(t)(t')([(x, t), P]$ exists $\&\ \text{Ct}(\text{loc}(x, t),$
$\text{loc}(y, t')) \rightarrow [(y, t'), Q]$ exists$)$.

We define 'contiguous cause' in terms of the ancestral of direct contiguous causation:

e is a contiguous cause of e' if and only if $e \neq e'$ and e bears to e' the ancestral of the relation of direct contiguous causation—that

[21] Hume writes: "Tho' distant objects may sometimes seem productive of each other, they are commonly found upon examination to be link'd by a chain of causes, which are contiguous among themselves, and to the distant objects; and when in any particular instance we cannot discover this connexion, we still presume it to exist." *Treatise of Human Nature*, bk. I, pt. III, sec. II.

is to say, $(S)(e' \epsilon S \ \& \ (f)(g)(f \epsilon S \ \& \ g$ is a direct contiguous cause of $f \supset g \epsilon S) \supset e \epsilon S)$.

Whether contiguity in this sense ought to be required of causal relations as a matter of definition is a debatable issue; in particular, the verification of the existence of a causal chain of the required sort may in practice be an impossible task in many areas of science in which causal attributions are regularly made; and the belief that such a chain must exist may be only metaphysical faith. But these are the questions we must leave aside.[22] Let us now turn to the last of the three ways of dealing with the pairing problem distinguished earlier.

Recall the example of two rifle shots causing two simultaneous deaths. We raised the question how each shot is to be paired with the death it causes. Causal chains will probably help us here, but there seems to be another, perhaps more natural and simpler, way of handling it. It may be said that the cause of a death here is not a rifle shot simpliciter, but rather the rifle shot cum the event (state) of the rifle's being in such-and-such spatiotemporal relationship to the man whose death it causes. Thus, the cause of the man's death is the set of events: the rifle's being fired and its being in a certain relation R to the man (at the time it was fired); we could perhaps speak of a single "compound" or "composite event" of the rifle's being fired and being in relation R to the man. In either case, the man, who is the constitutive object in the effect event, figures in the cause as a constitutive object. Again restricting ourselves essentially to monadic cases, we may capture this idea as follows:

The set of events, $[(a, T), F]$ and $[(a, b, T), R]$, is a cause of the event $[(b, T'), G]$ provided:
  (i) $[(a, T), F]$, $[(a, b, T), R]$, and $[(b, T'), G]$ exist, and
  (ii) $(x)(y)(t)([(x, t), F]$ exists $\& \ [(x, y, t), R]$ exists
              $\rightarrow [(y, t + \Delta t), G]$ exists), where $\Delta t = T' - T$.
  (iii) The law in (ii) does not hold if one or the other of its antecedent clauses is deleted.

We should be wary of speaking of "composite events" before a precise characterization of them is at hand. But at least we can say this: if $[(a,T), P]$ and $[(a,b,T), R]$ exist, then, by the existence condition, the event $[(a,b,T), R^*]$ exists, where $R^*(x,y)$ at t just in case $P(x)$ at $t \ \& \ R(x,y)$ at t, on the assumption that R^* is a generic event. Also, conversely, if this dyadic event exists, the two former

[22] For a brief discussion of these problems see Patrick Suppes, *A Probabilistic Theory of Causality* (Amsterdam: North Holland, 1970), pp. 30–32, 82–91.

events exist. This is the intuitive content of the concept of "conjunctive event" in a simple case of this kind; but a general formulation of this concept is yet to be worked out. In any case, if we allow ourselves conjunctive events of at least this simple sort, we can simplify the preceding formulation of Humean causation:

$[(a, b, T), P]$ is a cause of $[(b, T'), Q]$ provided:
  (i) $[(a, b, T), P]$ and $[(b, T'), Q]$ exist, and
  (ii) $(x)(y)(t)([(x, y, t), P]$ exists $\rightarrow [(y, t + \Delta t), Q]$ exists),
      where $\Delta t = T' - T$.

(There is of course no simple way of stating (iii) of the preceding formulation; but when the definition is stated for composite events, (iii) doesn't seem needed.) In special cases, $a = b$, and the cause event as well as the effect event would be monadic. But generally the cause event will be a dyadic or higher-place event involving, as one of its constitutive objects, the constitutive object of the effect event; and the first term of a constant conjunction will in general be a relational generic event rather than a monadic one.[23]

Let us briefly note here how the problem of parasitic constant conjunctions arises for direct contiguous causation as formulated above. What happens is this: suppose $[(a,t), F]$ is a direct contiguous cause of $[(b,t), G]$, where for simplicity we have assumed $t = t'$. Let c be any object such that b is the only object with which c is contiguous (for simplicity we drop t) and b is the only object with which both a and c are contiguous. We can then construct a property H such that c has H and $[(c,t), H]$ is a direct contiguous cause of $[(b,t), G]$; letting R be some relation such that $R(c,a)$, we can let H be the property that belongs to an object x just in case $(\exists w)(R(x,w)$ & $F(w)$ & $(\exists ! z)(\text{Cont}(x,z))$ & $(\exists ! z)(\text{Cont}(x,z)$ & $\text{Cont}(w,z)))$, where again for simplicity we have deleted reference to time and where 'Cont' is used as a contiguity predicate applicable to objects simpliciter. But notice that the conditions on the object c here are severer than for Foster's definition; and there seems to be no general argument to show that our definition of 'contiguous causation' succumbs generally to this sort of difficulty. In this way, the dif-

[22] The causal relation defined here is, in many respects, weaker than the relation of contiguous causation earlier defined, and is open to the following sort of difficulty. Jones has terminal cancer, and there is a law that any human being having cancer (of the kind and stage Jones has) is dead within two years. And in two years Jones is dead. However, Jones actually died in a traffic accident. The present definition of the causal relation will erroneously certify Jones's cancer as a cause of his being dead, whereas contiguous causation avoids cases of this sort in a natural way.

ficulty of parasitic constant conjunctions is somewhat mitigated for the relation of contiguous causation.

It is easily seen that our last formulation of Humean causation is also open to the difficulty of parasitic constant conjunctions; however, we omit the details.

Apart from this difficulty of parasitic constant conjunctions, I find the preceding two accounts of Humean causation (contiguous causation and the account that takes cause as essentially a relational event) attractive; on the other hand, the first account borrowed from Foster is somewhat unintuitive, and, even with the suggested alteration of the condition (ii), the last condition (iii) on the pairing relation appears somewhat ad hoc. In any event, various refinements can be attempted on these definitions. In particular, there is the problem of building temporal asymmetry into them, if this is desired. Also, according to these definitions, all correlated properties in the same object, e.g., thermal and electrical conductivity in metals (at constant temperature), turn out to be symmetrically related by the causal relation. (I assume that we would not want to attribute a causal relation directly between electrical and thermal conductivity; the correlation is to be explained by reference to the microstructure of metals.) It seems likely that clues to a correct account of these cases will be found not at the level of analysis in this paper but at a deeper metaphysical level involving such concepts as substance, power, and accident, or at a pragmatic level involving the concept of controlling one parameter by controlling another.[24]

These refinements, as well as others which are necessary to account for some of the well-known difficulties for Humean causation,[25] are beyond the scope of the present paper and must await another occasion. It is best, therefore, to look upon the tentative accounts of Humean causation in this section not as full-fledged analyses of causation, but rather as approximations to the broader notion of subsumption of events under a law, an idea that forms the foundation of the Humean, or nomological, approach to causation. In any event, my aim here has been to outline a uniform and coherent

[24] Georg H. von Wright has recently worked out an account of causation on the basis of the concept of an agent's bringing about some state of affairs by doing a certain action, in *Explanation and Understanding* (Ithaca, N.Y.: Cornell, 1971).

[25] One such refinement would consist in taking account of the common observation that what we ordinarily take as a cause is seldom by itself a necessary or sufficient condition for the event it is said to have caused. For a plausible treatment of this problem, see J. L. Mackie, "Causes and Conditions," *American Philosophical Quarterly*, II, 4 (October 1965): 245–264; and my "Causes and Events: Mackie on Causation," this JOURNAL, LXVIII, 14 (July 22, 1971): 426–441.

For an interesting treatment of other important problems, see Ernest Sosa, "On Causation," forthcoming.

ontological framework of events adequate for formulation of Humean causation rather than to resolve substantive issues traditionally associated with the Humean approach. These issues must of course ultimately be handled within the suggested framework if it is to prove its worth. It is hoped, however, that we have at least made a modest beginning and that we now have a clearer perception of the directions in which to explore and the problems and promises to be expected along the way.

JAEGWON KIM

The University of Michigan

THE JOURNAL OF PHILOSOPHY

VOLUME LXIV, NO. 21, NOVEMBER 9, 1967

CAUSAL RELATIONS*

WHAT is the logical form of singular causal statements like: 'The flood caused the famine', 'The stabbing caused Caesar's death', 'The burning of the house caused the roasting of the pig'? This question is more modest than the question how we know such statements are true, and the question whether they can be analyzed in terms of, say, constant conjunction. The request for the logical form is modest because it is answered when we have identified the logical or grammatical roles of the words (or other significant stretches) in the sentences under scrutiny. It goes beyond this to define, analyze, or set down axioms governing, particular words or expressions.

I

According to Hume, "we may define a cause to be an object, followed by another, and where all the objects similar to the first are followed by objects similar to the second." This definition pretty clearly suggests that causes and effects are entities that can be named or described by singular terms; probably events, since one can follow another. But in the *Treatise*, under "rules by which to judge of causes and effects," Hume says that "where several different objects produce the same effect, it must be by means of some quality, which we discover to be common amongst them. For as like effects imply like causes, we must always ascribe the causation to the circumstances, wherein we discover the resemblance." Here it seems to be the "quality" or "circumstances" of an event that is the cause rather than the event itself, for the event itself is the same as others in some respects

* To be presented in APA symposium of the same title, December 28, 1967.

I am indebted to Harry Lewis and David Nivison, as well as to other members of seminars at Stanford University to whom I presented the ideas in this paper during 1966/67, for many helpful comments. I have profited greatly from discussion with John Wallace of the questions raised here; he may or may not agree with my answers. My research was supported in part by the National Science Foundation.

and different in other respects. The suspicion that it is not events, but something more closely tied to the descriptions of events, that Hume holds to be causes, is fortified by Hume's claim that causal statements are never necessary. For if events were causes, then a true description of some event would be 'the cause of b', and, given that such an event exists, it follows logically that the cause of b caused b.

Mill said that the cause "is the sum total of the conditions positive and negative taken together . . . which being realized, the consequent invariably follows." Many discussions of causality have concentrated on the question whether Mill was right in insisting that the "real Cause" must include all the antecedent conditions that jointly were sufficient for the effect, and much ingenuity has been spent on discovering factors, pragmatic or otherwise, that guide and justify our choice of some "part" of the conditions as the cause. There has been general agreement that the notion of cause may be at least partly characterized in terms of sufficient and (or) necessary conditions.[1] Yet it seems to me we do not understand how such characterizations are to be applied to particular causes.

Take one of Mill's examples: some man, say Smith, dies, and the cause of his death is said to be that his foot slipped in climbing a ladder. Mill would say we have not given the whole cause, since having a foot slip in climbing a ladder is not always followed by death. What we were after, however, was not the cause of death in general but the cause of Smith's death: does it make sense to ask under what conditions Smith's death invariably follows? Mill suggests that part of the cause of Smith's death is "the circumstance of his weight," perhaps because if Smith had been light as a feather his slip might not have injured him. Mill's explanation of why we don't bother to mention this circumstance is that it is too obvious to bear mention, but it seems to me that if it was Smith's fall that killed him, and Smith weighed twelve stone, then Smith's fall was the fall of a man who weighed twelve stone, whether or not we know it or mention it. How could Smith's actual fall, with Smith weighing, as he did, twelve stone, be any more efficacious in killing him than Smith's actual fall?

The difficulty has nothing to do with Mill's sweeping view of the cause, but attends any attempt of this kind to treat particular causes as necessary or sufficient conditions. Thus Mackie asks, "What is the exact force of [the statement of some experts] that this short-circuit caused this fire?" And he answers, "Clearly the experts are not saying that the short-circuit was a necessary condition for this house's catch-

[1] For a recent example, with reference to many others, see J. L. Mackie, "Causes and Conditions," *American Philosophical Quarterly*, II, 4 (October 1965): 245–264.

ing fire at this time; they know perfectly well that a short-circuit somewhere else, or the overturning of a lighted oil stove . . . might, if it
had occurred, have set the house on fire" (*ibid.*, 245). Suppose the
experts know what they are said to; how does this bear on the question whether the short circuit was a necessary condition of this
particular fire? For a short circuit elsewhere could not have caused
this fire, nor could the overturning of a lighted oil stove.

To talk of particular events as conditions is bewildering, but perhaps causes aren't events (like the short circuit, or Smith's fall from
the ladder), but correspond rather to sentences (perhaps like the fact
that this short circuit occurred, or the fact that Smith fell from the
ladder). Sentences can express conditions of truth for others—hence
the word 'conditional'.

If causes correspond to sentences rather than singular terms, the
logical form of a sentence like:

(1)   The short circuit caused the fire.

would be given more accurately by:

(2)   *The fact that* there was a short circuit *caused it to be the case that*
       there was a fire.

In (2) the italicized words constitute a sentential connective like
'and' or 'if . . . then . . .'. This approach no doubt receives support
from the idea that causal laws are universal conditionals, and singular
causal statements ought to be instances of them. Yet the idea is not
easily implemented. Suppose, first that a causal law is (as it is usually
said Hume taught) nothing but a universally quantified material
conditional. If (2) is an instance of such, the italicized words have
just the meaning of the material conditional, 'If there was a short
circuit, then there was a fire'. No doubt (2) entails this, but not conversely, since (2) entails something stronger, namely the conjunction
'There was a short circuit *and* there was a fire'. We might try treating (2) as the conjunction of the appropriate law and 'There was
a short circuit and there was a fire'—indeed this seems a possible interpretation of Hume's definition of cause quoted above—but then (2)
would no longer be an instance of the law. And aside from the inherent implausibility of this suggestion as giving the logical form of (2) (in
contrast, say, to giving the grounds on which it might be asserted)
there is also the oddity that an inference from the fact that there was a
short circuit and there was a fire, and the law, to (2) would turn out
to be no more than a conjoining of the premises.

Suppose, then, that there is a non-truth-functional causal connective, as has been proposed by many.[2] In line with the concept of a cause as a condition, the causal connective is conceived as a conditional, though stronger than the truth-functional conditional. Thus Arthur Pap writes, "The distinctive property of causal implication as compared with material implication is just that the falsity of the antecedent is no ground for inferring the truth of the causal implication" (212). If the connective Pap had in mind were that of (2), this remark would be strange, for it is a property of the connective in (2) that the falsity of either the "antecedent" or the "consequent" is a ground for inferring the falsity of (2). That treating the causal connective as a kind of conditional unsuits it for the work of (1) or (2) is perhaps even more evident from Burks' remark that "p is causally sufficient for q is logically equivalent to $\sim q$ is causally sufficient for $\sim p$" (369). Indeed, this shows not only that Burks' connective is not that of (2), but also that it is not the subjunctive causal connective 'would cause'. My tickling Jones would cause him to laugh, but his not laughing would not cause it to be the case that I didn't tickle him.

These considerations show that the connective of (2), and hence by hypothesis of (1), cannot, as is often assumed, be a conditional of any sort, but they do not show that (2) does not give the logical form of singular causal statements. To show this needs a stronger argument, and I think there is one, as follows.

It is obvious that the connective in (2) is not truth-functional, since (2) may change from true to false if the contained sentences are switched. Nevertheless, substitution of singular terms for others with the same extension in sentences like (1) and (2) does not touch their truth value. If Smith's death was caused by the fall from the ladder and Smith was the first man to land on the moon, then the fall from the ladder was the cause of the death of the first man to land on the moon. And if the fact that there was a fire in Jones's house caused it to be the case that the pig was roasted, and Jones's house is the oldest building on Elm street, then the fact that there was a fire in the oldest building on Elm street caused it to be the case that the pig was roasted. We must accept the principle of extensional substitution, then. Surely also we cannot change the truth value of the likes of (2) by substituting logically equivalent sentences for sentences in it. Thus (2) retains its truth if for 'there was a fire' we substitute the logically equivalent '$\hat{x}\ (x = x\ \&\ \text{there was a fire}) = \hat{x}\ (x = x)$'; retains it still

[2] For example by: Mackie, op. cit., p. 254; Arthur Burks, "The Logic of Causal Propositions," Mind, LX, 239 (July 1951): 363–382; and Arthur Pap, "Disposition Concepts and Extensional Logic," in Minnesota Studies in the Philosophy of Science, II, ed. by H. Feigl, M. Scriven, and G. Maxwell (Minneapolis: Univ. of Minnesota Press, 1958), pp. 196–224.

if for the left side of this identity we write the coextensive singular term '$\hat{x}$ ($x = x$ & Nero fiddled)'; and still retains it if we replace '$\hat{x}$ ($x = x$ & Nero fiddled) $= \hat{x}$ ($x = x$)' by the logically equivalent 'Nero fiddled'. Since the only aspect of 'there was a fire' and 'Nero fiddled' that matters to this chain of reasoning is the fact of their material equivalence, it appears that our assumed principles have led to the conclusion that the main connective of (2) is, contrary to what we supposed, truth-functional.[3]

Having already seen that the connective of (2) cannot be truth-functional, it is tempting to try to escape the dilemma by tampering with the principles of substitution that led to it. But there is another, and, I think, wholly preferable way out: we may reject the hypothesis that (2) gives the logical form of (1), and with it the ideas that the 'caused' of (1) is a more or less concealed sentential connective, and that causes are fully expressed only by sentences.

II

Consider these six sentences:

(3) *It is a fact that* Jack fell down.
(4) Jack fell down *and* Jack broke his crown.
(5) Jack fell down *before* Jack broke his crown.
(6) Jack fell down, *which caused it to be the case that* Jack broke his crown.
(7) *Jones forgot the fact that* Jack fell down.
(8) *That* Jack fell down *explains the fact that* Jack broke his crown.

Substitution of equivalent sentences for, or substitution of coextensive singular terms or predicates in, the contained sentences, will not alter the truth value of (3) or (4): here extensionality reigns. In (7) and (8), intensionality reigns, in that similar substitution in or for the contained sentences is not guaranteed to save truth. (5) and (6) seem to fall in between; for in them substitution of coextensive singular terms preserves truth, whereas substitution of equivalent sentences does not. However this last is, as we just saw with respect to (2), and hence also (6), untenable middle ground.

Our recent argument would apply equally against taking the 'before' of (5) as the sentential connective it appears to be. And of course we don't interpret 'before' as a sentential connective, but

[3] This argument is closely related to one spelled out by Dagfinn Føllesdal [in "Quantification into Causal Contexts" in *Boston Studies in the Philosophy of Science*, II, ed. R. S. Cohen and M. W. Wartofsky (New York: Humanities, 1966), pp. 263–274] to show that unrestricted quantification into causal contexts leads to difficulties. His argument is in turn a direct adaptation of Quine's [*Word and Object* (Cambridge, Mass.: MIT Press, 1960), pp. 197–198] to show that (logical) modal distinctions collapse under certain natural assumptions. My argument derives directly from Frege.

rather as an ordinary two-place relation true of ordered pairs of times; this is made to work by introducing an extra place into the predicates ('x fell down' becoming 'x fell down at t') and an ontology of times to suit. The logical form of (5) is made perspicuous, then, by:

(5′)   There exist times t and t′ such that Jack fell down at t, Jack broke his crown at t′, and t preceded t′.

This standard way of dealing with (5) seems to me essentially correct, and I propose to apply the same strategy to (6), which then comes out:

(6′)   There exist events e and e′ such that e is a falling down of Jack, e′ is a breaking of his crown by Jack, and e caused e′.

Once events are on hand, an obvious economy suggests itself: (5) may as well be construed as about events rather than times. With this, the canonical version of (5) becomes just (6′), with 'preceded' replacing 'caused'. Indeed, it would be difficult to make sense of the claim that causes precede, or at least do not follow, their effects if (5) and (6) did not thus have parallel structures. We will still want to be able to say when an event occurred, but with events this requires an ontology of pure numbers only. So 'Jack fell down at 3 P.M.' says that there is an event e that is a falling down of Jack, and the time of e, measured in hours after noon, is three; more briefly, $(\exists e)\,(F\,(\text{Jack}, e)\,\&\,t\,(e) = 3)$.

On the present plan, (6) means some fall of Jack's caused some breaking of Jack's crown; so (6) is not false if Jack fell more than once, broke his crown more than once, or had a crown-breaking fall more than once. Nor, if such repetitions turned out to be the case, would we have grounds for saying that (6) referred to one rather than another of the fracturings. The same does not go for 'The short circuit caused the fire' or 'The flood caused the famine' or 'Jack's fall caused the breaking of Jack's crown'; here singularity is imputed. ('Jack's fall', like 'the day after tomorrow', is no less a singular term because it may refer to different entities on different occasions.) To do justice to 'Jack's fall caused the breaking of Jack's crown' what we need is something like 'The one and only falling down of Jack caused the one and only breaking of his crown by Jack'; in some symbols of the trade, '$(\imath e)\,F\,(\text{Jack}, e)$ caused $(\imath e)\,B\,(\text{Jack's crown}, e)$'.

Evidently (1) and (2) do not have the same logical form. If we think in terms of standard notations for first-order languages, it is (1) that more or less wears its form on its face; (2), like many existentially quantified sentences, does not (witness 'Somebody loves somebody').

The relation between (1) and (2) remains obvious and close: (1) entails (2), but not conversely.[4]

III

The salient point that emerges so far is that we must distinguish firmly between causes and the features we hit on for describing them, and hence between the question whether a statement says truly that one event caused another and the further question whether the events are characterized in such a way that we can deduce, or otherwise infer, from laws or other causal lore, that the relation was causal. "The cause of this match's lighting is that it was struck.—Yes, but that was only *part* of the cause; it had to be a dry match, there had to be adequate oxygen in the atmosphere, it had to be struck hard enough, etc." We ought now to appreciate that the "Yes, but" comment does not have the force we thought. It cannot be that the striking of this match was only part of the cause, for this match was in fact dry, in adequate oxygen, and the striking was hard enough. What is partial in the sentence "The cause of this match's lighting is that it was struck" is the *description* of the cause; as we add to the description of the cause, we may approach the point where we can deduce, from this description and laws, that an effect of the kind described would follow.

If Flora dried herself with a coarse towel, she dried herself with a towel. This is an inference we know how to articulate, and the articulation depends in an obvious way on reflecting in language an ontology that includes such things as towels: if there is a towel that is coarse and was used by Flora in her drying, there is a towel that was used by Flora in her drying. The usual way of doing things does not, however, give similar expression to the similar inference from 'Flora dried herself with a towel on the beach at noon' to 'Flora dried herself with a towel', or for that matter, from the last to 'Flora dried herself'. But if, as I suggest, we render 'Flora dried herself' as about an event, as well as about Flora, these inferences turn out to be quite parallel to the more familiar ones. Thus if there was an event that was a drying by Flora of herself and that was done with a towel, on the beach, at noon, then clearly there was an event that was a drying by Flora of herself—and so on.

---

[4] A familiar device I use for testing hypotheses about logical grammar is translation into standard quantificational form; since the semantics of such languages is transparent, translation into them is a way of providing a semantic theory (a theory of the logical form) for what is translated. In this employment, canonical notation is not to be conceived as an improvement on the vernacular, but as a comment on it.

For elaboration and defense of the view of events sketched in this section, see my "The Logical Form of Action Sentences" in *The Logic of Action and Preference*, ed. Nicholas Rescher (Pittsburgh: University Press, 1967).

The mode of inference carries over directly to causal statements. If it was a drying she gave herself with a coarse towel on the beach at noon that caused those awful splotches to appear on Flora's skin, then it was a drying she gave herself that did it; we may also conclude that it was something that happened on the beach, something that took place at noon, and something that was done with a towel, that caused the tragedy. These little pieces of reasoning seem all to be endorsed by intuition, and it speaks well for the analysis of causal statements in terms of events that on that analysis the arguments are transparently valid.

Mill, we are now in better position to see, was wrong in thinking we have not specified the whole cause of an event when we have not wholly specified it. And there is not, as Mill and others have maintained, anything elliptical in the claim that a certain man's death was caused by his eating a particular dish, even though death resulted only because the man had a particular bodily constitution, a particular state of present health, and so on. On the other hand Mill was, I think, quite right in saying that "there certainly is, among the circumstances that took place, some combination or other on which death is invariably consequent . . . the whole of which circumstances perhaps constituted in this particular case the conditions of the phenomenon . . ." (*A System of Logic*, book III, chap. v, § 3.) Mill's critics are no doubt justified in contending that we may correctly give the cause without saying enough about it to demonstrate that it was sufficient; but they share Mill's confusion if they think every deletion from the description of an event represents something deleted from the event described.

The relation between a singular causal statement like 'The short circuit caused the fire' and necessary and sufficient conditions seems, in brief, to be this. The fuller we make the description of the cause, the better our chances of demonstrating that it was sufficient (as described) to produce the effect, and the worse our chances of demonstrating that it was necessary; the fuller we make the description of the effect, the better our chances of demonstrating that the cause (as described) was necessary, and the worse our chances of demonstrating that it was sufficient. The symmetry of these remarks strongly suggests that in whatever sense causes are correctly said to be (described as) sufficient, they are as correctly said to be necessary. Here is an example. We may suppose there is some predicate '$P(x,y,e)$' true of Brutus, Caesar, and Brutus's stabbing of Caesar and such that any stab (by anyone of anyone) that is P is followed by the death of the stabbed. And let us suppose further that this law meets Mill's requirements of being *unconditional*—it supports counterfactuals of the form 'If Cleopatra had received a stab that was P, she would have died'. Now we can prove (assuming a man dies only once) that Brutus's stab was sufficient for Caesar's death. Yet it was not the cause of Caesar's death, for Caesar's death was the death of a man with more wounds than Brutus inflicted, and such a death could not have been caused by an event that was P ('P' was chosen to apply only to stabbings administered by a single hand). The trouble here is not that the description of the cause is partial, but that the event described was literally (spatio-temporally) only part of the cause.

Can we then analyze 'a caused b' as meaning that a and b may be described in such a way that the existence of each could be demonstrated, in the light of causal laws, to be a necessary and sufficient condition of the existence of the other? One objection, foreshadowed in previous discussion, is that the analysandum does, but the analysans does not, entail the existence of a and b. Suppose we add, in remedy, the condition that either a or b, as described, exists. Then on the proposed analysis one can show that the causal relation holds between any two events. To apply the point in the direction of sufficiency, imagine some description '$(\imath x)\, Fx$' under which the existence of an event a may be shown sufficient for the existence of b. Then the existence of an arbitrary event c may equally be shown sufficient for the existence of b: just take as the description of c the following: '$(\imath y)\, (y = c \,\&\, (\exists!x)\, Fx)$'.[5] It seems unlikely that any simple and natural restrictions on the form of allowable descriptions would meet this difficulty, but since I have abjured the analysis of the causal relation, I shall not pursue the matter here.

There remains a legitimate question concerning the relation between causal laws and singular causal statements that may be raised independently. Setting aside the abbreviations successful analysis might authorize, what form are causal laws apt to have if from them, and a premise to the effect that an event of a certain (acceptable) description exists, we are to infer a singular causal statement saying that the event caused, or was caused by, another? A possibility I find attractive is that a full-fledged causal law has the form of a conjunction:

$$(\text{L}) \begin{cases} (\text{S}) & (e)(n)((Fe \,\&\, t(e) = n) \rightarrow \\ & \qquad\qquad (\exists!f)(Gf \,\&\, t(f) = n + \epsilon \,\&\, C(e, f))) \text{ and} \\ (\text{N}) & (e)(n)((Ge \,\&\, t(e) = n + \epsilon) \rightarrow \\ & \qquad\qquad (\exists!f)(Ff \,\&\, t(f) = n \,\&\, C(f, e))) \end{cases}$$

5 Here I am indebted to Professor Carl Hempel, and in the next sentence to John Wallace.

Here the variables 'e' and 'f' range over events, 'n' ranges over numbers, F and G are properties of events, '$C(e, f)$' is read 'e causes f', and 't' is a function that assigns a number to an event to mark the time the event occurs. Now, given the premise:

(P) $(\exists\, !e)\,(Fe\ \&\ t(e) = 3)$

(C) $(\imath e)\,(Fe\ \&\ t(e) = 3)$ caused $(\imath e)\,(Ge\ \&\ t(e) = 3 + \epsilon)$

It is worth remarking that part (N) of (L) is as necessary to the proof of (C) from (P) as it is to the proof of (C) from the premise '$(\exists\, !e)\ (Ge\ \&\ t(e) = 3 + \epsilon))$'. This is perhaps more reason for holding that causes are, in the sense discussed above, necessary as well as sufficient conditions.

Explaining "why an event occurred," on this account of laws, may take an instructively large number of forms, even if we limit explanation to the resources of deduction. Suppose, for example, we want to explain the fact that there was a fire in the house at 3:01 P.M. Armed with appropriate premises in the form of (P) and (L), we may deduce: that there was a fire in the house at 3:01 P.M.; that it was caused by a short circuit at 3:00 P.M.; that there was only one fire in the house at 3:01 P.M.; that this fire was caused by the one and only short circuit that occurred at 3:00 P.M. Some of these explanations fall short of using all that is given by the premises; and this is lucky, since we often know less. Given only (S) and (P), for example, we cannot prove there was only one fire in the house at 3:01 P.M., though we can prove there was exactly one fire in the house at 3:01 P.M. that was caused by the short circuit. An interesting case is where we know a law in the form of (N), but not the corresponding (S). Then we may show that, given that an event of a particular sort occurred, there must have been a cause answering to a certain description, but, given the same description of the cause, we could not have predicted the effect. An example might be where the effect is getting pregnant.

If we explain why it is that a particular event occurred by deducing a statement that there is such an event (under a particular description) from a premise known to be true, then a simple way of explaining an event, for example the fire in the house at 3:01 P.M., consists in producing a statement of the form of (C); and this explanation makes no use of laws. The explanation will be greatly enhanced by whatever we can say in favor of the truth of (C); needless to say, producing the likes of (L) and (P), if they are known true, clinches the matter. In most cases, however, the request for explanation will describe the event in terms that fall under no full-fledged law. The device to which we will then resort, if we can, is apt to be redescription of the event. For we can explain the occurrence of any event a if we know (L), (P), and the further fact that $a = (\imath e)\,(Ge\ \&\ t(e) = 3 + \epsilon)$. Analogous remarks apply to the redescription of the cause, and to cases where all we want, or can, explain is the fact that there was *an* event of a certain sort.

The great majority of singular causal statements are not backed, we may be sure, by laws in the way (C) is backed by (L). The relation in general is rather this: if 'a caused b' is true, then there are descriptions of a and b such that the result of substituting them for 'a' and 'b' in 'a caused b' is entailed by true premises of the form of (L) and (P); and the converse holds if suitable restrictions are put on the descriptions.[6] If this is correct, it does not follow that we must be able to dredge up a law if we know a singular causal statement to be true; all that follows is that we know there must be a covering law. And very often, I think, our justification for accepting a singular causal statement is that we have reason to believe an appropriate causal law exists, though we do not know what it is. Generalizations like 'If you strike a well-made match hard enough against a properly prepared surface, then, other conditions being favorable, it will light' owe their importance not to the fact that we can hope eventually to render them untendentious and exceptionless, but rather to the fact that they summarize much of our evidence for believing that full-fledged causal laws exist covering events we wish to explain.[7]

If the story I have told is true, it is possible to reconcile, within limits, two accounts thought by their champions to be opposed. One account agrees with Hume and Mill to this extent: it says that a singular causal statement 'a caused b' entails that there is a law to the effect that "all the objects similar to a are followed by objects similar to b" and that we have reason to believe the singular statement only in so far as we have reason to believe there is such a law. The second

---

[6] Clearly this account cannot be taken as a definition of the causal relation. Not only is there the inherently vague quantification over expressions (of what language?), but there is also the problem of spelling out the "suitable restrictions."

[7] The thought in these paragraphs, like much more that appears here, was first adumbrated in my "Actions, Reasons, and Causes," this JOURNAL, LX, 23 (Nov. 7, 1963): 685–700, especially pp. 696–699; reprinted in *Free Will and Determinism*, ed. Bernard Berofsky (New York: Harper & Row, 1966). This conception of causality was subsequently discussed and, with various modifications, employed by Samuel Gorovitz, "Causal Judgments and Causal Explanations," this JOURNAL, LXII, 23 (Dec. 2, 1965): 695–711, and by Bernard Berofsky, "Causality and General Laws," this JOURNAL, LXIII, 6 (Mar. 17, 1966): 148–157.

account (persuasively argued by C. J. Ducasse [8]) maintains that singular causal statements entail no law and that we can know them to be true without knowing any relevant law. Both of these accounts are entailed, I think, by the account I have given, and they are consistent (I therefore hope) with each other. The reconciliation depends, of course, on the distinction between knowing there is a law "covering" two events and knowing what the law is: in my view, Ducasse is right that singular causal statements entail no law; Hume is right that they entail there is a law.

IV

Much of what philosophers have said of causes and causal relations is intelligible only on the assumption (often enough explicit) that causes are individual events, and causal relations hold between events. Yet, through failure to connect this basic *aperçu* with the grammar of singular causal judgments, these same philosophers have found themselves pressed, especially when trying to put causal statements into quantificational form, into trying to express the relation of cause to effect by a sentential connective. Hence the popularity of the utterly misleading question: can causal relations be expressed by the purely extensional material conditional, or is some stronger (non-Humean) connection involved? The question is misleading because it confuses two separate matters: the logical form of causal statements and the analysis of causality. So far as form is concerned, the issue of nonextensionality does not arise, since the relation of causality between events can be expressed (no matter how "strong" or "weak" it is) by an ordinary two-place predicate in an ordinary, extensional first-order language. These plain resources will perhaps be outrun by an adequate account of the form of causal laws, subjunctives, and counterfactual conditionals, to which most attempts to analyze the causal relation turn. But this is, I have urged, another question.

This is not to say there are no causal idioms that directly raise the issue of apparently non-truth-functional connectives. On the contrary, a host of statement forms, many of them strikingly similar, at least at first view, to those we have considered, challenge the account just given. Here are samples: 'The failure of the sprinkling system caused the fire', 'The slowness with which controls were applied caused the rapidity with which the inflation developed', 'The col-

[8] See his "Critique of Hume's Conception of Causality," this JOURNAL, LXIII, 6 (Mar. 17, 1966): 141–148; *Causation and the Types of Necessity* (Seattle: University of Washington Press, 1924); *Nature, Mind, and Death* (La Salle, Ill.: Open Court, 1951), part II. I have omitted from my "second account" much that Ducasse says that is not consistent with Hume.

lapse was caused, not by the fact that the bolt gave way, but by the fact that it gave way so suddenly and unexpectedly', 'The fact that the dam did not hold caused the flood'. Some of these sentences may yield to the methods I have prescribed, especially if failures are counted among events, but others remain recalcitrant. What we must say in such cases is that in addition to, or in place of, giving what Mill calls the "producing cause," such sentences tell, or suggest, a causal story. They are, in other words, rudimentary causal explanations. Explanations typically relate statements, not events. I suggest therefore that the 'caused' of the sample sentences in this paragraph is not the 'caused' of straightforward singular causal statements, but is best expressed by the words 'causally explains'.[9]

A final remark. It is often said that events can be explained and predicted only in so far as they have repeatable characteristics, but not in so far as they are particulars. No doubt there is a clear and trivial sense in which this is true, but we ought not to lose sight of the less obvious point that there is an important difference between explaining the fact that there was *an* explosion in the broom closet and explaining the occurrence of *the* explosion in the broom closet. Explanation of the second sort touches the particular event as closely as language can ever touch any particular. Of course this claim is persuasive only if there are such things as events to which singular terms, especially definite descriptions, may refer. But the assumption, ontological and metaphysical, that there are events, is one without which we cannot make sense of much of our most common talk; or so, at any rate, I have been arguing. I do not know any better, or further, way of showing what there is.

DONALD DAVIDSON

Princeton University

9 Zeno Vendler has ingeniously marshalled the linguistic evidence for a deep distinction, in our use of 'cause', 'effect', and related words, between occurrences of verb-nominalizations that are fact-like or propositional, and occurrences that are event-like. [See Zeno Vendler, "Effects, Results and Consequences," in *Analytic Philosophy*, ed. R. J. Butler (New York: Barnes & Noble, 1962), pp. 1–15.] Vendler concludes that the 'caused' of 'John's action caused the disturbance' is always flanked by expressions used in the propositional or fact-like sense, whereas 'was an effect of' or 'was due to' in 'The shaking of the earth was an effect of (was due to) the explosion' is flanked by expressions in the event-like sense. My distinction between essentially sentential expressions and the expressions that refer to events is much the same as Vendler's and owes much to him, though I have used more traditional semantic tools and have interpreted the evidence differently.

My suggestion that 'caused' is sometimes a relation, sometimes a connective, with corresponding changes in the interpretation of the expressions flanking it, has much in common with the thesis of J. M. Shorter's "Causality, and a Method of Analysis," in *Analytic Philosophy*, II, 1965, pp. 145–157.

CAUSATION *

H UME defined causation twice over. He wrote "we may define a cause to be *an object followed by another, and where all the objects, similar to the first, are followed by objects similar to the second.* Or, in other words, *where, if the first object had not been, the second never had existed.*"[1]

Descendants of Hume's first definition still dominate the philosophy of causation: a causal succession is supposed to be a succession that instantiates a regularity. To be sure, there have been improvements. Nowadays we try to distinguish the regularities that count—the "causal laws"—from mere accidental regularities of succession. We subsume causes and effects under regularities by means of descriptions they satisfy, not by over-all similarity. And we allow a cause to be only one indispensable part, not the whole, of the total situation that is followed by the effect in accordance with a law. In present-day regularity analyses, a cause is defined (roughly) as any member of any minimal set of actual conditions that are jointly sufficient, given the laws, for the existence of the effect.

More precisely, let C be the proposition that c exists (or occurs) and let E be the proposition that e exists. Then c causes e, according to a typical regularity analysis,[2] iff (1) C and E are true; and (2) for some nonempty set $\mathcal{L}$ of true law-propositions and some set $\mathcal{F}$ of true propositions of particular fact, $\mathcal{L}$ and $\mathcal{F}$ jointly imply $C \supset E$, although $\mathcal{L}$ and $\mathcal{F}$ jointly do not imply E and $\mathcal{F}$ alone does not imply $C \supset E$.[3]

Much needs doing, and much has been done, to turn definitions like this one into defensible analyses. Many problems have been overcome. Others remain: in particular, regularity analyses tend to confuse causation itself with various other causal relations. If c belongs to a minimal set of conditions jointly sufficient for e, given the laws,

* To be presented in an APA symposium on Causation, December 28, 1973; commentators will be Bernard Berofsky and Jaegwon Kim; see this JOURNAL, this issue, pp. 568–569 and 570–572, respectively.

I thank the American Council of Learned Societies, Princeton University, and the National Science Foundation for research support.

[1] *An Enquiry concerning Human Understanding*, Section VII.

[2] Not one that has been proposed by any actual author in just this form, so far as I know.

[3] I identify a *proposition*, as is becoming usual, with the set of possible worlds where it is true. It is not a linguistic entity. Truth-functional operations on propositions are the appropriate Boolean operations on sets of worlds; logical relations among propositions are relations of inclusion, overlap, etc. among sets. A sentence of a language *expresses* a proposition iff the sentence and the proposition are true at exactly the same worlds. No ordinary language will provide sentences to express all propositions; there will not be enough sentences to go around.

then *c* may well be a genuine cause of *e*. But *c* might rather be an effect of *e*: one which could not, given the laws and some of the actual circumstances, have occurred otherwise than by being caused by *e*. Or *c* might be an epiphenomenon of the causal history of *e*: a more or less inefficacious effect of some genuine cause of *e*. Or *c* might be a preempted potential cause of *e*: something that did not cause *e*, but that would have done so in the absence of whatever really did cause *e*.

It remains to be seen whether any regularity analysis can succeed in distinguishing genuine causes from effects, epiphenomena, and preempted potential causes—and whether it can succeed without falling victim to worse problems, without piling on the epicycles, and without departing from the fundamental idea that causation is instantiation of regularities. I have no proof that regularity analyses are beyond repair, nor any space to review the repairs that have been tried. Suffice it to say that the prospects look dark. I think it is time to give up and try something else.

A promising alternative is not far to seek. Hume's "other words" —that if the cause had not been, the effect never had existed—are no mere restatement of his first definition. They propose something altogether different: a counterfactual analysis of causation.

The proposal has not been well received. True, we do know that causation has something or other to do with counterfactuals. We think of a cause as something that makes a difference, and the difference it makes must be a difference from what would have happened without it. Had it been absent, its effects—some of them, at least, and usually all—would have been absent as well. Yet it is one thing to mention these platitudes now and again, and another thing to rest an analysis on them. That has not seemed worth while.[4] We have learned all too well that counterfactuals are ill understood, wherefore it did not seem that much understanding could be gained by using them to analyze causation or anything else. Pending a better understanding of counterfactuals, moreover, we had no way to fight seeming counterexamples to a counterfactual analysis.

But counterfactuals need not remain ill understood, I claim, unless we cling to false preconceptions about what it would be like to understand them. Must an adequate understanding make no reference to unactualized possibilities? Must it assign sharply determinate truth conditions? Must it connect counterfactuals rigidly to covering laws? Then none will be forthcoming. So much the worse for those standards of adequacy. Why not take counterfactuals at face value:

[4] One exception: Aardon Lyon, "Causality," *British Journal for Philosophy of Science*, XVIII, 1 (May 1967): 1–20.

as statements about possible alternatives to the actual situation, somewhat vaguely specified, in which the actual laws may or may not remain intact? There are now several such treatments of counterfactuals, differing only in details.[5] If they are right, then sound foundations have been laid for analyses that use counterfactuals.

In this paper, I shall state a counterfactual analysis, not very different from Hume's second definition, of some sorts of causation. Then I shall try to show how this analysis works to distinguish genuine causes from effects, epiphenomena, and preempted potential causes.

My discussion will be incomplete in at least four ways. Explicit preliminary settings-aside may prevent confusion.

1. I shall confine myself to causation among *events*, in the everyday sense of the word: flashes, battles, conversations, impacts, strolls, deaths, touchdowns, falls, kisses, and the like. Not that events are the only things that can cause or be caused; but I have no full list of the others, and no good umbrella-term to cover them all.

2. My analysis is meant to apply to causation in particular cases. It is not an analysis of causal generalizations. Presumably those are quantified statements involving causation among particular events (or non-events), but it turns out not to be easy to match up the causal generalizations of natural language with the available quantified forms. A sentence of the form "c-events cause E-events," for instance, can mean any of

(a) For some c in c and some e in E, c causes e.

(b) For every e in E, there is some c in c such that c causes e.

(c) For every c in c, there is some e in E such that c causes e.

not to mention further ambiguities. Worse still, 'Only c-events cause E-events' ought to mean

(d) For every c, if there is some e in E such that c causes e, then c is in c.

if 'only' has its usual meaning. But no; it unambiguously means (b) instead! These problems are not about causation, but about our idioms of quantification.

3. We sometimes single out one among all the causes of some event and call it "the" cause, as if there were no others. Or we single out a few as the "causes," calling the rest mere "causal factors" or "causal conditions." Or we speak of the "decisive" or "real" or "principal" cause. We may select the abnormal or extraordinary

---

[5] See, for instance, Robert Stalnaker, "A Theory of Conditionals," in Nicholas Rescher, ed., *Studies in Logical Theory* (Oxford: Blackwell, 1968); and my *Counterfactuals* (Oxford: Blackwell, 1973).

causes, or those under human control, or those we deem good or bad, or just those we want to talk about. I have nothing to say about these principles of invidious discrimination.[6] I am concerned with the prior question of what it is to be one of the causes (unselectively speaking). My analysis is meant to capture a broad and nondiscriminatory concept of causation.

4. I shall be content, for now, if I can give an analysis of causation that works properly under determinism. By determinism I do not mean any thesis of universal causation, or universal predictability-in-principle, but rather this: the prevailing laws of nature are such that there do not exist any two possible worlds which are exactly alike up to some time, which differ thereafter, and in which those laws are never violated. Perhaps by ignoring indeterminism I squander the most striking advantage of a counterfactual analysis over a regularity analysis: that it allows undetermined events to be caused.[7] I fear, however, that my present analysis cannot yet cope with all varieties of causation under indeterminism. The needed repair would take us too far into disputed questions about the foundations of probability.

COMPARATIVE SIMILARITY

To begin, I take as primitive a relation of *comparative over-all* similarity among possible worlds. We may say that one world is *closer to actuality* than another if the first resembles our actual world more than the second does, taking account of all the respects of similarity and difference and balancing them off one against another.

(More generally, an arbitrary world w can play the role of our actual world. In speaking of our actual world without knowing just which world is ours, I am in effect generalizing over all worlds. We really need a three-place relation: world w_1 is closer to world w than world w_2 is. I shall henceforth leave this generality tacit.)

I have not said just how to balance the respects of comparison against each other, so I have not said just what our relation of comparative similarity is to be. Not for nothing did I call it primitive. But I have said what *sort* of relation it is, and we are familiar with relations of that sort. We do make judgments of comparative over-all similarity—of people, for instance—by balancing off many re-

[6] Except that Morton G. White's discussion of causal selection, in *Foundations of Historical Knowledge* (New York: Harper & Row, 1965), pp. 105–181, would meet my needs, despite the fact that it is based on a regularity analysis.

[7] That this ought to be allowed is argued in G. E. M. Anscombe, *Causality and Determination: An Inaugural Lecture* (Cambridge: University Press, 1971); and in Fred Dretske and Aaron Snyder, "Causal Irregularity," *Philosophy of Science*, XXXIX, 1 (March 1972): 69–71.

spects of similarity and difference. Often our mutual expectations about the weighting factors are definite and accurate enough to permit communication. I shall have more to say later about the way the balance must go in particular cases to make my analysis work. But the vagueness of over-all similarity will not be entirely resolved. Nor should it be. The vagueness of similarity does infect causation, and no correct analysis can deny it.

The respects of similarity and difference that enter into the over-all similarity of worlds are many and varied. In particular, similarities in matters of particular fact trade off against similarities of law. The prevailing laws of nature are important to the character of a world; so similarities of law are weighty. Weighty, but not sacred. We should not take it for granted that a world that conforms perfectly to our actual laws is *ipso facto* closer to actuality than any world where those laws are violated in any way at all. It depends on the nature and extent of the violation, on the place of the violated laws in the total system of laws of nature, and on the countervailing similarities and differences in other respects. Likewise, similarities or differences of particular fact may be more or less weighty, depending on their nature and extent. Comprehensive and exact similarities of particular fact throughout large spatiotemporal regions seem to have special weight. It may be worth a small miracle to prolong or expand a region of perfect match.

Our relation of comparative similarity should meet two formal constraints. (1) It should be a weak ordering of the worlds: an ordering in which ties are permitted, but any two worlds are comparable. (2) Our actual world should be closest to actuality, resembling itself more than any other world resembles it. We do *not* impose the further constraint that for any set A of worlds there is a unique closest A-world, or even a set of A-worlds tied for closest. Why not an infinite sequence of closer and closer A-worlds, but no closest?

COUNTERFACTUALS AND COUNTERFACTUAL DEPENDENCE

Given any two propositions A and C, we have their *counterfactual* $A \,\square\!\!\rightarrow C$: the proposition that if A were true, then C would also be true. The operation $\square\!\!\rightarrow$ is defined by a rule of truth, as follows. $A \,\square\!\!\rightarrow C$ is true (at a world w) iff either (1) there are no possible A-worlds (in which case $A \,\square\!\!\rightarrow C$ is *vacuous*), or (2) some A-world where C holds is closer (to w) than is any A-world where C does not hold. In other words, a counterfactual is nonvacuously true iff it takes less of a departure from actuality to make the consequent true along with the antecedent than it does to make the antecedent true without the consequent.

We did not assume that there must always be one or more closest A-worlds. But if there are, we can simplify: $A \;\square\!\!\rightarrow C$ is nonvacuously true iff C holds at all the closest A-worlds.

We have not presupposed that A is false. If A is true, then our actual world is the closest A-world, so $A \;\square\!\!\rightarrow C$ is true iff C is. Hence $A \;\square\!\!\rightarrow C$ implies the material conditional $A \supset C$; and A and C jointly imply $A \;\square\!\!\rightarrow C$.

Let $A_1, A_2, \ldots$ be a family of possible propositions, no two of which are compossible; let $C_1, C_2, \ldots$ be another such family (of equal size). Then if all the counterfactuals $A_1 \;\square\!\!\rightarrow C_1$, $A_2 \;\square\!\!\rightarrow C_2$, $\ldots$ between corresponding propositions in the two families are true, we shall say that the C's *depend counterfactually* on the A's. We can say it like this in ordinary language: whether C_1 or C_2 or $\ldots$ depends (counterfactually) on whether A_1 or A_2 or $\ldots$.

Counterfactual dependence between large families of alternatives is characteristic of processes of measurement, perception, or control. Let $R_1, R_2, \ldots$ be propositions specifying the alternative readings of a certain barometer at a certain time. Let $P_1, P_2, \ldots$ specify the corresponding pressures of the surrounding air. Then, if the barometer is working properly to measure the pressure, the R's must depend counterfactually on the P's. As we say it: the reading depends on the pressure. Likewise, if I am seeing at a certain time, then my visual impressions must depend counterfactually, over a wide range of alternative possibilities, on the scene before my eyes. And if I am in control over what happens in some respect, then there must be a double counterfactual dependence, again over some fairly wide range of alternatives. The outcome depends on what I do, and that in turn depends on which outcome I want.[3]

CAUSAL DEPENDENCE AMONG EVENTS

If a family $C_1, C_2, \ldots$ depends counterfactually on a family A_1, $A_2, \ldots$ in the sense just explained, we will ordinarily be willing to speak also of causal dependence. We say, for instance, that the barometer reading depends causally on the pressure, that my visual impressions depend causally on the scene before my eyes, or that the outcome of something under my control depends causally on what I do. But there are exceptions. Let $G_1, G_2, \ldots$ be alternative possible laws of gravitation, differing in the value of some numerical constant. Let $M_1, M_2, \ldots$ be suitable alternative laws of planetary motion. Then the M's may depend counterfactually on the G's, but

---

[3] Analyses in terms of counterfactual dependence are found in two papers of Alvin I. Goldman: "Toward a Theory of Social Power," *Philosophical Studies*, XXIII (1972): 221–268; and "Discrimination and Perceptual Knowledge," presented at the 1972 Chapel Hill Colloquium.

we would not call this dependence causal. Such exceptions as this, however, do not involve any sort of dependence among distinct particular events. The hope remains that causal dependence among events, at least, may be analyzed simply as counterfactual dependence.

We have spoken thus far of counterfactual dependence among propositions, not among events. Whatever particular events may be, presumably they are not propositions. But that is no problem, since they can at least be paired with propositions. To any possible event e, there corresponds the proposition $O(e)$ that holds at all and only those worlds where e occurs. This $O(e)$ is the proposition that e occurs.[9] (If no two events occur at exactly the same worlds—if, that is, there are no absolutely necessary connections between distinct events—we may add that this correspondence of events and propositions is one to one.) Counterfactual dependence among events is simply counterfactual dependence among the corresponding propositions.

Let $c_1, c_2, \ldots$ and $e_1, e_2, \ldots$ be distinct possible events such that no two of the c's and no two of the e's are compossible. Then I say that the family $e_1, e_2, \ldots$ of events *depends causally* on the family $c_1, c_2, \ldots$ iff the family $O(e_1), O(e_2), \ldots$ of propositions depends counterfactually on the family $O(c_1), O(c_2), \ldots$. As we say it: whether e_1 or e_2 or $\ldots$ occurs depends on whether c_1 or c_2 or $\ldots$ occurs.

We can also define a relation of dependence among single events rather than families. Let c and e be two distinct possible particular

[9] Beware: if we refer to a particular event e by means of some description that e satisfies, then we must take care not to confuse $O(e)$, the proposition that e itself occurs, with the different proposition that some event or other occurs which satisfies the description. It is a contingent matter, in general, what events satisfy what descriptions. Let e be the death of Socrates—the death he actually died, to be distinguished from all the different deaths he might have died instead. Suppose that Socrates had fled, only to be eaten by a lion. Then e would not have occurred, and $O(e)$ would have been false; but a different event would have satisfied the description 'the death of Socrates' that I used to refer to e. Or suppose that Socrates had lived and died just as he actually did, and afterwards was resurrected and killed again and resurrected again, and finally became immortal. Then no event would have satisfied the description. (Even if the temporary deaths are real deaths, neither of the two can be *the* death.) But e would have occurred, and $O(e)$ would have been true. Call a description of an event e *rigid* iff (1) nothing but e could possibly satisfy it, and (2) e could not possibly occur without satisfy it. I have claimed that even such common-place descriptions as 'the death of Socrates' are nonrigid, and in fact I think that rigid descriptions of events are hard to find. That would be a problem for anyone who needed to associate with every possible event e a sentence $\phi(e)$ true at all and only those worlds where e occurs. But we need no such sentences—only propositions, which may or may not have expressions in our language.

events. Then *e depends causally* on *c* iff the family $O(e)$, $\sim O(e)$ depends counterfactually on the family $O(c)$, $\sim O(c)$. As we say it: whether *e* occurs or not depends on whether *c* occurs or not. The dependence consists in the truth of two counterfactuals: $O(c) \;\square\!\!\rightarrow O(e)$ and $\sim O(c) \;\square\!\!\rightarrow \sim O(e)$. There are two cases. If *c* and *e* do not actually occur, then the second counterfactual is automatically true because its antecedent and consequent are true: so *e* depends causally on *c* iff the first counterfactual holds. That is, iff *e* would have occurred if *c* had occurred. But if *c* and *e* are actual events, then it is the first counterfactual that is automatically true. Then *e* depends causally on *c* iff, if *c* had not been, *e* never had existed. I take Hume's second definition as my definition not of causation itself, but of causal dependence among actual events.

CAUSATION

Causal dependence among actual events implies causation. If *c* and *e* are two actual events such that *e* would not have occurred without *c*, then *c* is a cause of *e*. But I reject the converse. Causation must always be transitive; causal dependence may not be; so there can be causation without causal dependence. Let *c*, *d*, and *e* be three actual events such that *d* would not have occurred without *c* and *e* would not have occurred without *d*. Then *c* is a cause of *e* even if *e* would still have occurred (otherwise caused) without *c*.

We extend causal dependence to a transitive relation in the usual way. Let *c*, *d*, *e*, . . . be a finite sequence of actual particular events such that *d* depends causally on *c*, *e* on *d*, and so on throughout. Then this sequence is a *causal chain*. Finally, one event is a *cause* of another iff there exists a causal chain leading from the first to the second. This completes my counterfactual analysis of causation.

COUNTERFACTUAL VERSUS NOMIC DEPENDENCE

It is essential to distinguish counterfactual and causal dependence from what I shall call *nomic dependence*. The family $C_1, C_2, \ldots$ of propositions depends nomically on the family $A_1, A_2, \ldots$ iff there are a nonempty set $\mathcal{L}$ of true law-propositions and a set $\mathcal{F}$ of true propositions of particular fact such that $\mathcal{L}$ and $\mathcal{F}$ jointly imply (but $\mathcal{F}$ alone does not imply) all the material conditionals $A_1 \supset C_1, A_2 \supset C_2,$ . . . between the corresponding propositions in the two families. (Recall that these same material conditionals are implied by the counterfactuals that would comprise a counterfactual dependence.) We shall say also that the nomic dependence holds *in virtue of* the premise sets $\mathcal{L}$ and $\mathcal{F}$.

Nomic and counterfactual dependence are related as follows. Say that a proposition *B* is *counterfactually independent* of the family A_1,

A_3, . . . of alternatives iff B would hold no matter which of the A's were true—that is, iff the counterfactuals $A_1 \;\square\!\!\rightarrow B$, $A_2 \;\square\!\!\rightarrow B$, . . . all hold. If the C's depend nomically on the A's in virtue of the premise sets $\mathcal{L}$ and $\mathcal{F}$, and if in addition (all members of) $\mathcal{L}$ and $\mathcal{F}$ are counterfactually independent of the A's, then it follows that the C's depend counterfactually on the A's. In that case, we may regard the nomic dependence in virtue of $\mathcal{L}$ and $\mathcal{F}$ as explaining the counterfactual dependence. Often, perhaps always, counterfactual dependences may be thus explained. But the requirement of counterfactual independence is indispensable. Unless $\mathcal{L}$ and $\mathcal{F}$ meet that requirement, nomic dependence in virtue of $\mathcal{L}$ and $\mathcal{F}$ does not imply counterfactual dependence, and, if there is counterfactual dependence anyway, does not explain it.

Nomic dependence is reversible, in the following sense. If the family C_1, C_2, . . . depends nomically on the family A_1, A_2, . . . in virtue of $\mathcal{L}$ and $\mathcal{F}$, then also A_1, A_2, . . . depends nomically on the family AC_1, AC_2, . . . , in virtue of $\mathcal{L}$ and $\mathcal{F}$, where A is the disjunction $A_1 \vee A_2 \vee$ . . . . Is counterfactual dependence likewise reversible? That does not follow. For, even if $\mathcal{L}$ and $\mathcal{F}$ are independent of A_1, A_2, . . . and hence establish the counterfactual dependence of the C's on the A's, still they may fail to be independent of AC_1, AC_2, . . . , and hence may fail to establish the reverse counterfactual dependence of the A's on the AC's. Irreversible counterfactual dependence is shown below: @ is our actual world, the dots are the other worlds, and distance on the page represents similarity "distance."

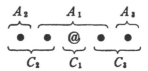

The counterfactuals $A_1 \;\square\!\!\rightarrow C_1$, $A_2 \;\square\!\!\rightarrow C_2$, and $A_3 \;\square\!\!\rightarrow C_3$ hold at the actual world; wherefore the C's depend on the A's. But we do not have the reverse dependence of the A's on the AC's, since instead of the needed $AC_2 \;\square\!\!\rightarrow A_2$ and $AC_3 \;\square\!\!\rightarrow A_3$ we have $AC_2 \;\square\!\!\rightarrow A_1$ and $AC_3 \;\square\!\!\rightarrow A_1$.

Just such irreversibility is commonplace. The barometer reading depends counterfactually on the pressure—that is as clear-cut as counterfactuals ever get—but does the pressure depend counterfactually on the reading? If the reading had been higher, would the pressure have been higher? Or would the barometer have been malfunctioning? The second sounds better: a higher reading would have been an incorrect reading. To be sure, there are actual laws and cir-

368

cumstances that imply and explain the actual accuracy of the barometer, but these are no more sacred than the actual laws and circumstances that imply and explain the actual pressure. Less sacred, in fact. When something must give way to permit a higher reading, we find it less of a departure from actuality to hold the pressure fixed and sacrifice the accuracy, rather than vice versa. It is not hard to see why. The barometer, being more localized and more delicate than the weather, is more vulnerable to slight departures from actuality.[10]

We can now explain why regularity analyses of causation (among events, under determinism) work as well as they do. Suppose that event c causes event e according to the sample regularity analysis that I gave at the beginning of this paper, in virtue of premise sets $\mathcal{L}$ and $\mathcal{F}$. It follows that $\mathcal{L}$, $\mathcal{F}$, and $\sim O(c)$ jointly do not imply $O(e)$. Strengthen this: suppose further that they do imply $\sim O(e)$. If so, the family $O(e)$, $\sim O(e)$, depends nomically on the family $O(c)$, $\sim O(c)$ in virtue of $\mathcal{L}$ and $\mathcal{F}$. Add one more supposition: that $\mathcal{L}$ and $\mathcal{F}$ are counterfactually independent of $O(c)$, $\sim O(c)$. Then it follows according to my counterfactual analysis that e depends counterfactually and causally on c, and hence that c causes e. If I am right, the regularity analysis gives conditions that are almost but not quite sufficient for explicable causal dependence. That is not quite the same thing as causation; but causation without causal dependence is scarce, and if there is inexplicable causal dependence we are (understandably!) unaware of it.[11]

EFFECTS AND EPIPHENOMENA

I return now to the problems I raised against regularity analyses, hoping to show that my counterfactual analysis can overcome them.

The *problem of effects*, as it confronts a counterfactual analysis, is as follows. Suppose that c causes a subsequent event e, and that e does not also cause c. (I do not rule out closed causal loops a priori, but this case is not to be one.) Suppose further that, given the laws and some of the actual circumstances, c could not have failed to

---

[10] Granted, there are contexts or changes of wording that would incline us the other way. For some reason, "If the reading had been higher, that would have been because the pressure was higher" invites my assent more than "If the reading had been higher, the pressure would have been higher." The counterfactuals from readings to pressures are much less clear-cut than those from pressures to readings. But it is enough that some legitimate resolutions of vagueness give an irreversible dependence of readings on pressures. Those are the resolutions we want at present, even if they are not favored in all contexts.

[11] I am not here proposing a repaired regularity analysis. The repaired analysis would gratuitously rule out inexplicable causal dependence, which seems bad. Nor would it be squarely in the tradition of regularity analyses any more. Too much else would have been added.

cause e. It seems to follow that if the effect e had not occurred, then its cause c would not have occurred. We have a spurious reverse causal dependence of c on e, contradicting our supposition that e did not cause c.

The *problem of epiphenomena*, for a counterfactual analysis, is similar. Suppose that e is an epiphenomenal effect of a genuine cause c of an effect f. That is, c causes first e and then f, but e does not cause f. Suppose further that, given the laws and some of the actual circumstances, c could not have failed to cause e; and that, given the laws and others of the circumstances, f could not have been caused otherwise than by c. It seems to follow that if the epiphenomenon e had not occurred, then its cause c would not have occurred and the further effect f of that same cause would not have occurred either. We have a spurious causal dependence of f on e, contradicting our supposition that e did not cause f.

One might be tempted to solve the problem of effects by brute force: insert into the analysis a stipulation that a cause must always precede its effect (and perhaps a parallel stipulation for causal dependence). I reject this solution. (1) It is worthless against the closely related problem of epiphenomena, since the epiphenomenon e does precede its spurious effect f. (2) It rejects a priori certain legitimate physical hypotheses that posit backward or simultaneous causation. (3) It trivializes any theory that seeks to define the forward direction of time as the predominant direction of causation.

The proper solution to both problems, I think, is flatly to deny the counterfactuals that cause the trouble. If e had been absent, it is not that c would have been absent (and with it f, in the second case). Rather, c would have occurred just as it did but would have failed to cause e. It is less of a departure from actuality to get rid of e by holding c fixed and giving up some or other of the laws and circumstances in virtue of which c could not have failed to cause e, rather than to hold those laws and circumstances fixed and get rid of e by going back and abolishing its cause c. (In the second case, it would of course be pointless not to hold f fixed along with c.) The causal dependence of e on c is the same sort of irreversible counterfactual dependence that we have considered already.

To get rid of an actual event e with the least over-all departure from actuality, it will normally be best not to diverge at all from the actual course of events until just before the time of e. The longer we wait, the more we prolong the spatiotemporal region of perfect match between our actual world and the selected alternative. Why diverge sooner rather than later? Not to avoid violations of laws of nature.

Under determinism *any* divergence, soon or late, requires some violation of the actual laws. If the laws were held sacred, there would be no way to get rid of e without changing all of the past; and nothing guarantees that the change could be kept negligible except in the recent past. That would mean that if the present were ever so slightly different, then all of the past would have been different—which is absurd. So the laws are not sacred. Violation of laws is a matter of degree. Until we get up to the time immediately before e is to occur, there is no general reason why a later divergence to avert e should need a more severe violation than an earlier one. Perhaps there are special reasons in special cases—but then these may be cases of backward causal dependence.

PREEMPTION

Suppose that c_1 occurs and causes e; and that c_2 also occurs and does not cause e, but would have caused e if c_1 had been absent. Thus c_2 is a potential alternate cause of e, but is preempted by the actual cause c_1. We may say that c_1 and c_2 overdetermine e, but they do so asymmetrically.[12] In virtue of what difference does c_1 but not c_2 cause e?

As far as causal dependence goes, there is no difference: e depends neither on c_1 nor on c_2. If either one had not occurred, the other would have sufficed to cause e. So the difference must be that, thanks to c_1, there is no causal chain from c_2 to e; whereas there is a causal chain of two or more steps from c_1 to e. Assume for simplicity that two steps are enough. Then e depends causally on some intermediate event d, and d in turn depends on c_1. Causal dependence is here intransitive: c_1 causes e via d even though e would still have occurred without c_1.

So far, so good. It remains only to deal with the objection that e does *not* depend causally on d, because if d had been absent then c_1 would have been absent and c_2, no longer preempted, would have caused e. We may reply by denying the claim that if d had been absent then c_1 would have been absent. That is the very same sort of spurious reverse dependence of cause on effect that we have just rejected in simpler cases. I rather claim that if d had been absent, c_1 would somehow have failed to cause d. But c_1 would still have been there to interfere with c_2, so e would not have occurred.

DAVID LEWIS

Princeton University

---

[12] I shall not discuss symmetrical cases of overdetermination, in which two overdetermining factors have equal claim to count as causes. For me these are useless as test cases because I lack firm naive opinions about them.

PROBABILISTIC CAUSAL INTERACTION*

ELLERY EELLS†

Department of Philosophy
University of Wisconsin–Madison

It is possible for a causal factor to raise the probability of a second factor in some situations while *lowering* the probability of the second factor in *other* situations. Must a genuine cause *always* raise the probability of a genuine effect of it? When it does not always do so, an "interaction" with some third factor may be the reason. I discuss causal interaction from the perspectives of Giere's counterfactual characterization of probabilistic causal connection (1979, 1980) and the "contextual unanimity" model developed by, among others, Cartwright (1979) and Skyrms (1980). I argue that the contextual unanimity theory must exercise care, in a new way that seems to have gone unnoticed, in order to adequately accommodate the phenomenon, and that the counterfactual theory must be substantially revised; although it will still, pending clarification of a second kind of revision, be unable to accommodate a kind of interaction exemplified in cases like those described by Sober (1982).

Common to theories of probabilistic causality recently discussed by philosophers is the idea that a cause raises the probability of its effect. In order to accommodate well-known examples of various kinds of "spurious correlation," however, it is necessary to be careful in formulating the idea. Effects often raise the probability of their causes; joint effects of a common cause often raise each others' probabilities; and if a cause is strongly correlated with a factor that prevents its effect, then the cause may actually *lower* the probability of its effect, on average. According to the theory advanced by Nancy Cartwright (1979), therefore, a factor C is a cause of a factor E (in a given population) if and only if C raises the probability of E given each way of *holding fixed* (positively or negatively) all and only those factors—other than C and C's effects—that are themselves causes of E or of $\sim E$. I call this the "contextual unanimity" model.[1] Let us call the relation Cartwright has defined that of C's

---

*Received October 1984; revised November 1984.

†I thank Elliott Sober and an anonymous referee of this journal for useful comments on earlier drafts of this paper, and the University of Wisconsin–Madison Graduate School for financial support.

[1]For other versions of the same kind of model, see, e.g., Suppes (1970) and Skyrms (1980). One variant worth noting is Skyrms's suggestion of the weaker, Pareto-like condition that a cause must raise the probability of its effect in *at least one* (as opposed to every) constellation of relevant background factors and must not lower it in any other. The discussion below applies to this version as well. It is also worth noting that this kind of

Philosophy of Science, 53 (1986) pp. 52–64.
Copyright © 1986 by the Philosophy of Science Association.

being a *positive causal factor for E*. Then, *negative causal factorhood* can be defined by replacing "raises" by "lowers," and *causal irrelevance* (or *neutrality*) can be defined by replacing "raises" by "leaves unchanged." Note that on this model, the three possibilities of a given factor's being causally *positive, negative,* or *neutral* for another are mutually exclusive but *not* collectively exhaustive.

According to the counterfactual approach to probabilistic causal connection that Ronald Giere advocates (1979, 1980),[2] *C* is a *positive (negative) causal factor for E in population P* if and only if there *would* be a higher (lower) frequency of *E* if every individual in *P* had *C* than there would be if no individual in *P* had *C*, where the individuals in the two *hypothetical populations* (one in which every individual has *C* and one in which no individual has *C*) are exactly like individuals in the actual population except (at most) with respect to the presence of *C*. *C* is *causally irrelevant to* (or *neutral for*) *E* when the two hypothetical populations exhibit the same frequency of *E*. On this model, it seems that the three possibilities of one factor's being causally *positive, negative,* or *neutral* for another are both mutually exclusive *and* collectively exhaustive.[3]

The question naturally arises of whether a cause really must, for *every* kind of individual, raise the probability of its effect, as the contextual unanimity model requires, but Giere's seems not to. Cartwright gives a reason why a cause may seem not to be "unanimous" for its effect: there

---

analysis of "*C* causes *E*" does not assume that there actually are individuals for whom *C* causes *E*. Under a propensity, or a hypothetical relative frequency, interpretation of probability, the probabilistic criterion merely describes how *C would* effect the probability of *E* in causally homogeneous subpopulations. Thus, it could be true (under the analysis) that *C* causes *E* even if no individual exhibited *C*—and it could be true even if no individuals with *C* actually have *E*. See Eells and Sober (1983, pp. 37–38) for elaboration of this point.

[2] It should be noted that Giere's approach is not intended to provide a *reductive analysis of the meaning* of causal statements in terms of counterfactuals; rather, his models are intended to help in understanding causal hypotheses *in terms of the ways in which they are tested and applied, using also the ideas of physical necessity and propensity* (see Giere 1984). Of course, these two kinds of approaches to understanding causal hypotheses are not unrelated, for (as I think Giere would allow) what we understand causal hypotheses to assert helps in the understanding of methodology, and understanding that such and such a methodology is appropriate for the confirmation, disconfirmation or application of a causal hypothesis helps us understand the meaning of the hypothesis. The distinctions involving interaction that I shall advance in this paper are relevant to both kinds of approaches: both conceptual analysis and models of causation in terms of methodology, if adequate, must be sensitive to them.

[3] Actually, the question of whether or not this holds for Giere's theory is complicated by another condition that Giere sometimes includes, the condition that for *C* to be a positive causal factor for *E* in population *P*, there must be at least one individual in *P* for whom the presence or absence of *C* makes a difference with respect to *E*. It is unclear to me what the analogue of this would be for the case of causal *neutrality*. Also, the model presented here is the version for populations of deterministic systems. Giere also advances a model for stochastic systems. The discussion below applies to both.

may be an "interaction," and "two causal factors are interactive if in combination they act like a single causal factor whose effects are different from at least one of the two acting separately" (1979, 427–28). Her example involves the two factors of ingesting an acid poison and ingesting an alkali poison: the presence of one without the other causes death, but the presence of neither or both results in survival. Should we deny that ingesting acid poison causes death just because there is a kind of individual (those who have ingested alkali poison) for whom ingesting acid *decreases* the probability of death?[4]

I shall discuss two general ways in which a theory of probabilistic causality may deal with examples of this sort, which initially may seem to come to quite the same thing—but don't. *First*, in Cartwright's example, we might say that neither of the factors *separately* have a definite causal role with respect to survival and death, and that we should focus instead on the four *combined, conjunctive* factors specified by saying *both* whether or not the acid is ingested *and* whether or not the alkaline is ingested. According to this approach, the causal truth will be captured by four causal statements, each to the effect that one of the four combined factors causes, or does not cause, death. This approach is adopted by Cartwright, and she claims that it "accords" with her analysis of probabilistic causality, but adds that "considerably more has to be said about interactions" (1979, p. 428). In the sequel, this approach to causal interaction will be called 'treatment I'. A *second* approach would make use of the fact that probabilistic causality is a relation between three things: a cause, an effect, and a *population* within which the former is a cause of the latter. On this approach, the causal truth in Cartwright's example would be expressed as follows: *among* individuals who have ingested an alkaline (acid) poison, ingesting an acid (alkaline) poison will cause survival, and *among* people who have not ingested the alkaline (acid), ingesting acid (alkaline) will cause death. I will call this second approach to causal interaction 'treatment II'.

I want to consider the merits of these two approaches to causal interaction. But first, I want to be more precise about what causal interaction amounts to. The question with which we began, recall, was: "Must a cause increase the probability of its effect in *every* causally fixed situation?" *Thus, let us understand causal interaction to be the (not always symmetrical) relation on which a factor C interacts with a factor F, with respect to E as the effect, if and only if C raises the probability of E in the presence of F and lowers the probability of E in the absence of F,*

---

[4]Note the reference to *kinds* of individuals; for interaction, as well as for simple causation, it is not assumed that there actually are individuals of all the relevant kinds. See n. 1.

or vice versa.[5] This is somewhat different from Cartwright's character-ization of interaction, and it would, of course, be unfair to hold Cart-wright to her approach to causal interaction given this recharacterization of the idea—that is why I refer to the two approaches to interaction sketched above simply as 'treatment I' and 'treatment II'. I will argue, by way of examples, that: (i) treatment I of causal interaction is not generally ad-equate (either for Giere's counterfactual theory or for the contextual un-animity model), (ii) treatment II has more promise of being generally adequate, and (iii) both the counterfactual theory and Cartwright's for-mulation of the contextual unanimity theory must be revised to deal ad-equately with the phenomenon of causal interaction (though I will note that even the suggested revision of Giere's theory will still, pending clar-ification of a *second* kind of revision discussed below, not be able to accommodate a kind of interaction first brought to bear on Giere's theory by Elliott Sober [1982]).

The examples I will use will be versions of the following. Suppose smoking (S) caused heart attacks (H) in individuals who do not exercise and that, for exercisers (X), smoking prevents heart attacks. Thus, smok-ing interacts with exercising, with respect to getting a heart attack. Let us first see how Giere's theory deals with this example, and then turn to the contextual unanimity model.

Does smoking cause heart attacks in the general population on Giere's model? On this model, we must compare the frequency of heart attack victims in two hypothetical populations, one just like the actual popula-tion except that everyone smokes, the other just like the actual population except that no one smokes. Now if the *actual frequency of exercisers* is high enough, then the frequency of heart attack victims should be *lower* were everyone to smoke than were nobody to smoke—since, by hypoth-esis, smoking prevents heart attacks among exercisers. But if the actual frequency of exercisers is *low* enough, then the *reverse* should hold true. *Thus, in this example, whether smoking is causally positive or negative for heart attacks depends on the actual frequency of exercisers, on Giere's theory.*

But it seems clear that it is (at least) incomplete or misleading to char-acterize the causal role of smoking for heart attacks simply as positive or as negative, when smoking's impact on one's chances of getting a heart attack depends on whether or not one exercises. Indeed, it would seem *false* to characterize smoking's role as univocally positive or negative in

[5]A more general approach here, and in the sequel, would be to define causal interaction as a relation between C and a *partition* of factors. For a discussion of the connection of the idea of interaction used here with the more general idea of interaction used in the analysis of variance (ANOVA), see Sober (1984a, chap. 7). In particular, I am not count-ing *changes* of probability that are not *reversals* as cases of interaction.

this kind of case; for regardless of the actual frequency of exercisers, smoking has, by hypothesis, *two* causal roles for heart attacks, *both* positive *and* negative, positive for nonexercisers and negative for exercisers. Indeed, it seems that the causal role of smoking with respect to heart attacks is the same *regardless of the frequency of exercisers:* for nonexercising individuals smoking causes heart attacks, and for exercising individuals smoking prevents attacks.

Thus, one thing I think this example shows is that the proposition "*C* causes *E* in population *P*" is not equivalent to the proposition that an increase in the frequency of *C* (say to 1) in population *P* would result in a higher frequency of *E* than would result from a decrease (say to 0) of the frequency of *C*. Because of the role of interactive causal factors (like exercising in the example), it may be true that, for a given population, *an increase in the frequency of C to 1* is a positive causal factor for *an overall increase in the frequency of E,* and that *a decrease in the frequency of C to 0* is a positive causal factor for *an overall decrease in the frequency of E,* even though it is misleading, incomplete, even false, to say that the factor *C* itself is simply causally positive (and not negative) for *E*.

But perhaps if we can appropriately accommodate the phenomenon of causal interaction, then Giere's characterization of causation in terms of *relations between certain changes in the frequencies* of factors would agree with the causal facts *relating to the factors themselves.* Let us first try treatment I of causal interaction. According to this approach, it is the combined factor of smoking and exercising that is a negative causal factor for heart attacks and the combined factor of smoking and not exercising that is a positive causal factor for heart attacks. Does Giere's theory yield this result? And what about the causal role of not smoking but exercising and of neither smoking nor exercising?

Figures 1(a) and 1(b) exhibit two of the many ways in which it could be true that smoking lowers the probability of a heart attack among ex-

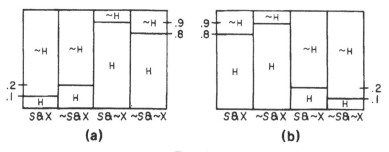

Figure 1

ercisers and raises it among nonexercisers. (The spacing of the vertical lines is not meant to indicate the frequencies of the four combined factors.) In the case of 1(a), it is clear that Giere's theory will imply that *S&X* is negative, and *S&~X* positive, for *H*; and in case 1(b) it is clear that this theory will imply that *~S&X* is positive, and *~S&~X* negative, for *H*. But what does Giere's theory tell us about the roles of *~S&X* and *~S&~X* in the 1(a) case and of *S&X* and *S&~X* in the 1(b) case? Take, for example, the factor *~S&X* in case 1(a). Can treatment I capture the causal truth about its causal relation to *H*? Although it seems clear what the frequency of *H* is in the hypothetical population in which everyone exhibits *~S&X* (namely, just the actual frequency of *H* in the presence of this factor: 0.2), we cannot say, without further elaboration of the example, what the frequency of *H* would be in the hypothetical population where no one has *~S&X*. That depends on how the actual *~S&X*'s distribute themselves among the three alternatives (*S&X*, *S&~X*, and *~S&~X*) in the hypothetical population in which no one has *~S&X*.

There are, presumably, certain individuals who, *were* they to behave differently with respect to *S* or *X* from the way they *do* behave, *would* be *S&X*'s, some who *would* be *S&~X*'s, others who *would* be *~S&X*'s, and others who *would* be *~S&~X*'s. Let us call these four types of individuals types *A*, *B*, *C*, and *D*, respectively. Clearly, the combined factors determined by whether or not one smokes and whether or not one exercises interact with the four factors *A*, *B*, *C*, and *D*. For example, in the 1(a) case, among individuals of type *A*, *~S&X* is positive for *H*, for $Pr(H/A\&{\sim}S\&X) = 0.2 > Pr(H/A\&{\sim}({\sim}S\&X)) = Pr(H/S\&X) = 0.1$ (assuming that, given what you *actually do* in terms of smoking and exercise, what you *would* do were you to behave differently does not affect the probability of a heart attack). Also, among individuals of types *B* or *D*, *~S&X* is negative for *H*. Thus, according to treatment I of causal interaction, the causal truth is captured in terms of more complex combined factors, such as *~S&X&A*, *~S&X&B*, and *~S&X&D*. But again, *absence* of any one of *these* factors in every individual in a hypothetical population does not yield a unique causally significant frequency for *H*, for there are seven alternative factors for each such factor, some of the alternatives conferring a higher probability on *H* than the original, and others a lower probability. And this phenomenon will recur (again) if further interacting modal features are included in combined causal factors. I conclude that combining Giere's counterfactual approach with treatment I of causal interaction cannot capture the whole causal truth.

Thus, let us turn to treatment II. We now consider two populations separately, one consisting of the exercisers and the other consisting of the nonexercisers. Then Giere's theory leads straight to the correct conclusions: in the population of exercisers smoking is negative for heart attacks, and in the population of nonexercisers, smoking is positive for heart attacks. It thus appears that the theory must be modified by the inclusion of some such clause as this, which overrides the application of the rest of the theory:

> If there is a factor F such that when population P is divided into F's and non-F's we get two populations across which the causal role of C for E reverses, then conclude that, in P, C is neither causally positive, negative, nor neutral, for E in P.[6]

If the rule is applicable, then, of course, an investigator should seek the causal truth in the two subpopulations. Note also that this suggested revision makes Giere's theory agree with the contextual unanimity theory that the three possibilities of a given factor's being causally positive, negative, or neutral for a second factor in a given population are *not* collectively exhaustive. It is also interesting to note that Collier (1983) and Giere (1984) have suggested that Giere's approach can avoid difficulties of a different kind (discussed below) by the choice of appropriate subpopulations to which the model should be applied. The explicit incorporation of the above clause into Giere's approach is perhaps a step in the direction of a general characterization of the relevant subpopulations. I will return to this question below.

Examples of the kind under discussion have a bearing on the contextual unanimity model of probabilistic causal connection as well. This model (at least Cartwright's formulation of it) must also be modified in order to get the right answer in cases of causal interaction. Consider figure 2, where the frequency of each of the combined factors is equal to .25. In this case, on the contextual unanimity model, X is neither positive nor negative for H: $Pr(H/X) = Pr(H/{\sim}X) = 0.5$. Even holding S fixed, positively or negatively, X cannot count as a cause of H or of ${\sim}H$, for lack of unanimity. Thus, according to Cartwright's theory, X should *not* be held fixed in specifications of background contexts. (Recall that, for Cartwright, all *and only* those factors, other than the putative cause and its effects, that are causes of the putative effect or of its absence are to be held fixed.[7]) Thus, since $Pr(H/S) = 0.35 < 0.65 = Pr(H/{\sim}S)$, the Cartwright formulation of the contextual unanimity model implies that,

---

[6]As usual, a more adequate and general principle could be formulated in terms of a *partition* of factors F.

[7]It is worth emphasizing that this condition is part of *Cartwright's formulation* of the contextual unanimity theory, where other formulations have been advanced (see n. 1 above, for example) in which the background contexts are described differently, e.g., more vaguely as simply specifying all relevant background factors. In Eells and Sober (1983), we argue (contra Cartwright) that it *cannot hurt* to specify factors beyond those which Cartwright's theory insists we *must* specify (as long as we do not specify the putative cause or any of its effects). Here I am arguing that there is a further kind of factor that *must* be held fixed.

in the population of figure 2, S causes $\sim H$, where the *right answer* is that S is neither simply causally positive, negative, nor neutral, for H in the whole population.

This suggests the following revision of the characterization of background contexts:

> For evaluating the causal impact of factor C on factor E in population P, hold fixed, positively or negatively, not only factors—other than C and its effects—that are causes of E or of $\sim E$, but also factors with which C interacts.

This will make the contextual unanimity model give the right answer in the case of figure 2: namely, that S is neither (simply) causally positive, (simply) negative, nor (simply) neutral, for H in the population "mapped" by figure 2. Again, of course, the investigator applying this revision of the contextual unanimity model should, in cases like this, seek the causal truth in the subpopulations determined by the presence or absence of the factor with which C interacts, in accordance with treatment II of causal interaction.

There remains to be considered, however, the possibility of combining treatment I of interaction with the Cartwright version of probabilistic causes. Incorporating this treatment would go something like this, I suppose:

> If C interacts with some factor F, with respect to E as the effect, then it is the four combined factors of the presence or absence of C together with the presence or absence of F that have univocal causal roles for E.

This approach is flawed for much the same reasons as its incorporation in Giere's theory was flawed. In the example of figure 2, although the

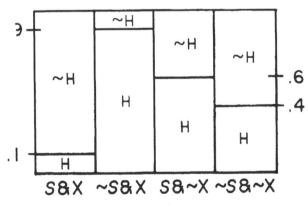

Figure 2

factors $S\&X$ and $\sim S\&X$ have univocal causal roles of preventing and causing heart attacks, respectively, the factors $S\&\sim X$ and $\sim S\&\sim X$ clearly have no such roles. Also, whether the latter two raise or lower the probability of heart attacks depends on the frequencies of the four combined factors, but the question of whether these two factors cause or prevent heart attacks evidently should not turn on this.

Let us finally compare the two revised theories of probabilistic causality. If C raises the probability of E in every *subpopulation* determined by a maximal specification of causally relevant background factors, then should not the *universal introduction* of C result in a higher frequency of E in the entire population than would result from its *universal elimination*? Also, shouldn't we always be able to find interaction-free (sub)populations in which the following will hold: if the universal introduction of C would result in a higher population frequency of E than would result from the universal elimination of C, then this would also have to be true within each background context (for otherwise there would be an interaction)? In fact, the revision suggested above of Giere's theory does *not* always give the same answer as the revised contextual unanimity theory. This is because there is a kind of interaction which Giere's "all or none" counterfactual approach cannot appropriately accommodate in the form given above, but which the contextual unanimity theory apparently does: interactions, exemplified in cases which Elliott Sober (1982) first brought to bear on Giere's theory, in which a causal factor *interacts with its own frequency* in the relevant population.

The answer to both of the questions with which the previous paragraph began is no. Rather than showing this explicitly (basically, by repeating examples of Sober, some pertaining to the necessity and others to the sufficiency of Giere's characterization of probabilistic causal connection), I shall merely present an example that illustrates the (perhaps more general) idea of a causal factor's interacting with certain of its *subpopulation* frequencies. I shall also briefly discuss the responses to such examples given by Collier (1983) and Giere (1984).

Consider the population of all students now attending some lecture somewhere. If a student raises a hand, and nobody else at the same lecture does, then this student will (probably) be noticed by the professor and asked what's on his mind. In this case (nobody else at the lecture raising a hand), raising one's hand is a positive causal factor for being noticed. But if everyone else at that lecture is raising a hand, *this* student's raising hand may actually be a *negative* factor for getting noticed, and an *unraised* would be more apt to be noticed. Thus, in general, if the frequency of hand-raisers at a given lecture is low, raising one's hand is positive for getting noticed, and if this frequency is high, then raising one's hand is negative for getting noticed. Let us first consider the question of whether raising a hand is positive for getting noticed in the *general* population (including all the lectures). In cases like this, there is no hypothetical (general) population with 100 percent *C*'s that is exactly like the actual population even only with respect to all other factors that are causally relevant to the putative effect *E*, for the *frequencies* of *C* in the different lecture rooms are themselves causally relevant. Here, we cannot have 100 percent hand-raisers and at the same time keep all the lecture-room frequencies of hand-raisers the same as they are in the actual population. And no matter what these actual frequencies are, it seems that Giere's theory will tell us that raising a hand is (simply) causally *irrelevant* to getting noticed, for (plausibly) there would be the same frequency of noticed students if they were all to raise a hand as there would be if none were to raise a hand.

But in this example, the causal factor of having a raised hand clearly interacts with the lecture-room frequencies of hand-raisers, with respect to getting noticed as the effect (with a low frequency, the causal factor is positive, and with a high frequency, the causal factor is negative). This suggests that we take into account the suggested revision of Giere's theory: that is, the overriding clause according to which, if the role of a causal factor reverses itself in subpopulations determined by the presence or absence of an interacting factor, then we should conclude that the putative causal factor is neither causally positive, negative, nor neutral, for the effect under investigation in the general population—the causal investigator then being directed to seek the causal truth in the relevant *subpopulations*. But in examples of the kind under discussion, Giere's theory, as presented at the beginning of this paper (a revision will be considered below), *does not imply that the causal role of the putative causal factor reverses itself in subpopulations determined by the interacting factor (i.e., the frequency of the causal factor)*. Consider any subpopulation of students who are in lecture rooms that have some given, fixed, frequency of hand-raisers (thus "holding fixed" the interacting causal factor in this example), and consider two hypothetical subpopulations just like the subpopulation except that the frequency of hand-raisers is 0 in one case and 1 in the other. Then the frequency of noticed students should, plausibly, be the *same* (low) in the two hypothetical subpopulations. Thus, even the suggested revision of Giere's theory, according to which we should consider subpopulations determined by interacting causal factors, gives the intuitively incorrect answer that raising a hand is (simply) causally neutral with respect to getting noticed.

The suggested revision of Giere's theory given above thus does not accommodate all kinds of causal interaction. However, Collier (1983) and Giere (1984) have advanced revisions of the original formulation of Giere's theory intended to accommodate examples of the kind just described. Basically, the approach attempts to *both* hold fixed the (interacting) frequency of the putative causal factor *and* still maintain that we should understand causal hypotheses in terms of comparisons of subpopulations exhibiting 100 percent and 0 percent frequencies of the causal factor under investigation. In applying their approach to the hand-raising example described above, it seems that they would propose that we consider *a very small subpopulation* of any given subpopulation of students for which the frequency of hand-raisers in their lecture rooms is the same. Consider the case of this frequency being .95, and consider the subpopulation P' (taken from the general population P of all students in a lecture room now) of all students whose lecture room frequency of hand-raisers is .95. Now consider a very small subpopulation, P'', of the subpopulation P'. The desired result is that if *all* of P'' were to raise their hands, then there would be a *lower* frequency of getting noticed in P'' than there would be in P' if *none* of the individuals in P'' were to raise their hands—that being because the population P' within which P'' is scattered has a high frequency of hand-raisers. Given this result, we conclude (correctly) that, in P'', raising a hand is negative for getting noticed.

There are three points that deserve mention with regard to this kind of elaboration of the original formulation of Giere's "all or none" counterfactual approach to causation, and the basic idea retained. First, in the example, we get the right answer for the very small subpopulation P''. But what about the causal role of raising a hand in the more inclusive subpopulation P'? Collier says, "It is quite possible for some factor to be a positive causal factor in a subpopulation without being causal in a more inclusive population" (1983, p. 625), adding that while this may disagree with intuition, there is no good reason to rule it out a priori. In any case, we have here, I believe, a *genuine disagreement* between the counterfactual approach and the contextual unanimity model. As explained below, the contextual unanimity model implies that in P', raising a hand is negative for getting noticed, while it seems that the counterfactual approach does not have this consequence, directing us instead to find the causal truth in very small subpopulations of P'. Thus, it seems that cases like this are worthy of further investigation: because the two theories give different answers in them, they provide a focal point for assessing the relative merits of the two approaches. Although my intuitions tell me that raising a hand is negative for being noticed in P', I shall not pursue this line of thought here. (On the other hand, it is open to advocates of the "all or none" theory to further elaborate the approach, in such a way that we may assess the causal role of a factor in a given population itself in terms of hypothetical versions of small subpopulations of it.)

Second, we are given little direction concerning *which* small subpopulations of a given population are appropriate for dealing adequately with cases of frequency-dependent causation. A characterization of appropriate small subpopulations is quite important. In the hand-raising example, for example, if the small subpopulation P'' of P' consisted *solely* of *all* the students from *some particular lecture room in which the frequency of hand-raisers is .95*, then the revised version of Giere's approach again becomes insensitive to frequency dependence. On the other hand, if P'' consisted of just one student from each lecture room in which the frequency of hand-raisers is .95, then the revision should yield the correct answer. Other relevant considerations would presumably include such factors as where the students in the small subpopulation are seated in their lecture rooms, how long their arms are, and so on. What would be ideally desired here is an adequate general characterization of the appropriate small subpopulations.

And third, it seems that there can be no single, simple revision of the counterfactual, all or none, approach that will handle all cases of causal interaction. One suggested revision handles cases like the one described above in which smoking interacts with exercising, with respect to a heart attack as the effect; and it seems that a quite different kind of elaboration is required for the special case of causal interaction in which a causal factor interacts with its own frequency.

The (revised) contextual unanimity model, however, is appropriately sensitive, in a simple and unified way, to both kinds of interaction discussed above. We have already seen how the revised contextual unanimity theory deals appropriately with the smoking/exericising/heart attack example. In the hand-raising example, since hand-raising interacts with its own lecture-room frequency (with respect to getting noticed), that frequency must, according to the revised theory, be held fixed in specifications of background contexts. This leads to the conclusion that there is *no single causal role of hand-raising for getting noticed in the general population:* in background contexts in which the frequency of hand-raisers is low, hand-raisers are more frequently noticed than non-hand-raisers, and in background contexts in which the frequency of hand-raisers is high, hand-raisers are less frequently noticed than the non-hand-raisers. Also, for the contextual unanimity theory, treatment II of causal interaction directs us to subpopulations of students who have the *same* lecture-room frequency of hand-raisers. And in subpopulations of students in which this frequency is low (high), the frequency of noticed students will be higher (lower) for the hand-raisers than for the non-hand-raisers. The (revised) contextual unanimity theory thus seems to be able both to *give the right answers* for the relevant subpopulations and to *adequately identify appropriate subpopulations.*

The phenomenon of causal interaction requires the exercise of care on the part of *both* kinds of theories of probabilistic causality. The revised contextual unanimity model handles both kinds of causal interaction discussed in this paper in a simple and unified way. But since a causal factor can interact *with its own frequency,* as Sober has shown, it seems that an *"all or none"* kind of counterfactual approach such as Giere's must involve complexities and subtleties not required of the contextual unanimity model in order to give the right answers in all cases of causal interaction.

REFERENCES

Cartwright, N. (1979), "Causal Laws and Effective Strategies", *Noûs 13*: 419–37.
Collier, J. (1983), "Frequency-Dependent Causation: A Defense of Giere", *Philosophy of Science 50*: 618–25.
Eells, E., and Sober, E. (1983), "Probabilistic Causality and the Question of Transitivity", *Philosophy of Science 50*: 35–57.
Giere, R. (1979), *Understanding Scientific Reasoning.* New York: Holt, Rinehart and Winston.
———. (1980), "Causal Systems and Statistical Hypotheses", in *Applications of Inductive Logic,* L. Jonathan Cohen (ed.). New York: Oxford University Press.
———. (1984), "Causal Models with Frequency Dependence", *Journal of Philosophy 81*: 384–91.
Skyrms, B. (1980), *Causal Necessity.* New Haven and London: Yale University Press.
Sober, E. (1982), "Frequency-Dependent Causation", *Journal of Philosophy 79*: 247–53.
———. (1984a), *The Nature of Selection.* Cambridge: The MIT Press/A Bradford Book.
———. (1984b), "Discussion: What Would Happen If Everyone Did It?", *Philosophy of Science 52*: 141–50.
Suppes, P. (1970), *A Probabilistic Theory of Causality.* Amsterdam: North-Holland.

Acknowledgments

Hempel, Carl G., and Paul Oppenheim. "Studies in the Logic of Explanation."
Philosophy of Science 15 (1948): 135–75. Reprinted with the permission of the
University of Chicago Press.

van Fraassen, Bas C. "The Pragmatics of Explanation." *American Philosophical Quarterly*
14 (1977): 143–50. Reprinted with the permission of the *American
Philosophical Quarterly*.

Kitcher, Philip. "Explanatory Unification." *Philosophy of Science* 48 (1981): 507–31.
Reprinted with the permission of the University of Chicago Press.

Friedman, Michael. "Explanation and Scientific Understanding." *Journal of Philosophy*
71 (1974): 5–19. Reprinted with the permission of the Journal of
Philosophy, Inc., Columbia University, and the author.

Lewis, David. "Causal Explanation." In *Philosophical Papers*, Vol. II (New York: Oxford
University Press, 1986): 214–40. Reprinted with the permission of Oxford
University Press.

Salmon, Wesley C. "Theoretical Explanation." In *Explanation*, edited by Stephan
Körner (New Haven: Yale University Press, 1975): 118–45. Reprinted with
the permission of Blackwell Publishers.

Railton, Peter. "A Deductive-Nomological Model of Probabilistic Explanation."
Philosophy of Science 45 (1978): 206–26. Reprinted with the permission of the
University of Chicago Press.

Salmon, Wesley C. "Statistical Explanation and Its Models." In *Scientific Explanation
and the Causal Structure of the World*, Chapter 2 (Princeton: Princeton
University Press, 1984): 24–47. Reprinted with the permission of Princeton
University Press.

Ayer, A.J. "What is a Law of Nature." In *The Concept of a Person*, Chapter 8 (London:
Macmillan, 1963): 209–34. Reprinted with the permission of Macmillan
Publishers.

Chisholm, Roderick M. "Law Statements and Counterfactual Inference." *Analysis* 15
(1955): 97–105. Reprinted with the permission of the author.

Goodman, Nelson. "The Problem of Counterfactual Conditionals." In *Fact, Fiction, and
Forecast*, Chapter 1 (Cambridge: Harvard University Press, 1955): 13–34.
Reprinted with the permission of Harvard University Press.

Lewis, David K. "The Metalinguistic Theory: Laws of Nature." In *Counterfactuals*, Section 3.3 (Cambridge: Harvard University Press, 1973): 72–77. Reprinted with the permission of the author.

Tooley, Michael. "The Nature of Laws." *Canadian Journal of Philosophy* 7 (1977): 667–98. Reprinted with the permission of the University of Calgary Press.

Cartwright, Nancy. "Fundamentalism vs. The Patchwork of Laws." *Proceedings of the Aristotelian Society* 94 (1994): 279–92. Reprinted by courtesy of the Editor of the Aristotelian Society. Copyright 1994.

Mackie, J.L. "Causes and Conditions." *American Philosophical Quarterly* 2 (1965): 245–64. Reprinted with the permission of the *American Philosophical Quarterly*.

Kim, Jaegwon. "Causation, Nomic Subsumption, and the Concept of Event." *Journal of Philosophy* 70 (1973): 217–36. Reprinted with the permission of the Journal of Philosophy, Inc., Columbia University, and the author.

Davidson, Donald. "Causal Relations." *Journal of Philosophy* 64 (1967): 691–703. Reprinted with the permission of the Journal of Philosophy, Inc., Columbia University, and the author.

Lewis, David. "Causation." *Journal of Philosophy* 70 (1973): 556–67. Reprinted with the permission of the Journal of Philosophy, Inc., Columbia University, and the author.

Eells, Ellery. "Probabilistic Causal Interaction." *Philosophy of Science* 53 (1986): 52–64. Reprinted with the permission of the University of Chicago Press.

.